NEW CONCEPTS IN LIPID RESEARCH

In Honour of Stina and Einar Stenhagen

Editor

RALPH T. HOLMAN

Executive Director and Professor of Biochemistry,
Hormel Institute of the University of Minnesota,
Austin, Minnesota 55912

PERGAMON PRESS

OXFORD · NEW YORK · TORONTO · SYDNEY
PARIS · FRANKFURT

U.K.	Pergamon Press Ltd., Headington Hill Hall, Oxford OX3 0BW, England
U.S.A.	Pergamon Press Inc., Maxwell House, Fairview Park, Elmsford, New York 10523, U.S.A.
CANADA	Pergamon of Canada Ltd., 75 The East Mall, Toronto, Ontario, Canada
AUSTRALIA	Pergamon Press (Aust.) Pty. Ltd., 19a Boundary Street, Rushcutters Bay, N.S.W. 2011, Australia
FRANCE	Pergamon Press SARL, 24 rue des Ecoles, 75240 Paris, Cedex 05, France
FEDERAL REPUBLIC OF GERMANY	Pergamon Press GmbH, 6242 Kronberg-Taunus, Pferdstrasse 1, Federal Republic of Germany

First edition 1978

British Library Cataloguing in Publication Data

Progress in the chemistry of fats and other lipids.

Vol. 16: New concepts in lipid research.
1. Lipids
I. Stenhagen, Stina   II. Stenhagen, Einar
III. Holman, Ralph Theodore   IV. New concepts in lipid research
574.1'9247   QP751   77-30703
ISBN   0 08 022663 9

Published as a commemorative volume of the Journal *Progress in the Chemistry of Fats and other Lipids* (Volume 16) and supplied to subscribers as their 1978 subscription.

Printed in Great Britain by A. Wheaton & Co. Ltd., Exeter

NEW CONCEPTS IN LIPID RESEARCH

In Honour of Stina and Einar Stenhagen

PROGRESS IN THE CHEMISTRY
OF FATS AND OTHER LIPIDS

Editor

RALPH T. HOLMAN

Executive Director and Professor of Biochemistry,
The Hormel Institute, University of Minnesota,
Austin, Minnesota 55912, U.S.A.

EDITORIAL ADVISORY BOARD

DEDICATION

Photos : Gunnar Stenhagen

Professors STINA STÄLLBERG-STENHAGEN and EINAR STENHAGEN have left a permanent mark upon the record of lipid science. Although their accomplishments were monumental, their humility was greater, and they did nothing to attract attention except to do serious scientific work. It is left to us who remain to call proper attention to their accomplishments.

Einar and Stina Stenhagen were beset by difficulties with health which, on more than one occasion, interrupted their education and their careers. They were unique in the modern scientific community in that their formal education was never completed, yet they were able to pursue science at its highest levels and were both appointed professors. Their doctors' degrees were honorary. The Swedish scientific community and government are to be commended for flexibility and wisdom in recognizing the genius of the somewhat unorthodox Stenhagens.

Never in robust health, the Stenhagens sharpened their focus, conserved their energies and concentrated their dedication upon significant fundamental medical science. They taught by example, and a large group of students and colleagues now follow the examples of their beloved professors. We who had the rare opportunity to work beside and learn from such masters can think of no more fitting memorial to Einar and Stina than to dedicate our best efforts to their memory.

Ralph T. Holman

CONTENTS

Dedication   v

Preface   viii

Einar August Stenhagen (1911–1973);   1
Stina Lisa Ställberg-Stenhagen (1916–1973)

Quantitative Chemical Taxonomy based upon Composition of Lipids   9
RALPH T. HOLMAN

Biological and Chemical Aspects of the Aquatic Lipid Surface Microlayer   31
GÖRAN ODHAM, BÖRJE NORÉN, BIRGITTA NORKRANS, ANDERS SÖDERGREN and
HÅKAN LÖFGREN

Occurrence, Synthesis and Biological Effects of Substituted Glycerol Ethers   45
BO HALLGREN, GUNNEL STÄLLBERG and BERNT BOERYD

Trehalose-containing Glycolipids   59
C. ASSELINEAU and J. ASSELINEAU

Molecular Arrangement in Condensed Monolayer Phases   101
MONICA LUNDQUIST

Lateral Packing of Hydrocarbon Chains   125
SIXTEN ABRAHAMSSON, BIRGITTA DAHLÉN, HÅKAN LÖFGREN and IRMIN PASCHER

Liquid Crystalline Behavior in Lipid–Water Systems   145
KRISTER FONTELL

Stability of Emulsions Formed by Polar Lipids   163
KÅRE LARSSON

Dermatophyte Lipids   171
JAN VINCENT

Optically Active Higher Aliphatic Hydroxy Compounds Synthesized from   179
Chiral Precursors by Chain Extension
K. SERCK-HANSSEN

Perdeuteriated Normal-chain Saturated Mono- and Dicarboxylic Acids and   195
Methyl Esters
NGUYÊÑ DINH-NGUYÊN and AINO RAAL

Mass-spectrometric Sequence Studies of Lipid-linked Oligosaccharides, Blood-   207
group Fucolipids, Gangliosides and Related Cell-surface Receptors
KARL-ANDERS KARLSSON

Acyclic Diterpene Alcohols: Occurrence and Synthesis of Geranylcitronellol,   231
Phytol and Geranylgeraniol
LARS AHLQUIST, GUNNAR BERGSTRÖM and CONNY LILJENBERG

Functional Aspects of Odontocete Head Oil Lipids with Special Reference to   257
Pilot Whale Head Oil
JONAS BLOMBERG

Mass Spectrometry of Fatty Acid Pyrrolidides   279
BENGT Å. ANDERSSON

Index   309

Contents of Previous Volumes   315

PREFACE

This volume is a collection of works honoring Professors Einar and Stina Stenhagen contributed by their students and associates. The Professors Stenhagen entered many areas of science to solve problems they encountered with lipids, and the breadth of their interests is mimicked by the range of subject matter now offered by their successors. The contributions apply methods of synthetic organic chemistry, physical chemistry and computer analysis to structural problems with natural products, to study of monolayer and membrane lipids, to taxonomy and to biological and physical properties of lipids.

R. T. Holman

Prog. Chem. Fats other Lipids. Vol. 16, pp. 1–7. Pergamon Press, 1978. Printed in Great Britain

BIOGRAPHY

EINAR AUGUST STENHAGEN
(1911–1973)

STINA LISA STÄLLBERG-STENHAGEN
(1916–1973)

This volume of the "Progress" series is dedicated to the memory of Einar and Stina Stenhagen, whose untimely deaths were a great loss to family, friends, colleagues and to Science to which they had contributed so much of their lives.

Einar Stenhagen (ES) had a strong interest in technology even in his childhood and chose at the beginning of his academic career in 1929 to study electrical engineering at the Royal Institute of Technology in Stockholm. During the period of a severe attack of tuberculosis, which ever after degraded his health, he developed a keen interest in medicine and a dedication to study of the chemical composition of the human tubercle bacillus. He also considerably broadened and deepened his knowledge of chemistry, physics and engineering while confined in the hospital. In 1933, after returning from a Swiss sanatorium, he started medical studies at the Karolinska Institute, Stockholm, later continuing at the universities of Uppsala and Lund. He obtained a preclinical degree in Medicine at Uppsala in 1935. His lack of personal career ambitions and several years of ill health prevented him from obtaining further formal qualifications. A Rockefeller fellowship, awarded to ES in 1938, involved him for 1 year at Cambridge, United Kingdom, in the Department of Colloid Science, and was of great importance to the direction of his future work. Recognition of his talents came in 1943, when he was appointed docent in Medical Chemistry at the University of Uppsala. This was further confirmed in 1952 by the award of a personal chair as professor in Medical Biochemistry which he transferred from the University of Uppsala to the University of Göteborg in 1959.

Though Stina Ställberg-Stenhagen (SS) always had a basic interest in botany and entomology, she began academic studies in medicine at the University of Uppsala in 1936 and received her preclinical degree in 1939. She was appointed docent in 1951 and the following year she was made assistant professor at the University of Göteborg. It was during the years in Uppsala that she met ES, whom she later married, and that they began their cooperation on the structural problems of lipids, which was to continue throughout their lives. SS also suffered severely from tuberculosis, which struck her in 1951. In 1963 she was appointed full professor in Medical Chemistry at the University of Göteborg.

In the beginning, their goal was to determine the structure of the strange lipids in human tubercle bacillus. At that time they used, among other techniques, surface chemistry and X-ray crystallography, developing new methods in both techniques. They designed and built a surface balance, which even today is not bettered. As new methods, such as infrared spectroscopy, gas liquid chromatography (GLC) and mass spectrometry (MS), became available, they adopted them in their laboratory, and in all cases they made important contributions to the fundamental application and development of the techniques.

Model compounds constituted an important part of their activity, and this can be reflected in the long series of publications reporting synthesis of optically active aliphatic compounds. Even in this field, their contributions and methods of synthesis are important. Their collection of reference compounds is still of great value for many scientists, as they are of the highest purity, even by modern standard demands.

They soon became aware of the usefulness of mass spectrometry, and the introduction of this instrument into their arsenal of analytical tools involved a breakthrough regarding detailed solutions to the structures of organic molecules. They built their own MS apparatus, with which the structures of some wax constituents of human tubercle bacillus were solved and the first mass spectra of peptides were recorded. Their synthetic reference compounds now became exceedingly useful and their pioneering MS work in the field of lipids is remarkable. GLC had also been their tool since it appeared in the early fifties, and the idea to combine GLC and MS into one instrument quickly entered their minds. The basic problem in this idea was to remove the GLC carrier gas from the sample before it reached the ion source of the MS. The solution was the jet separator, which was designed in their laboratory. This device is now in common use throughout the world. With this improved instrument, their group solved such difficult problems as the composition of preen gland waxes of birds.

In the late sixties, techniques such as electron spectroscopy for chemical analysis (ESCA), microwave spectroscopy, and computers became available, and the Stenhagens soon adopted them in their research work. The GLC–MS was coupled to a computer to facilitate interpretation of spectra and, in collaboration with others, they compiled a library of around 20,000 spectra of organic compounds. Before the availability of commercial ESCA and microwave instruments, ES participated in groups constructing and building their own instruments.

As the synthesis program grew, large, optically active molecules such as mycoserosic acids and geranylcitronellol were synthesized, and their group became specialists at introducing deuterium into lipid molecules. Their last contribution in this area was the total synthesis of optically active phytol.

In 1950 they encountered the problem of identification of insect attractant substances occurring in orchids. With the techniques available at that time, this project was unsolvable. In the early sixties, when GLC–MS became a reality, their group continued to investigate these substances and also solved some structures of pheromone character occurring in bees. These studies on natural odoriferous compounds remained their main interest until their death.

Appreciation of their work was manifested in the series of honors and awards they received. Among these, the following can be mentioned: ES was made Honorary Doctor of Medicine (Uppsala, 1949), Fellow of the Royal Swedish Academy of Sciences (1961), Fellow of the Royal Swedish Engineering Academy of Science (1965), and received the John Ericson Medal (1970). ES was also a member of the Nobel Prize committee for Chemistry. SS was made an Honorary Doctor of Medicine (Göteborg, 1960) and shared with her husband the Swedish Medical Association Anniversary Prize (1951).

The success they achieved in their research, covering such a wide scientific spectrum, ranging from the development of instruments and their application to structural chemical analyses, chemical synthesis, biochemistry, ecology and medicine, had its foundations in the complementary characters of their nature, which made a perfectly well balanced combination. For their many friends, colleagues and those who in one way or another know their work, the Stenhagens will be remembered with admiration and respect.

Bengt Å. Andersson

LIST OF PUBLICATIONS

STINA STÄLLBERG-STENHAGEN (SS), EINAR STENHAGEN (ES)

1. Über die Beziehungen zwischen Kephalin und Neutralsalzen in wässrigem Medium. (ES) *Skand. Arch. Physiol.* **77** (1937).
2. Electrophoretic behaviour in nucleic acid–protein mixtures. (ES with T. TEORELL) *Nature*, **141**, 415 (1938).
3. Electrophoresis of human blood plasma. Electrophoretic properties of fibrinogen. (ES) *Biochem. J.*, **32**, 714 (1938).
4. Complex formation in lipoid films. (ES with J. H. SCHULMAN and E. K. RIDEAL) *Nature*, **141**, 785 (1938).

5. An apparatus for electrometric titrations of high precision. (ES) *Ind. Eng. Chem., Anal. Ed.,* **10,** 432 (1938).
6. Built-up films of esters. (ES) *Trans. Faraday Soc.,* **34,** 1328 (1938).
7. Molecular interaction in monolayers. III. Complex formation in lipoid monolayers. (ES with J. H. SCHULMAN) *Proc. R. Soc., B,* **126,** 356 (1938).
8. Ein Universalpuffer für den pH-Bereich 2.0 bis 12.0. (ES with T. TEORELL) *Biochem. Z.,* **299,** 416 (1938).
9. Deposition of protein multilayers. (ES with R. B. DEAN and O. GATTY) *Nature,* **143,** 721 (1939).
10. Electrophoretic properties of thymonucleic acid. (ES with T. TEORELL) *Trans. Faraday Soc.,* **35,** 743 (1939).
11. A new method of spreading monolayers of albumin and lipoid–albumin mixtures. (SS with T. TEORELL) *Trans. Faraday Soc.,* **35,** 1413 (1939).
12. Surface films of heat-denaturated serum albumin. (SS) *Trans. Faraday Soc.,* **35,** 1416 (1939).
13. The interaction between porphyrins and lipoid and protein monolayers. (ES with E. K. RIDEAL) *Biochem. J.,* **33,** 1591 (1939).
14. Monolayers of a long chain ester sulphate. (ES) *Trans. Faraday Soc.,* **36,** 496 (1940).
15. Monolayers of compounds with branched hydrocarbon chains. I. Disubstituted acetic acids. (ES) *Trans. Faraday Soc.,* **36,** 597 (1940).
16. Monolayers of compounds with branched hydrocarbon chains. II. Effects of substitution on the packing of long hydrocarbon chains. (SS, ES) *Sven. Kem. Tidskr.,* **52,** 223 (1940).
17. On the structure of multilayers and the relation between optical and mechanical thickness and X-ray spacing. (ES) *Ark. Kemi, Mineral. Geol.,* **14 A:11** (1940).
18. Monolayers of compounds with branched hydrocarbon chains. III. Evenly distributed methyl side chains. Phytol and phytanic acid. (SS, ES) *Sven. Kem. Tidskr.,* **53,** 44 (1941).
19. Monolayers of compounds with branched hydrocarbon chains. IV. Phthioic acid. (SS, ES) *J. Biol. Chem.,* **139,** 345 (1941).
20. Monolayers of 15-phenylpentadecanoic acid and 22-phenylbehenic acid. (SS, ES) *Sven. Kem. Tidskr.,* **53,** 355 (1941).
21. 2-Methyl-tetradecanol-l. (ES with K. LINDBLAD) *J. Am. Chem. Soc.,* **63,** 3539 (1941).
22. Monolayer studies on ketones. I. Effect of the displacement of the ketonic oxygen along a straight hydrocarbon chain of 17 carbon atoms. II. Effect of ring closure. (SS, ES) *Sven. Kem. Tidskr.,* **53,** 478 (1941).
23. Monolayers of compounds with branched hydrocarbon chains. V. Phthiocerol. (SS, ES) *J. Biol. Chem.,* **143,** 171 (1942).
24. The synthesis of 2-methyl-eicosanoic acid (methyl-*n*-octadecylacetic acid) and 2-methyl-tetracosanoic acid (methyl-*n*-docosylacetic acid). (ES with B. TÄGTSTRÖM) *Sven. Kem. Tidskr.,* **54,** 145 (1942).
25. Tuberkelbakteriens fettkemi. (ES with G. BLIX) *Nord. Med.,* **15,** 2577 (1942).
26. The state of bile salt solutions. (ES with O. MELLANDER) *Acta Physiol. Scand.,* **4,** 349 (1942).
27. 3,3-Dimethyl-$\Delta^{13:14}$-tetradecenoic acid. (ES with G. GUSTBÉE) *Sven. Kem. Tidskr.,* **54,** 243 (1942).
28. Monolayers of compounds with branched hydrocarbon chains. VI. 2-Methyl- and 10-methyl substituted carboxylic acids of high molecular weight. (SS, ES) *J. Biol. Chem.,* **148,** 685 (1943).
29. A new type of thermal expansion in monolayers. (SS, ES) *Sven. Kem. Tidskr.,* **55,** 63 (1943).
30. X-ray study of the hydrocarbon from phthicerol. (ES) *J. Biol. Chem.,* **148,** 695 (1943).
31. *n*-Nonatriacontane. (ES with B. TÄGTSTRÖM) *J. Am. Chem. Soc.,* **65,** 845 (1944).
32. Monolayers of optically active long chain compounds. I. D-Eicosanol-2 and DL-eicosanol-2. (SS, ES) *Ark. Kemi, Mineral. Geol.,* **18 A:19** (1944).
33. An easily assembled and adjusted mercury-sealed stirrer. (SS, ES) *Sven. Kem. Tidskr.,* **56** (1944).
34. A recording surface balance of the Wilhelmy–Dervichian type (SS, ES with K. J. I. ANDERSSON) In *The Svedberg,* p. 11. Almqvist & Wiksell, Uppsala, Sweden (1944).
35. The synthesis of long carbon chains using long chain β-keto esters. A new synthesis of *n*-tetratriacontanoic acid. (SS, ES) *Ark Kemi, Mineral. Geol.,* **19 A:1** (1944).
36. Long chain iso-acids. I. 16-Methylheptadecanoic (isostearic) acid and certain derivatives. (ES with B. TÄGTSTRÖM-EKETORP) *Ark. Kemi, Mineral. Geol.,* **19 A:8** (1944).
37. The synthesis of 15-methylheptadecanoic acid. (SS) *Ark. Kemi, Mineral. Geol.,* **19 A:28** (1945).
38. A monolayer and X-ray study of mycolic acid from the human tubercle bacillus. (SS, ES) *J. Biol. Chem.,* **159,** 255 (1945).
39. Phase transitions in condensed monolayers of normal chain carboxylic acids. (SS, ES) *Nature,* **156,** 239 (1945).
40. Synthesis of higher β-keto esters. Methyl esters of normal chain β-keto acids with 9 to 24 carbon atoms. (SS) *Ark. Kemi, Mineral. Geol.,* **20 A:19** (1945).
41. Branched long chain compounds. Introduction of tertiary carbon atoms using β-keto esters derived from disubstituted acetic acids. I. Methyl substituted carboxylic acids. (SS) *Ark. Kemi, Mineral. Geol.,* **22 A:19** (1946).
42. Optically active higher aliphatic compounds. I. The synthesis of (+)-2-methyldodecanoic and (+)-2-methylhexacosanoic acids. (SS) *Ark. Kemi, Mineral. Geol.,* **23 A:15** (1946).
43. On the nature of two carboxylic acids of high molecular weight obtained from the waxes of acid-fast bacteria. (SS, ES) *J. Biol. Chem.,* **165,** 599 (1946).
44. Virus. (Summary) (ES with S. SJÖLIN) *Tek. Tidskr.,* **41** (1946).
45. On the steric relations between certain optically active methyl substituted monocarboxylic acids and glyceric aldehyde. The configuration of isoleucine. (SS, ES) *Ark. Kemi, Mineral. Geol.,* **24 B:9** (1947).
46. The resolution of methyl hydrogen β-methylglutarate and the steric relations between the enantiomorphs and glyceric aldehyde. (SS) *Ark. Kemi. Mineral. Geol.,* **25 A:10** (1947).
47. Rearrangement in the preparation of ester acid chlorides. (SS) *J. Am. Chem. Soc.,* **69,** 2568 (1947).
48. A recording surface balance of the horizontal type. (SS, ES) *Nature,* **159,** 814 (1947).

49. Infra-red spectrum and molecular structure of phthiocerane. (SS, ES with N. Sheppard, G. B. B. M. Sutherland and A. Walsch) *Nature*, **160**, 580 (1947).

50. Optically active higher aliphatic compounds. II. The synthesis of D(+)- and L(−)-3-methyltetracosanoic acids. (SS) *Ark. Kemi, Mineral. Geol.*, **26 A:1** (1948).

51. Synthesis and X-ray investigation of methyl-substituted long chain hydrocarbons related to phthiocerane. (SS, ES) *J. Biol. Chem.*, **173**, 383 (1948).

52. Optically active higher aliphatic compounds. III. On the structure of tuberculostearic acid. Synthesis of D(−)- and L(+)-10-methyl- and DL-9-methyldecanoic acids. (SS) *Ark. Kemi, Mineral. Geol.*, **26 A:12** (1948).

53. Long chain iso-chains. II. Synthesis of acids with 13, 15, 17, 24, 25, 26 and 35 carbon atoms and an X-ray study of synthetic acids and amides. (ES with K. E. Arosenius, G. Ställberg and B. Tägt-ström-Eketorp) *Ark. Kemi, Mineral. Geol.*, **26 A :19** (1948–49).

54. Unsaturated higher carboxylic acids. The synthesis of $\Delta^{22:23}$-tricosenoic, $\Delta^{22:23}$-tricosynoic, and $\Delta^{21:22}$-tricosynoic acids. (ES) *Ark. Kemi*, **1**, 99 (1949).

55. Optically active higher aliphatic compounds. IV. The degradation of D(+)-3-methylhepatacosanoic acid to D(−)-2-methyl-hexanoic acid. (SS) *Ark. Kemi*, **1**, 153 (1949).

56. Optically active higher aliphatic compounds. V. Synthesis of the dextrorotatory diastereoisomers of 2,9-dimethyltetracosanoic acid. (SS) *Ark. Kemi*, **1**, 187 (1949).

57. Branched long chain compounds. The synthesis of 20,20-dimethylheneicosanoic and 14,14-dimethylpentadecanoic acids. (SS, ES with K. E. Arosenius) *Ark. Kemi*, **1**, 413 (1949).

58. On the use of monolayer phase diagrams for determining the composition of mixtures of homologous long chain compounds of high molecular weight. (SS, ES) *Acta Chem. Scand.*, **3**, 1035 (1949).

59. Studies on hydrocarbons structurally related to phthiocerol. Synthesis of the levorotatory enantiomorph of 4-methyltritriacontane. (SS, ES) *J. Biol. Chem.*, **183**, 223 (1950).

60. Optically active higher aliphatic compounds. VI. The synthesis of D(−)-21-methyltricosanoic, D(−)-21-methyltetracosanoic, and L(+)-21-methylpentacosanoic acids. (SS) *Ark. Kemi*, **2**, 95 (1950).

61. Optically active higher aliphatic compounds. VII. The synthesis of (−)-3 (L), 6 (D)-dimethyltetracosanoic and L(+)-6-methylpentacosanoic acids. (SS) *Ark. Kemi*, **2**, 431 (1950).

62. Sulla sintesi per elttrolisi degli antipodi ottici e della forma DL dell-acido LL-metil-nonadecanico. (SS with R. Cavanna) *Ric. Sci.*, **20**, 1710 (1950).

63. Ricerche su composti alifatici superiori otticamente attivi in rapporto con la struttura dell-acido fitomonico. (SS with R. Cavanna) *Atti Lincei* (VIII), **3**, 31 (1950).

64. Optically active higher aliphatic compounds. VIII. The synthesis of D(+)-4-methyltetracosanoic acid and D(+)-5-methyl-pentacosanoic acid. (SS) *Ark. Kemi*, **3**, 117 (1951).

65. Optically active higher aliphatic compounds. IX. The synthesis of the dextrorotatory diastereoisomers of 3,4-dimethyldocosanoic acid. (SS) *Ark. Kemi*, **3**, 249 (1951).

66. Optically active higher aliphatic compounds. X. Degradation of the dextrorotatory 3,4-dimethyldocosanoic acid derived from (−)-3,4-dimethyladipic acid to a dextrorotatory 2,3-dimethylheneicosanoic acid. (SS) *Ark. Kemi*, **3**, 267 (1951).

67. Optically active higher aliphatic compounds. XI. The synthesis of (−)-2-methyl-2-ethyleicosanoic acid. (SS) *Ark. Kemi*, **3**, 273 (1951).

68. Undersökningar över optiskt aktiva högre fettsyror med förgrenad kolkedja. (SS) Dissertation, University of Uppsala, Sweden, 1951.

69. Monolayers of diastereoisomeric long chain compounds. The behaviour of (+)-2(L), 9(L)-dimethyltetracosanoic acid, (+)-2(L), 9(D)-dimethyltetracosanoic acid, and the corresponding 7-keto acids. (SS, ES) *Acta Chem. Scand.*, **5**, 481 (1951).

70. X-ray camera for the continuous recording of diffraction pattern–temperature diagrams. (ES) *Acta Chem. Scand.*, **5**, 805 (1951).

71. Higher aliphatic β-keto acids. Preparation and investigation of the crystal behaviour of seventeen homologous normal chain β-keto acids with 8 to 24 carbon atoms. (ES) *Ark. Kemi*, **3**, 381 (1951).

72. Synthesis of (+)- and (−)-mucone. (SS) *Ark. Kemi*, **3**, 517 (1951).

73. The crystal structure of isopalmitic acid. (ES with V. Vand and A. Sim) *Acta Crystallogr.*, **5**, 695 (1952).

74. Very long hydrocarbon chains. I. The synthesis of *n*-dooctacontane and *n*-hectane. (SS, ES with G. Ställberg) *Acta Chem. Scand.*, **6**, 313 (1952).

75. On the determination of C-methyl groups in aliphatic long chain compounds. (ES with W. Kirsten) *Acta Chem. Scand.*, **6**, 682 1952.

76. The monoketo- and monohydroxyoctadecanoic acids. Preparation and characterization by thermal and X-ray methods. (ES with S. Bergström, G. Aulin-Erdtman, B. Rolander and S. Östling) *Acta Chem. Scand.*, **6**, 1157 (1952).

77. Optically active higher aliphatic hydroxy compounds. I. The resolution of 3-acetoxybutanoic acid and the synthesis of (+)- and (−)-eicosanol-2. (SS, ES with K. Serck-Hanssen) *Ark. Kemi*, **5**, 203 (1953).

78. On the phase transitions in normal chain carboxylic acids with 12 up to and including 29 carbon atoms between 30°C and the melting point. (ES with E. v. Sydow) *Ark. Kemi*, **6**, 309 (1953).

79. Optically active higher aliphatic compounds. XII. *trans*-2,5D-dimethyl-$\Delta^{2:3}$-heneicosenoic acid and *trans*-2,4D-di-methyl-$\Delta^{2:3}$-heneicosenic acid. (SS) *Ark. Kemi*, **6**, 537 (1954).

80. The crystal structure of *n*-dodecylammonium chloride and bromide. (ES with M. Grodon and V. Vand) *Acta Crystallogr.*, **6**, 739 (1953).

81. Surface films. (ES) In *The Determination of Organic Structures by Physical Methods*, Chapter 8, p. 325. Braude and Nachod, Eds. New York, 1955.

82. A general method for the synthesis of optically active β-hydroxy acids. (ES with K. Serck-Hanssen) *Acta Chem. Scand.*, **9**, 866 (1955).

83. Rapid high precision conductivity recorder. (ES with C.-O. Andersson and O. Mellander) *Acta Chem. Scand.*, **9**, 1044 (1955).

84. Inhibition of estrogen-induced proliferation of the vaginal epithelium in the rat by topical application of certain 4,4'-hydroxydiphenylalkanes and related compounds. (ES with E. BÁRÁNY, P. MORSING, W. MÜLLER and G. STÄLLBERG) *Acta Soc. Med. Ups.*, **60,** 68 (1955).

85. Constituents of tall oil. Part I. The nature of "carnauba acid" from pitch wood. (ES with H. BERGSTRÖM and R. RYHAGE) *Sven. Papperstidn*, **59,** 593 (1956).

86. Information about the structure of phthiocerol from mass spectral and X-ray crystallographic data. (ES with R. RYHAGE and E. V. SYDOW) *Acta Chem. Scand.*, **10,** 158 (1956).

87. *Masspektrometri som hjälpmedel vid organisk-kemiska strukturbestämningar*. p. 59 (ES) 9. Nordiske Kemikermöde i Aarhus, 1956.

88. Synthesis of racemic methyl phthienoate. (SS, ES with C. ASSELINEAU and J. ASSELINEAU) *Acta Chem. Scand.*, **10,** 478, (1956).

89. Synthesis of racemic methyl C_{27}-phthienoate. Part II. (SS, ES with C. ASSELINEAU and J. ASSELINEAU) *Acta Chem. Scand.*, **10,** 1035 (1956).

90. Fast-writing precision apparatus for continuous recording of electrolytic resistance. (ES with C.-O. ANDERSSON and O. MELLANDER) *Acta Chem. Scand.*, **10,** 1317 (1956).

91. Significance of bronchospirometric values. (SS with G. BIRATH and E. W. SWENSON) *Am. Rev. Tuberc. Pulm. Dis.*, **75,** 699 (1956).

92. Mass spectrometric determination of the structure of phthiocerol. (ES with R. RYHAGE and E. V. SYDOW) *Acta Chem. Scand.*, **11,** 180 (1957).

93. Mass spectrometric studies of long chain methyl esters. A determination of the molecular weight and structure of mycocerosic acid. (ES with J. ASSELINEAU and R. RYHAGE) *Acta Chem. Scand.*, **11,** 196 (1957).

94. Constituents of tall oil. Part II. The nature of a solid hydrocarbon fraction of the so-called B-oil. (ES with H. BERGSTRÖM and R. RYHAGE) *Sven. Papperstidn.*, **60,** 96 (1957).

95. Synthesis of the *cis*- and *trans*-isomers of methyl 2,4L,21,21-tetramethyl-$\Delta^{2:3}$-docosenoate. (ES with I. HEDLUND-STOLTZ) *Acta Chem. Scand.*, **11,** 405 (1957).

96. Quantitative mass spectrometric analysis of mixtures of unsaturated and saturated fatty acids. (ES with B. HALLGREN and R. RYHAGE) *Acta Chem. Scand.*, **11,** 1064 (1957).

97. Servo-controlled Wheatstone bridge arrangements for the simultaneous recording of resistance and capacitance. (ES with C.-O. ANDERSSON and F. MÖHL) *Acta Chem. Scand.*, **12,** 415 (1958).

98. Mass spectrometric evidence regarding the structural relations between dextropimaric, *iso*dextropimaric, and cryptopimaric acids. (ES with H. H. BRUUN and R. RYHAGE) *Acta Chem. Scand.*, **12,** 789 (1958).

99. Mass spectrometric studies on bile acids and other steroid derivatives. (ES with S. BERGSTRÖM and R. RYHAGE) *Acta Chem. Scand.*, **12,** 1349 (1958).

100. Mass spectrometric studies on esters of rosin esters. (ES with H. H. BRUUN and R. RYHAGE) *Acta Chem. Scand.*, **12,** 1355 (1958).

101. Synthesis of the *cis*- and *trans*-isomers of *erythro*-2,4,6-trimethyl-$\Delta^{2:3}$-tetracosenoic acid. (SS, ES with L. AHLQUIST, C. ASSELINEAU and J. ASSELINEAU) *Ark. Kemi*, **13,** 543 (1958).

102. Mass spectrometric studies. I. Methyl esters of saturated normal chain carboxylic acids. (ES with R. RYHAGE) *Ark. Kemi*, **13,** 523 (1959).

103. Structure of phthiocerol. (SS, ES with H. DEMARTEAU-GINSBURG, E. LEDERER and R. RYHAGE) *Nature*, **183,** 1117 (1959).

104. Synthesis and configuration of the 3,5-dimethylpimelic acid and the four optically active methyl hydrogen 3,5-dimethylpimelates. (SS, ES with L. AHLQUIST, J. ASSELINEAU, C. ASSELINEAU and K. SERCK-HANSSEN) *Ark. Kemi*, **14,** 171 (1959).

105. An infrared spectroscopic study of the structural relationships of dextropimaric, *iso*dextropimaric, and cryptopimaric acids. (ES with H. H. BRUUN and I. FISCHMEISTER) *Acta Chem. Scand.*, **13,** 379 (1959).

106. Studies on phthiocerol. I. Mass spectrometric investigation of phthiocerol and certain structurally related compounds. (ES with L. AHLGUIST, R. RYHAGE and E. V. SYDOW) *Ark. Kemi*, **14,** 211 (1959).

107. Studies on phthiocerol. II. The nature of the acidic products obtained on oxidation of phthiocerol by chromic acid. (SS, ES with R. RYHAGE) *Ark. Kemi*, **14,** 247 (1959).

108. Studies on phthiocerol. III. Identification of phthiocerane as a mixture of 4-methyldotriacontane and 4-methyltetratriacontane. (SS, ES with R. RYHAGE) *Ark. Kemi*, **14,** 259 (1959).

109. Synthesis of ($-$)-methyl 2D, 4D, 6D-trimethylnonacosanoate and identification of C_{32}-mycocerosic acid as a 2,4,6,8-tetramethyloctacosanoic acid. (SS, ES with C. ASSELINEAU, J. ASSELINEAU and R. RYHAGE) *Acta Chem. Scand.*, **13,** 822 (1959).

110. A surface-balance study of the structural relationships between cryptopimaric, detropimaric and *iso*dextropimaric acids. (ES with H. H. BRUUN) *Acta Chem. Scand.*, **13,** 832 (1959).

111. The mass spectra of methyl oleate, methyl linoleate and methyl linolenate. (ES with B. HALLGREN and R. RYHAGE) *Acta Chem. Scand.*, **13,** 845 (1959).

112. The carbon skeleton of lagosin (antibiotic A 246). (SS, ES with M. L. DHAR, V. THALLER, M. C. WHITING and R. RYHAGE) *Proc. Chem. Soc.*, 154, (1959).

113. Mass spectrometric studies. II. Saturated normal long chain esters of ethanol and higher alcohols. (ES with R. RYHAGE) *Ark. Kemi*, **14,** 483 (1959).

114. Mass spectrometric studies. III. Esters of saturated dibasic acids. (ES with R. RYHAGE) *Ark. Kemi*, **14,** 497 (1959).

115. On the chemistry of tobacco smoke. (ES) *Acta Soc. Med. Ups.*, **54,** 322 (1959).

116. Mass spectrometric studies. IV. Esters of monomethyl-substituted long chain carboxylic acids. (ES with R. RYHAGE) *Ark. Kemi*, **15,** 291 (1960).

117. Mass spectrometric studies. V. Methyl esters of monomethyl-substituted acids with ethyl or longer side chain and methyl esters of di- and polyalkyl-substituted acids. (ES with R. RYHAGE) *Ark. Kemi*, **15,** 333 (1959).

118. Fettsyrasammansättningen i några födoämnen. (SS with B. HALLGREN) *Nord. Med.*, **63,** 732 (1960).

119. Détermination de la position des doubles liaisons carbone–carbone par spectrométrie de masse. I. Double liaison dans les esters méthyliques des acides cis-pétrosélinique, oléique et élaidique. (SS with NG. DINH-NGUYEN and R. RYHAGE) Ark. Kemi, **15**, 433 (1960).

120. Gas chromatographic analysis of the fatty acid composition of the plasma lipid in normal and diabetic subjects. (SS with B. HALLGREN, A. SVANBORG and L. SVENNERHOLM) J. Clin. Invest., **39**, 1424 (1960).

121. Mass spectrometric studies. VI. Methyl esters of normal chain oxo-, hydroxy-, methoxy- and epoxy-acids. (ES with R. RYHAGE) Ark. Kemi, **15**, 545 (1960).

122. Mass spectrometry in lipid research. (ES with R. RYHAGE) J. Lipid Res., **1**, 361 (1960).

123. Mass spectrometric studies of hop bitter substances. (ES with S. BROHULT, R. RYHAGE and L.-O. SPETZIG) Eur. Brewery Convention, p. 121. Elsevier, Amsterdam, 1960.

124. Synthèses dans le domaine de l'acide mycocérosique. (SS, ES with L. AHLQUIST, C. ASSELINEAU and J. ASSELINEAU) Bull. Soc. Chim., 1316 (1960).

125. Massenspektrometrie als Hilfsmittel bei der Strukturbestimmung organischer Verbindungen, besonders bei Lipiden und Peptiden. (ES) Z. Anal. Chem., **181**, 462 (1961).

126. Mass spectrometric studies. VII. Methyl esters of α,β-unsaturated long chain acids. On the structure of C_{27}-phthienoic acid. (SS, ES with R. RYHAGE) Ark. Kemi, **18**, 179 (1961).

127. Arterial oxygen, carbon dioxide, and pH levels in patients undergoing pulmonary resection. (SS with E. W. SWENSON and M. BECK) J. Thorac. Cardiovasc. Surg., **42**, 179 (1961).

128. Mass spectrometric studies. VIII. A study of the fragmentation of normal long chain methyl esters and hydrocarbons under electron impact with the aid of deuterium-substituted compounds. (SS, ES with NG. DINH-NGUYEN and R. RYHAGE) Ark. Kemi, **18**, 393 (1961).

129. Mass spectrometric studies on sterols and bile acids. (ES with S. BERGSTRÖM and R. RYHAGE) Sven. Kem. Tidskr., **73**, 566 (1961).

130. Grindelic and oxygrindelic acids. (ES with T. BRUNN and L. M. JACKMAN) Acta Chem. Scand., **16**, 1675 (1962).

131. Mass spectrometric studies. IX. Methyl and ethyl esters of some aliphatic α-amino acids. (SS, ES with C.-O. ANDERSSON and R. RYHAGE) Ark. Kemi, **19**, 405 (1962).

132. Mass spectrometric studies. X. Methyl and ethyl esters of N-acetylamino acids. (ES with C.-O. ANDERSSON and R. RYHAGE) Ark. Kemi, **19**, 417 (1962).

133. Mass spectrometric demonstration of the absence of $N \rightarrow O$ acyl shift in N-acetylserine treated with anhydrous formic acid. (ES with G. FÖLSCH and R. RYHAGE) Ark. Kemi, **20**, 55 (1962).

134. Om tobaksrökningens kemi (On the chemistry of tobacco smoke). (ES) In Svensk Naturvetenskap, NFR, p. 350. B. LUNDHOLM Ed. 1962.

135. Mass spectrometry of long chain esters. (ES) In Mass Spectrometry of Organic Ions, Chapter 9, p. 399. F. W. MCLAFFERTY, Ed., Academic Press, New York, N.Y., 1963.

136. The higher saturated branched chain fatty acids. (SS, ES with S. ABRAHAMSSON) Progress in the Chemistry of Fats and Other Lipids, Part I, Vol. 7, R. T. HOLMAN, Ed. Pergamon Press, Oxford, 1963.

137. Mass spectrometric studies. XI. On the nature of the ions of m/e $84 + n \times 14$ $(n = 0,1,2...)$ present in mass spectra of esters of dibasic acids. (ES with R. RYHAGE) Ark. Kemi, **23**, 167 (1964).

138. Jetziger Stand der Massenspektrometrie in der organischen Analyse. (ES) Z. Anal. Chem., **205**, 109 (1964).

139. The chemistry and physical chemistry of unsaturated fatty acids. (ES) Symposia of the Swedish Nutrition Foundation **IV**, p. 9, 1934.

140. Location of double bonds by mass spectrometry. (ES with G. W. KENNER) Acta Chem. Scand., **18**, 1551 (1964).

141. Biokemi (Biochemistry). (ES) In Svensk Naturvetenskap, p. 382. NFR, B. LUNDHOLM Ed., 1964.

142. A device for analysis of organic compounds with the use of a "jet separator". (ES) Swed. Pat. Appl. 2009 (1964).

143. Mass spectrometry in organic structure determination. A problem of storage and identification. (SS, ES with S. ABRAHAMSSON) Biochem. J., **92**, 2 (1964).

144. Fatty acid composition of human brain sphingomyelins: normal variation with age and changes during myelin disorders. (SS with L. SVENNERHOLM) J. Lipid Res., **6**, 146 (1965).

145. Deuteriation of organic compounds. (ES with NG. DINH-NGUYEN) Swed. Pat. Appl. 1280 (1965); Fr. Pat. 1466088; Swiss Pat. 1201; Jap Pat. 5255; Br. Pat. 1103607; W. Ger. Pat. D. 49213; Can. Pat. 950514; U.S. Pat. 522030.

146. Strukturbestämning av organiska föreningar med hjälp av masspektrometri. (ES with R. RYHAGE) In 20 års medicinsk forskning, p. 228 The Swedish Medical Research Council, Stockholm (1966).

147. Studies on natural odoriferous components. I. Identification of macrocyclic lactones as odoriferous components of the scent of the solitary bees Halictus calceatus Scop. and Halictus albipes F. (SS with C.-O. ANDERSSON, G. BERGSTRÖM and B. KULLENBERG) Ark. Kemi, **26**, 191 (1966).

148. Fatty acid composition of sphingomyelins in blood, spleen, placenta, liver, lung and kidney. (SS with E. SVENNERHOLM and L. SVENNERHOLM) Biochim. Biophys. Acta, **125**, 60 (1966).

149. A convenient process for the synthesis of organic compounds of high deuterium content. (ES with NG. DINH-NGUYEN) Acta Chem. Scand., **20**, 1423 (1966).

150. Photoelectron spectroscopy of fatty acid multilayers. (ES with K. LARSSON, C. NORDLING and K. SIEGBAHN) Acta Chem. Scand., **20**, 2880 (1966).

151. Mass spectrometry in biochemical research. (ES) Chimia, **20**, 346 (1966).

152. Kemisk kommunikation inom insektsvärlden. (SS, ES) Sven. Kem. Tidskr., **80**, 178 (1968).

153. Changes in the fatty acid composition of cerebrosides and sulfatides of human nervous tissue with age. (SS with L. SVENNERHOLM) J. Lipid Res., **9**, 215 (1968).

154. Studies on natural odoriferous compounds. II. Identification of a 2,3-dihydrofarnesol as the main component of the marking perfume of male bumble bees of the species Bombus terrestris L. (SS, ES with G. BERGSTRÖM and B. KULLENBERG) Ark. Kemi, **28**, 453 (1968).

155. Atlas of Mass Spectral Data, Eds: ES with S. ABRAHAMSSON and F. W. MCLAFFERTY, Wiley-Interscience, New York, 1969.

156. Some recent studies on the structural arrangement of lipids in surface layers and interphases. (SS, ES with K. LARSSON and M. LUNDQUIST) *J. Colloid Interface Sci.*, **29**, 268 (1969).

157. The absolute configuration of terrestrol. (SS) *Acta Chem. Scand.*, **24**, 358 (1970).

158. Volatile components of the cephalic marking secretion of male bumble bees. (SS with B. KULLENBERG and G. BERGSTRÖM) *Acta Chem. Scand.*, **24**, 1481 (1970).

159. Gas liquid chromatography–mass spectrometry combination. (SS, ES) *Topics in Organic Mass Spectrometry*, p. 167, A. L. BURLINGAME, Wiley, 1970.

160. Stereospecific total synthesis of mycocerosic acids. (ES with K. WAERN and G. ODHAM) *Ark. Kemi*, **31**, 533 (1970).

161. On the chemistry of preen gland waxes of waterfowl. (ES with G. ODHAM) *Acc. Chem. Res.*, **4**, 121 (1971).

162. Specific protium labelling of perdeuteriated compounds. (ES with NG. DINH-NGUYEN and A. RAAL) *Chem. Scr.*, **1**, 117 (1971).

163. Studies on natural odoriferous compounds. III. Synthesis of (+)- and (−)-3,7,11-trimethyl-6-*trans*, 10 dodecadien-1-ol. (SS with L. AHLQVIST) *Acta Chem. Scand.*, **25**, 1685 (1971).

164. Studies on natural odoriferous compounds. IV. The synthesis of (−)-3,7,11,15-tetramethylhexadeca-6-*trans*, 10-*trans*, 14-trien-l-ol and its enantiomer. (SS with L. AHLQVIST. B. OLSSON and A.-B. STÅHL) *Chem. Scr.*, **1**, 237 (1971).

165. Some early examples of the application of mass spectrometry to the elucidation of the structure of complex molecules of biological origin. (ES) in Chapter I "A historical survey of mass spectrometry" of *Biochemical Applications of Mass Spectrometry*, pp. 11–19, G. WALLER Ed. Wiley, 1972.

166. Fatty acids. (ES) In *Biochemical Applications of Mass Spectrometry*, Chapter 8, pp. 211–228, G. WALLER Ed. Wiley, 1972.

167. Complex lipids (ES with G. ODHAM) In *Biochemical Applications of Mass Spectrometry*, Chapter 9, pp. 229–249. G. WALLER Ed. Wiley, 1972.

168. Perdeuteriated organic compounds. I. Normal long chain saturated deuteriocarbons, monocarboxylic acids and methyl esters. (ES with NG. DINH-NGUYEN and A. RAAL) *Chem. Scr.*, **2**, 171 (1972).

169. Studies on natural odoriferous compounds. V. Splitter-free all glass intake system for glass capillary gas chromatography of volatile compounds from biological material. (SS) *Chem. Scr.*, **2**, 97 (1972).

170. The Ecological Station of Uppsala University on Öland 1963–1973. (ES with B. KULLENBERG) *Zoon*, Suppl. 1, 5 (1973).

171. Analytical techniques in pheromone studies. (SS, ES with G. BERGSTRÖM) *Zoon*, Suppl. 1, 77. (1973).

172. Very long hydrocarbon chains. II. The synthesis of normal chain fatty acids with 41, 50, 60 and 69 carbon atoms and a mass spectrometric study of these acids and related compounds. (ES with G. STÄLLBERG and A. RAAL) *Chem. Scr.*, **3**, 125 (1973).

173. The use of optically active half-esters of methoxy-substituted succinic acids in the stereospecific synthesis of long chain oxygenated compounds. (ES with G. ODHAM and B. PETTERSSON, *Acta Chem. Scand.*, **B28**, 36 (1974).

174. Studies on natural odoriferous compounds. VII. Recognition of two forms of *Bombus lucorum* L. (Hymenoptera, Apidae) by analysis of the volatile marking secretion from individual males. (SS with G. BERGSTRÖM and B. KULLENBERG) *Chem. Scr.*, **4**, 174 (1974).

175. *Registry of Mass Spectral Data*. Eds: ES with S. ABRAHAMSSON and F. W. MCLAFFERTY, Wiley-Interscience, New York, 1974.

176. The chemical composition of the free-flowing secretion of the preen gland of the dipper. (*Cinclus cinclus*). (ES with O. BERTELSEN, B. ELIASSON, G. ODHAM) *Chem. Scr.*, **8**, 5 (1975).

177. Perdeuteriated organic compounds. II. Normal chain saturated deuteriocarbons. (ES with NG. DINH-NGUYEN and A. RAAL) *Chem. Scr.*, in preparation.

Prog. Chem. Fats other Lipids. Vol. 16, pp. 9–29. Pergamon Press, 1978. Printed in Great Britain

QUANTITATIVE CHEMICAL TAXONOMY BASED UPON COMPOSITION OF LIPIDS

RALPH T. HOLMAN

The Hormel Institute, University of Minnesota, 801 16th Avenue N.E., Austin, Minnesota 55912

CONTENTS

I. DEVELOPMENT OF THE METHOD 10
   A. Preliminary approach 10
   B. Pseudo three-dimensional matrix 11
   C. Equivalence of characters 12
   D. Index of relationship, *R* 13
   E. Display of relationships in *Magnolia* by the new procedure 13
   F. Variable threshold values 14
   G. Dissimilarity index, $1 - R$ 14
   H. Rearrangement of sequential order 17
   I. Population studies 17
   J. Environmental and nutritional conditions 18
   K. Other uses of indices of relationship 19

II. TAXONOMIC RELATIONSHIPS BASED UPON LIPIDS 20
   A. Hydrocarbons and terpenoids 20
   B. Fatty acids 20
   C. Waxes from preen glands of birds 25
   D. Phospholipids 25
   E. Cyanolipids 25

III. CONCLUSION 28

ACKNOWLEDGEMENTS 28

REFERENCES 28

Here and there (especially in vegetable seed fats), some natural fats, or fats from a group of biologically related organisms, are found to contain in combination some particular fatty acid which has been found in no other instance in nature; but this is on the whole decidedly exceptional. The general case is that a number of higher fatty acids occur continually throughout nature. The consequence is that the differences between one fat and another depend very largely on the varying proportions of the fatty acids in combination in the different fats, as well as upon the particular acids which happen to be components. The study of the natural fats is therefore somewhat differently placed from that of many other groups of naturally occurring organic compounds in that it must be conducted on a quantitative basis rather than solely on a qualitative basis.[17]

T. P. HILDITCH

This quotation taken from the first chapter of the first edition of the *Chemical Constitution of Natural Fats* published in 1940 was intended to state the case for chemical taxonomy based upon fat analysis. Although Hilditch arranged his monumental work according to the traditional taxonomic classification system, composition of fats has not been an especially useful tool in taxonomy. Nearly a generation later in a masterful treatment of biochemical systematics, Alston and Turner[4] commented that they knew of no data concerning the fat composition of species applied to the solution of a taxonomic or phylogenetic problem. This still appears to be the condition, for although there is a plethora of analytical data on lipid composition arranged comparatively, the leap from comparative tabulation to taxonomic meaning has not been made.

One contributory reason is that the emphasis has been to discover unique compounds which can be used as taxonomic labels for each species. Unfortunately, there are not enough possible fatty acids or even unique lipid entities to serve as unique markers for all the species which occur in nature! The general case, as pointed out by Hilditch, is that the same fatty acids occur repeatedly in the lipids of fats of plants, animals and microorganisms. In the evolution of species, the change in metabolism and therefore composition has not been so much the development of unique metabolic pathways and molecular structures in each instance as it has been to develop control mechanisms which govern proportions of a limited number of common molecules. If the fats and lipids which are major structural and functional entities in nature are to find use in taxonomic arrangements, analyses must be done on a quantitative basis, and the *patterns of composition must be the basis for quantitative comparisons rather than reliance only upon the unique and occasional occurrences of marking compounds.* This requires a mathematical or quantitative treatment of large bodies of complex data, and the taxonomic phase of a study, as well as the chemical analysis, "must be conducted on a quantitative basis rather than solely on a qualitative basis".

Before any analytical or conceptual method can be used to solve a problem, it must be tested against previous methods or current concepts. Any system of numerical taxonomy based upon lipid analysis must thus be compared with traditional taxonomic classifications. It must be tested in the general case before it can be extrapolated to the solution of a unique problem. It will be the purpose of this brief review to offer a method of treatment and presentation of analytical data based upon lipids, and to test it on several bodies of data available from the literature, with the intent to deduce whether or not the system and concept may have any general applicability. Examples will be shown from a wide range of lipid types and from a wide range of taxa. From a chemist's point of view, these presentations seem to reveal relationships which are obscured in tabular presentations. However, it must be left to those trained in taxonomy to decide whether the approach is valid and useful.

I. DEVELOPMENT OF THE METHOD

A. *Preliminary Approach*

In a recent study of the composition of floral odors of species of *Magnolia*, gas chromatographic analyses and mass spectrometric identifications of most components were made.[43] The compounds encountered included methyl esters of common fatty acids, normal hydrocarbons and a group of terpenes and aromatic compounds, all of which are volatile lipids of low molecular weight. The partial relationships of species to species based upon similar patterns of these compounds were considered to be equal to the sum of coincidences of compounds in the two species and weighting was given for proportions of substances in the two species. Some of the substances could be identified only as belonging to the terpene, hydrocarbon, or methyl ester groups. Inasmuch as compounds of the same class occurring in two species must bear some metabolic relationship to each other,[42] terms were added to the equation for total quantities of terpenes, hydrocarbons and methyl esters. The equation used in that investigation was as follows:

$$R_{x,y} = \left(1 + \frac{C_x}{C_y}\right)_a + \left(1 + \frac{C_x}{C_y}\right)_b \cdots + \left(1 + \frac{C_x}{C_y}\right)_n + \left(\frac{C_x}{C_y}\right)_h + \left(\frac{C_x}{C_y}\right)_m + \left(\frac{C_x}{C_y}\right)_t.$$

In this equation, R is relationship index between two species x and y, C is concentration and a through n are individual compounds. For each compound, a value of 1 was assigned for coincidence of the compound in the two species and a factor was added equal to the minor ratio of concentrations of that compound in the two species. Terms were also added for the minor ratios of total hydrocarbon content, h, total methyl ester content, m, and total terpenes, t, in the two species. Indices so calculated were used to express degree of relationship between species based upon floral odors alone, and a traditional form of graphic presentation attempted to show via broadness of lines between pairs of species the relative relationships between them. This chart is shown in Fig. 1. Although for a total of nine species, the chart is clear and informative, it is apparent that if the number of species rises much above this number, this mode of presentation would become confusing and useless. In subsequent consideration of the problem of presenting quantitative data for taxonomic purposes, we have come to realize that both the calculation of index of similarity and the mode of graphic presentation of the data were inadequate.

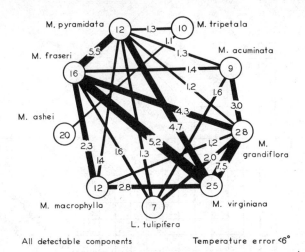

FIG. 1. Graphic presentation of quantitative relationships between species of *Magnolia* and *Liriodendron*. Values on lines connecting two species indicate the degree of relationship as does the width of the line. Values within the circle for each species indicates its composite relationship to the group.[43]

During the investigation of species of *Magnolia*, it became apparent that correlations could be constructed between species even when similar or identical components of the floral odors were unidentified. Preliminary correlations were made by comparing the five major peaks in the mass spectrum of an odor constituent plus the time of emergence of the components from the gas chromatograph with comparable data from a constituent of another species. Similarities thus calculated in a preliminary fashion on unidentified components gave approximately the same interrelationships between species as were ultimately obtained through mass spectrometric identification of the compounds.[43]

B. *Pseudo Three-dimensional Matrix*

When this latter approach was applied to the components of the fragrances of the genera *Epidendrum* and *Encyclia* in Orchidaceae, the number of species treated was much larger than in the study of *Magnolia* requiring that the traditional square matrix be used for graphic presentation of the results. The data for this study[19] are shown in Fig. 2 in which the species have been arranged in order of their taxonomic relationships according to R. L. Dressler. Open circles indicate similarity of components of the odors from two species, deduced from coincidence of four of the five mass spectral peaks from each, with an average error of 10–20% in intensity of the five mass spectral peaks. Solid circles indicate probable identity based upon coincidence of all five spectral peaks with an average error in intensity of less than 10%. The presentation of such data in half of a traditional square matrix, whether the values for identity or relationship are symbols or numbers, is in reality a quasi three-dimensional presentation, the dimensions of which are species × species × intensity.

Even though data for species are displayed in a matrix, most viewers have difficulty in translating enough numbers or symbols into vectors at one time to be able to perceive

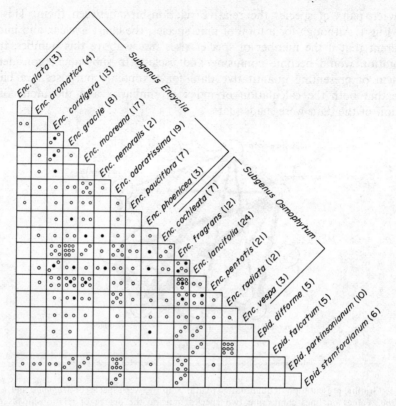

FIG. 2. Presentation of degrees of relationship between species of *Epidendrum* (*Encyclia*) in traditional matrix form.[19] Open circles indicate similarity of mass spectra of a component common to two species and hence similarity of structure. Solid circles indicate coincidence of 5 major peaks of the mass spectra of a component common to the two species, indicating probable identity.

the inherent pattern in the data. Therefore, a means was sought to make the intensity value more readily apparent as a line or bar in the third dimension. Using the PDP-12 computer and the FOCAL language, programs were written to graphically present a pseudo three-dimensional bar graph with the intensity value as the vertical dimension.

C. *Equivalence of Characters*

During the development of programs for the handling of gas chromatographic data and the calculation of the indices of relationship between all individuals or species in a group, the formula for calculation of the index, *R*, was re-evaluated. The concept of unequal weighting of characters in the calculation of similarities was questioned, and search of the controversial literature on numerical taxonomy[1,39] confirmed our opinion that characters belonging to a group of chemical compounds to be used in numerical formulations should not be weighted initially but should be given equal importance.[34] There is no *a priori* reason why one member of a group of substances should be assigned more importance than another, but perhaps *a posteriori* weighting of characters may be justified.[1] This is seemingly contrary to much previous practice in morphological taxonomy in which many characters common to a group of species are ignored and others which are more unique are emphasized in the ranking of relationship. Of necessity, something must be ignored or left out of consideration because all microscopic and macroscopic morphological characters and all chemical substances cannot be evaluated by a single observer even on a single specimen. Linnaeus[35] chose to base his classification of higher plants largely upon floral characters, and his system is still the basis of modern taxonomy. In chemical taxonomy, selection of data must likewise be made because a total chemical analysis of all molecules present in a specimen is patently impossible. Investigators are limited to analysis of one class of compounds at a time and must select the most relevant class to study. However, within that group, no one compound is more important than another until a reason is found for assigning greater importance to one. In the examples which are to follow, lipids are being examined to the exclusion of carbohydrates, minerals, proteins and small water-soluble molecules, but within each group of lipids examined or analyzed, all components are given equal weight, for there is no logical reason at the outset to assign greater importance to some than to others. We do not imply that lipids should have more taxonomic importance than other classes of substances, but we explore the possibility that lipids too may contribute to interspecific relationships.[42] We do not imply that the relationship indices calculated in this study are *true* relationships, but only suggest that these indices based upon lipid composition may give insight into understanding true relationships. The approaches presented here should be applicable to any class of substances and to any characters which can be expressed numerically. True relationships obviously must be based upon all characters morphological and compositional, and true compositional chemical taxonomic expressions must be based upon composition of total substance of the organism. Calculated relationships based upon lipid composition are only one small window through which the reality of relationships may be perceived partially. It is one window which has heretofore been unused.

D. *Index of Relationship,* R

To achieve equal weight for each character, the index of relationship, *R*, was redefined. For each compound in the group under consideration, a comparison is made for a pair of species. If a compound is absent in one of the species, it cannot contribute to the quantified relationship between the two. If a compound is a minor component in both species, it can contribute only in a minor way toward the total relationship, but if it is a major component, its contribution to the relationship should be major. If a compound is a minor component in one species and a major component in the other, the coincidence of metabolic pathways for that compound is modulated by major differences in the regulation of that pathway in the two species, diminishing that relationship. Therefore, in the new equation, for each coincident compound in the two species, one term is included, and it is equal to the product of the minor ratio of the concentrations of that substance in the two species times the average percentage of that component in the two species. If the two samples consist of the same number of components in the same proportions, they are identical, and the sum of individual contributions to the relationship will be equal to one. Degrees of coincidence, intermediate between zero and 1, regardless of the number of components, are measures of degrees of relationship between the species. The formula developed is as follows:

$$R_{x,y} = \left(\frac{C_x}{C_y}\right)_1 \left(\frac{C_x + C_y}{200}\right)_1 + \left(\frac{C_x}{C_y}\right)_2 \left(\frac{C_x + C_y}{200}\right)_2 + \cdots + \left(\frac{C_x}{C_y}\right)_n \left(\frac{C_x + C_y}{200}\right)_n$$

in which *x* and *y* are the two species compared, *C* is concentration in percentage, and 1 through *n* are the components of the samples.

The values calculated by this means for the similarity of one species to all other species of a genus or for one individual to all others in a group are averaged to express the relationship to the genus or to the group as a whole. Values for $\Sigma R/n$ may be plotted beside the grid by the same FOCAL plotting program if desired.

E. *Display of Relationships in* Magnolia
by the New Procedure

The analytical data of the floral odors of the species of *Magnolia* and *Liriodendron* were recalculated using the new formula and were redrawn using the new graphic pro-

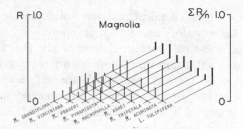

FIG. 3. Pseudo three-dimensional plot of R values for *Magnolia* and *Liriodendron* species using the same data from which Fig. 1 was generated.[43] Heavy bars to the right are average relationships to the group as a whole, $\Sigma R/n$.

cedure. The calculated indices, R, and the relationships to the group as a whole, $\Sigma R/n$, are depicted in Fig. 3. In this diagram, it is readily apparent that relationships between floral odors of some species are much more intense than for others. The composition of the floral odor of *M. acuminata* bears little relationship to that of any other *Magnolia* but bears strong relationship to that of *L. tulipifera*. In this sense, *M. acuminata* is least related to the group as a whole. *L. tulipifera* is strongly related to the *Magnolia* species and from point of view of floral odor as well as from floral structure, might well be included in the *Magnolia*. The increased clarity of perception of similarities through this mode of graphic presentation suggested that the method may also be successfully applied to other types of analytical or comparative information. It need not be limited to display of chemical data, but should be equally useful for display of morphological characters numerically expressed.

F. *Variable Threshold Values*

Recently a study has been made at The Hormel Institute of the unsaturated fatty acids occurring in the lipids of mosses.[5] In this study, seven individual unsaturated fatty acids were measured in twenty-two species of varieties of moss. Treatment of the published data by the formula given above and the computerized graphics procedure yielded the diagram shown in Fig. 4A. The interspecific indices of relationship for the several species of mosses were so similar that the perception of relationships and of group relationships was difficult. This example illustrates precisely the problem which has discouraged investigators from using fatty acid composition as a basis of comparing species. The occurrence of most of the measured components in most of the samples makes all the species seem to appear to be the same at first glance. Nevertheless, small variations in relationship exist, superimposed upon large similarities. By raising the threshold value for the plotting of the intensities of relationship, the pattern of relationship is easily perceived. In Fig. 4A, the entire range of values of R from 0 to 1 is plotted and the differences are obscured. In Fig. 4B, 0.3 has been subtracted from each value and the differences between species become more apparent. Thus, the general and equal relationships between species contributed by coincidence of many fatty acids in the patterns are deleted, leaving only the residual differences in relationships. This maneuver is equivalent to ignoring similarity and amplifying that portion of the composite character which differs. When this is done, it is more readily seen for example, that the *Polytrichum* species are indeed closely related to each other and are not well related to the *Rhacomitrium* species which in turn are strongly similar. By these criteria, *Mnium cuspidatum* and *M. medium* are closely related, but neither is closely related to *M. punctatum*, and the genus *Mnium* is poorly related to other genera. The setting of appropriate threshold values reveals differences between species in the midst of overwhelming similarities.

G. *Dissimilarity Index,* $1 - R$

Figure 5A shows the numerical relationships calculated between mammalian species on the basis of fatty acid composition of milk fats.[13,14] In this diagram, it is immediately

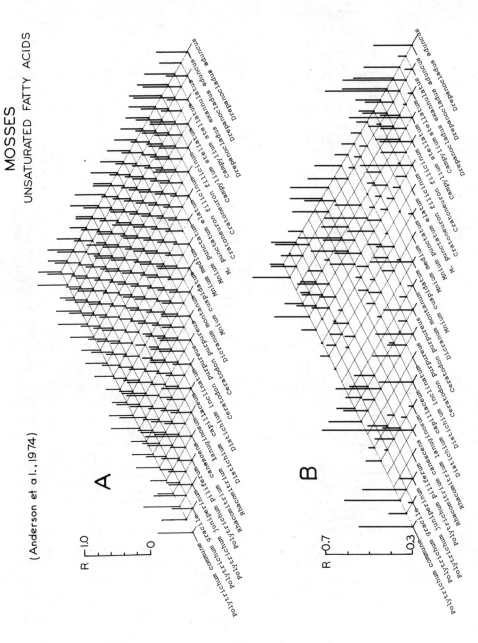

Fig. 4. Relationships between species of mosses based upon content of unsaturated fatty acids in total lipids.[5] (A) Index of relationship on a scale of 0 to 1. (B) Index of relationship on a scale of 0.3–0.7, revealing more clearly the differences between species.

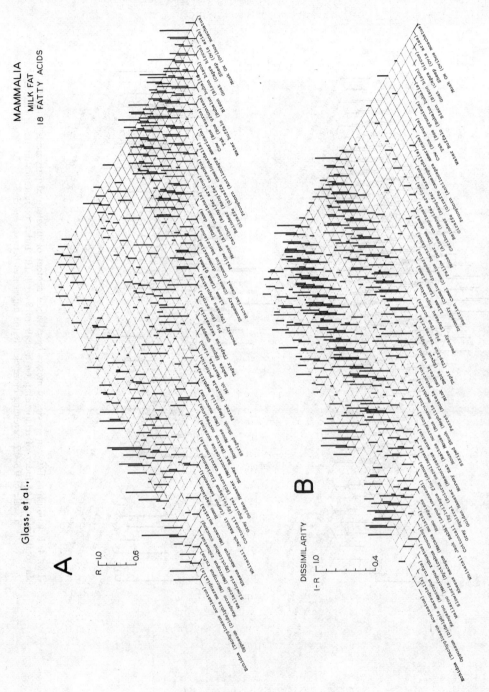

Fig. 5. Relationship between species of mammals based upon fatty acid composition of milk fats.[13,14] (A) Index of relationship on a scale of 0.6–1.0. (B) Index of dissimilarity, $1 - R$, on a scale of 0.4–1.0.

apparent that the marsupials at the far left corner of the triangle are rather well related to each other, but that in general they are poorly related to the ruminants which occur at the far right of the diagram, and which generally show strong relationship to each other. The two species of rabbits are closely related but show poor relationship to any other group. By plotting $1 - R$ rather than R in the vertical dimension in Fig. 5B, the inverse image of the upper figure is obtained. This presentation makes it possible to perceive at a glance dissimilarities between species and to identify the most remotely related species by proper manipulation of the threshold value.

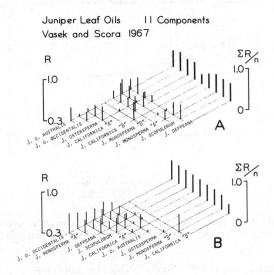

FIG. 6. Relationships between species of *Juniperus* based upon composition of leaf oils.[44] (A) Arranged in order as published by Vasek and Scora. (B) Arranged in order of decreasing relationship to genus.

H. *Rearrangement of Sequential Order*

In the pseudo three-dimensional plots thus far discussed, the arrangements of species were in the same order as were tabulated in the original publications, presumably according to known taxonomic relationships. The relationships between *Juniperus* species deduced from analyses of the leaf oils are shown in Fig. 6A.[44] In this chart, *Juniperus californica* B is placed beside *J. californica* A, because they are regarded as varieties of the same species. Their leaf oils do not bear much relationship to each other, nor do they bear equal relationships to the group as a whole. To regroup the species on the basis of relationships, a computer program was written which rearranged the file of data in decreasing order of $\Sigma R/n$ values. When the chart was redrawn with the rearranged data, *J. californica* B occurred at the extreme right, rather remote from *J. californica* A (Fig. 6B). This rearrangement also has the effect that within the triangular matrix, the species having the greatest relationships to each other lie in the left half. A variety of other rearrangements of species within the matrix are, of course, possible and can be performed as needed. Regrouping data by this and other means should lead to additional insights of quantified relationships.

I. *Population Studies*

Perhaps the most obvious extension of the index of relationships and the pseudo three-dimensional plot is for comparison of individuals within a population. Moreover, the inherent variability within a population ideally should be known before comparisons between populations or species are made, and the same techniques are applicable to populations. There is considerable current difference of opinion concerning advisability of making a single analysis as characteristic of a species, for in many cases, biological variation is significant. In many of the studies used as examples in this review, samples were composites of many individuals from the same population, and might be taken as representative of a population. In a study of *Picea* (Fig. 7), individual plants were sampled to test variation of individuals at different altitudes.[45] Of course, different altitudes may involve different populations developed because of different environments.

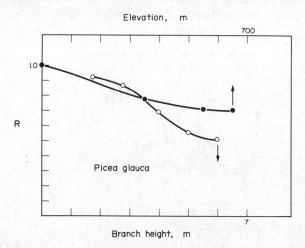

FIG. 7. Index of relationship used to relate samples of leaves from *Picea glauca* as a function of height in a single tree (open circles) or as a function of elevation (solid circles).[45]

In several of the studies cited here, the number of analyses or of samples of each species were not specified, and in these cases one must assume that single analyses were made. Even though single samplings are made and populational variations cannot be assessed, it is striking that the relationships indicated by the lipid analyses in most cases parallel accepted relationships based upon morphological characters. These experiments suggest that individual variations within a species are less than are interspecific differences. Nevertheless, assessment of biological variation should be a part of comparisons between species whenever feasible. Such feasibility must be rare, for in search of studies of interspecific differences based upon lipids, no example was found in which primary data was given for whole populations of each of several species. Analysis of a single sample for taxonomic purposes is tantamount to describing a species from a single type specimen. In studies in breadth, analysis of populations of each species may be neither feasible nor necessary.

In a study currently under way in our laboratory, the serum lipids of more than 200 hospital patients whose diagnosis was not metabolic diseases have been analyzed for fatty acid composition. Serum phospholipid fatty acid compositions from a subgroup of this population, 46 males and females from 20 to 39 years of age, were analyzed by the same programs described above. Indices of relationship, R, were calculated for individual—individual similarities, and $\Sigma R/n$ values were calculated for each individual's similarity to the population. The R values varied from 0.617 to 0.935 with an average of 0.805 and a standard deviation of 0.057. This indicates the variability of the population with respect to pattern of fatty acids of serum phospholipids, and indicates that R values must be lower than 0.75 for the individual to be considered abnormal. This approach and these data are finding use in screening humans for metabolic disease which affects serum lipid composition.

J. *Environmental and Nutritional Conditions*

Classification of microorganisms and primitve plants by their chemical composition suffers limitation because these organisms are notoriously responsive to changes in conditions of culture. In some cases, conditions of culture can cause greater differences in the composition of the organism than the differences between species cultured under the same conditions. The investigator who wishes to use chemical taxonomy with such species, must be aware of these phenomena and must carefully control culture conditions. This limitation likewise applies to some degree for most organisms because even the higher plants and animals are affected by nutrient and environmental conditions. This is illustrated by Fig. 7 in which differences in elevation were related to differences in leaf oils,[45] and in Fig. 8 in which the rats were subjected to different nutrient conditions.[18] Despite such hazards in the way of chemical taxonomy, it is remarkable that

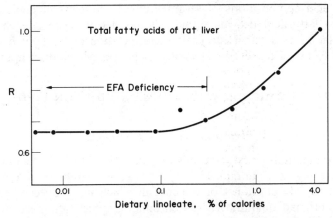

FIG. 8. Index of relationship used to relate fatty acids of liver lipids from rats fed different levels of the essential fatty acid, linoleic acid.[18]

in the case of mosses recognized taxa based upon morphology are distinguished by consideration of the chemical composition of the unsaturated fatty acids, and recognized taxa of bacteria were confirmed by fatty acid analyses. In Fig. 5, accepted taxonomy of mammalia is confirmed in many instances by the chemical differences in the milk fat composition, despite the known effect nutrition has upon composition of milk. Indeed, some of the differences between species of mammals are directly due to their different habitats and to their different eating habits, both of which influence the chemical composition of the milk fat as well as body composition. That is, the differences in lipid composition in response to nutritional and environmental influences are probably a significant component of the set of characters which distinguish species in their natural habitats.

K. *Other Uses of Indices of Relationship*

The index of relationship R is applicable to the comparison of any two bodies of data which have components in common. It could be applied to follow the total change in composition as a consequence of such variables as culture conditions, age,[2] nutrition, temperature, geography[11,12,27] or altitude,[45] season,[3,38,47] to name but a few. As an example, the study by von Rudloff of the effects of height of branch in a tree (age of tissue?) and of elevation of the tree upon the composition of the oil from the foliage of the white spruce, *Picea glauca*[45] was chosen. His primary data was available and has been reduced to indices of relationship and plotted against the relevant parameters in Fig. 7. In each case, a table of data has been reduced to one curve which demonstrates the degree of change of composition of the leaf oil. In the curve for branch height, the first point in the curve is the index of relationship between two samples from the same height of the same tree, and which bore a relationship to each other of 0.928. It appears from this comparison of data that the change in level within a single tree may cause greater changes in composition of leaf oils than does the effect of change in elevation.

Figure 8 shows the effect of changes in the level of dietary linoleate in rats upon index of relationship.[18] All values in this figure are compared with the value for animals fed the highest and adequate level of linoleate, and all values were calculated from analyses of several animals. Diminishing the level of linoleate in the diet changes the composition of the liver fatty acids, and at 0.1% of calories intake of linoleate, a plateau is reached at which the index of relationship is 0.66. Further decrease of the content of linoleate in the diet makes no further change in composition. In other words, the essential fatty acid deficiency is not worsened from a biochemical, compositional point of view.

In our laboratory, the index, R, is currently being used to detect aberrations from normal of fatty acid compositions of serum lipids. In humans, it depicts readily the onset of essential fatty acid deficiency, and should detect composite changes as the result of metabolic diseases which affect fatty acid metabolism. Plots of R values versus time also should be useful to observe physiologic changes which affect lipid metabolism.

II. TAXONOMIC RELATIONSHIPS BASED UPON LIPIDS

A. *Hydrocarbons and Terpenoids*

The literature on analysis of essential oils and the use of their terpenoid components in chemical taxonomy is voluminous,[1,4,46] but in our search for a typical body of data, we found a paucity of studies of terpenes in which many species were examined and from which original quantitative gas chromatographic data were available. Our study of *Magnolia*[43] (Fig. 3) is an example in which normal hydrocarbons, terpenoids and methyl esters of common fatty acids in floral odors found use in relating species. Figure 6 illustrates the use of essential oils from the foliage of plants for the quantification of relationships between species. The less volatile hydrocarbons of leaf waxes and the carotenoids have also been measured in comparative studies for purposes of classification.

Members of the Department of Chemistry of the University of Glasgow and their cooperators around the world have made a massive effort to use the composition of leaf hydrocarbons as a means of chemical classification of plants.[7-10] Figure 9A shows a comparison made of 54 species of Gymnosperms based upon the content of 30 hydrocarbons taken taken from a publication of this group.[7] This type of lipid analysis lends itself very well to taxonomic surveys, for leaf samples can be easily collected in the field and gas chromatographic analysis for these hydrocarbons are relatively easy to perform. Figure 9B reveals the calculated relationships between 48 species of algae based upon analysis for 47 carotenoids.[15,16,40,41] Three remotely related groups of algae are apparent. Within each group, relationships are strong and between the groups, the relationships are few and weak. These three groups of algae are clearly three different taxa.

B. *Fatty Acids*

A voluminous literature on fatty acid composition of microorganisms has been generated for use in description and classification of species. Figure 10 shows interspecific relationships calculated for two groups of microorganisms. Analysis of the lipids of 21 strains or species of spirochetes by Livermore reveals considerable differences between those strains.[26] The species of *Spirochaeta* listed at the left of Fig. 10A show strong relationships to each other. The *Treponema* species likewise show strong relationships to each other and little relationship to other species. The last six *Treponema* species listed bear lesser relationships to the group as a whole. Clearly the fatty acid compositions can distinguish between the *Spirochaeta* and the *Treponema* species. In Fig. 10B, the fatty acid compositions of 29 species of *Micrococci* and *Staphylococci* were used to calculate interspecific relationships.[22] The *Micrococci* group consists of at least three

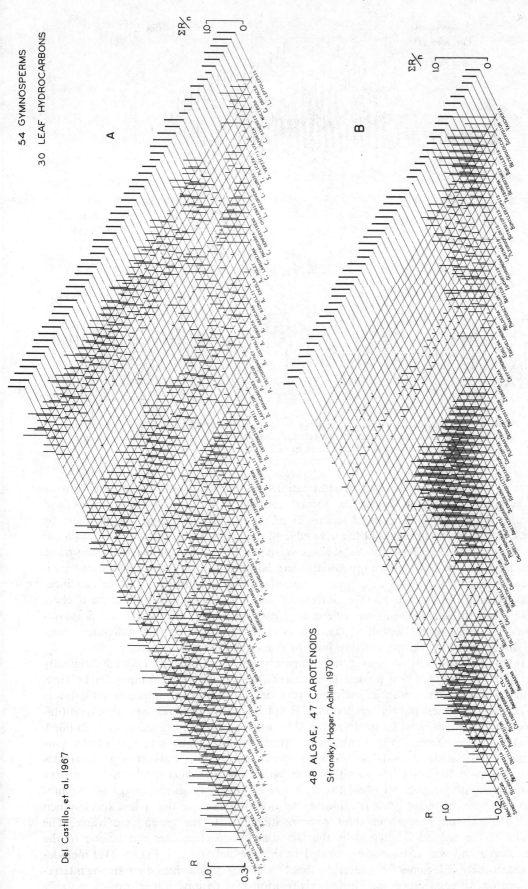

Fig. 9. (A) Relationships between 54 species of *Gymnosperms* on the basis of the content of 30 hydrocarbons in the leaf cuticular waxes.[7] (B) Relationships between 48 species of algae on the basis of their content of 47 carotenoids.

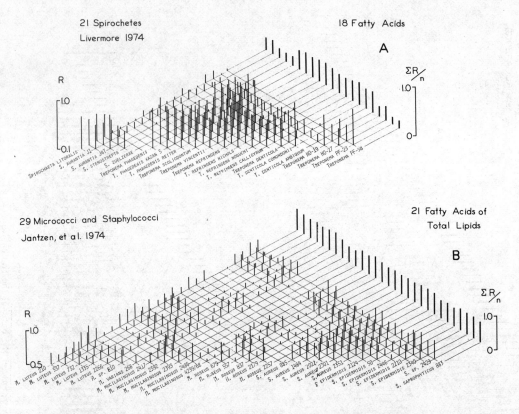

FIG. 10. (A) Relationships between 21 species or strains of spirochetes based upon content of 18 fatty acids in the total lipids.[26] (B) Relationships between 29 species of *Micrococci* and *Staphylococci* based upon content of 21 fatty acids in total lipids.[22]

distinct subgroups within which strong relationships occur. One of these groups consists of several strains of *M. luteus*, the second consists of *M. mucilaginosus* strains, and the third consists of the several strains of *M. roseus*. These groups are manifest by the clusters of tall bars lying at the near edge of the chart. Similarly, the *Staphylococcus aureus* strains, *S. epidermidis*, *Staphylocoeus* sp. 2429 and *S. saprophyticus* constitute one large group showing strong mutual interrelationships. These two illustrations indicate the usefulness of fatty acid composition of microorganisms as an aid to classification. Among the spirochetes, the differences are great enough so that even a chart with a low threshold reveals the differences clearly. With the *Micrococci* and *Staphylococci*, the relationships for all species are strong and similar so that differences must be distinguished by raising the threshold to approximately 0.5.

In Fig. 11, two sets of data quite comparable to the kind that Hilditch originally gathered on seed fats are treated to measure interspecific relationships. To be sure, modern gas chromatographic technique provides more precise information than was available to Hilditch, but the prediction of Hilditch that to reveal relationships, quantitative considerations must be made rather than only qualitative ones, has been fulfilled. Within the Cruciferae[29] (Fig. 11A), species from the same genus have very similar patterns of fatty acids in seed oils. *Brassica* species as a group show strong similarities in a cluster in the lower left portion of the figure. The *Lepidium* species (front center), likewise constitute a cluster of well related species. In the Cruciferae, most of the subgroups show moderate cross-relationship to other subgroups, but a few species such as *Lobularia* and *Tropaeolum* show poor relationships to the group as a whole. The utility of the method is attested by the fact that *Tropaeolum* does not belong to the Cruciferae and was erroneously included in the original study.[29] Figure 11B includes a wide variety of legumes from several tribes.[48] The *Caesalpinioideae* show strong interrelationships to each other and show cross-relationships to most other groups, notably

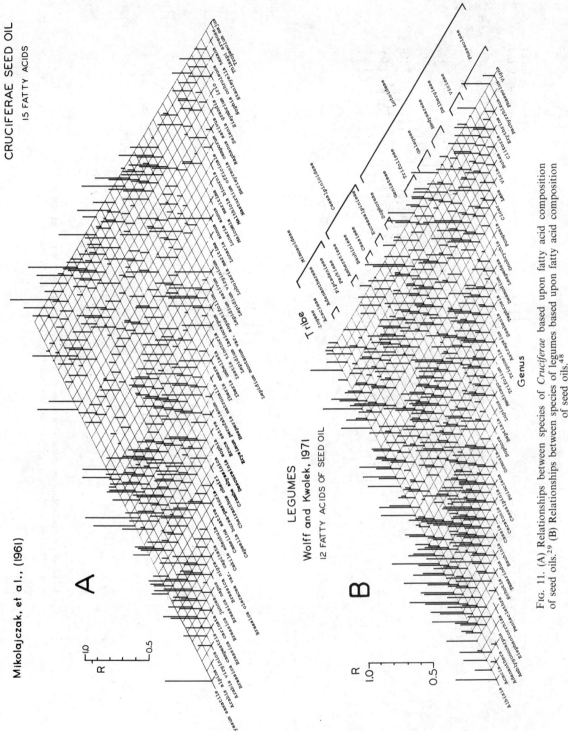

FIG. 11. (A) Relationships between species of *Cruciferae* based upon fatty acid composition of seed oils.[29] (B) Relationships between species of legumes based upon fatty acid composition of seed oils.[48]

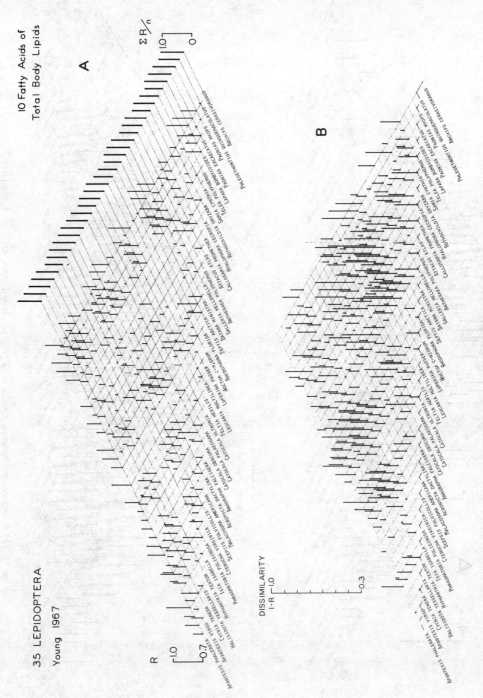

FIG. 12. (A) Relationships between 35 species of *Lepidoptera* based upon fatty acid composition of whole body lipids.[49] (B) Dissimilarity of species of *Lepidoptera*.

to the *Mimosoideae. Pongamia*, on the other hand, shows only a few strong similarities to other species, being rather poorly related to most. *Elephantorrhiza* likewise shows few and weak relationships to other groups.

The fatty acid composition of the whole body fats of 35 species of *Lepidoptera* were used as the basis for comparing and contrasting these species.[49] In Fig. 12A it is apparent that these species all bear strong and nearly equal relationships to the group as a whole. To display the differences that occur between species, it was necessary to raise the threshold to 0.7. This example involves no unusual marker fatty acids for any of the species, and provides an illustration that even with only the common fatty acids, relationships can be measured. This is also a study in which the use of the dissimilarity index, $1 - R$, revealed which species are the most unlike each other (Fig. 12B).

C. Waxes from Preen Glands of Birds

This subject as an illustration of quantitative chemical taxonomy is most appropriate because work in this field began in the laboratory of Professors Einar Stenhagen and Stina Stenhagen,[6,24,30–35] to whose memory this volume is dedicated. The uropygial glands of birds contain a wide variety of lipids of which wax esters are prominent. Odham and others in the Stenhagen laboratory undertook an investigation of the wax acids and wax alcohols in the uropygial glands of several birds, using a gas chromatograph coupled directly to a mass spectrometer. By this means, a large number of wax alcohols and wax acids were identified. In more recent years, Jacob and Poltz in Germany have continued this kind of investigation and have added a large number of birds to the list.[20,21,36,37] Figure 13 combines selected data from the Stenhagen laboratory and the laboratory of Jacob and Poltz. In Fig. 13A, the calculated interspecific relationships between 39 birds based upon their contents of 172 wax alcohols are presented in pseudo three-dimensional diagram. On this basis, the water birds including ducks, geese and swans, show strong relationship to each other but lesser relationships to other birds such as warblers and owls. In Fig. 13B, a similar presentation is made based upon the contents of 199 wax acids from the same wax ester fractions. In this diagram, three major groups of birds are apparent, even without elevation of the threshold value. These data illustrate the purpose of the original studies, that wax alcohols and wax acids of uropygial glands of birds should be useful as an adjunct to the classification of birds. Perhaps when the individual wax esters can be identified and measured, the interspecific interrelationships can be revealed even more clearly.

D. Phospholipids

Kaneko and his associates have measured the content of 12 different phospholipids in the total lipids of 31 species or strains of yeast[23] (Fig. 14A). Display of calculated interspecific relationships in the pseudo three-dimensional plot showed that on the basis of phospholipid composition most of the yeasts are very similar. To reveal any differences, the threshold value was raised to 0.7. In this topographic chart, the four species of *Saccharomyces* were very similar. *Cr. neoformans* and *Cr. laurentii* show lesser relationship to each other and few and weak relationships to the other species. *Rh. rubra*, strain AY–2 and strain 2–17–4B, show strong relationship to each other indicating near identity, but they show very weak relationship to all other species studied. These more obvious relationships suggest that detailed examination of the chart by taxonomists familiar with the yeasts may reveal yet other meaningful correlations.

E. Cyanolipids

The final illustration, Fig. 14B, involves an attempt at chemical taxonomy of 17 species of Sapindaceae based upon analysis of the seed oils for some newly discovered classes

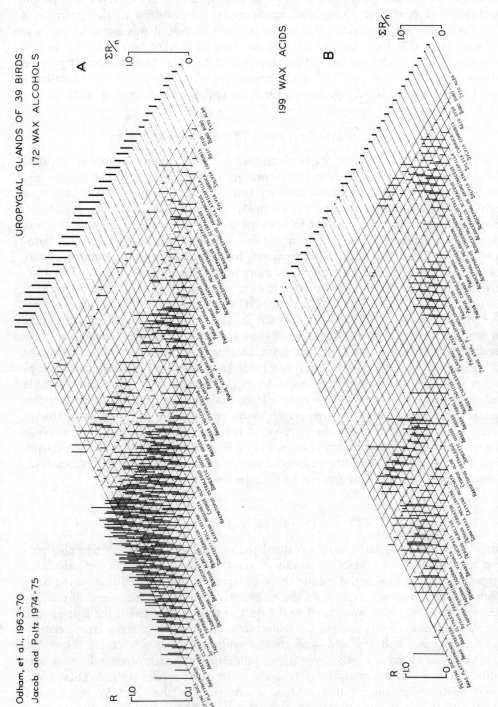

FIG. 13. Relationships between 39 species of birds based upon (A) 172 wax alcohols, and (B) 199 wax acids. 6,20,21,24,30-37

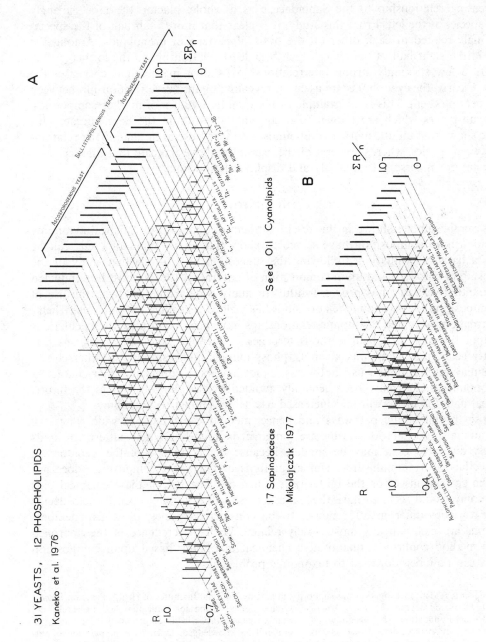

FIG. 14. (A) Relationships between 31 species and strains of yeast based upon their contents of 12 phospholipids.[23] (B) Relationships between 17 species of *Sapindaceae* calculated from the content of cyanolipids and triglycerides in the seed oils.[28]

of compounds, the cyanolipids.[28] Four different groups of cyanolipids occur, each containing a different nitrogen-containing moiety. The proportions of these compounds in seed oils of Sapindaceae range from traces to major proportions and triglycerides constitute the remainder. The data of Mikolajczak[28] have been used for calculation of R values between the 17 species thus far analyzed, and a three-dimensional plot of them is shown in Fig. 14B. In this plot, the species have been rearranged in order of decreasing relationship to the Sapindaceae as a whole, placing the most strongly related species at the left. From this study, it is clear that more than half of the species are strongly related to each other on the basis of pattern of cyanolipids. Among the species in the right half of the chart which bear least relationship to the former group, there are a few strikingly strong interrelationships, shown in the right corner of the triangle. Clearly, the cyanolipids are useful in revealing quantitative relationship between species bearing them. This is an example of the quantitative use of marker compounds, unique substances which are known to occur only thus far in the Sapindaceae. This example shows that unique marker compounds are indeed useful in quantifying relationship between species, whereas several of the examples above indicate that marker compounds are not necessary to quantify and display relationships.

III. CONCLUSION

The several examples shown in this brief treatment of quantitative chemical taxonomy based upon lipid composition have served to show that many types of lipids can be used as a basis of evaluation. Following the suggestion by Hilditch that study of the relationship of species based upon composition of natural fat (lipids) must be conducted on a quantitative basis, has proven fruitful. In many widely different taxa, the use of lipid composition for taxonomic relationships has yielded revealing and potentially helpful information. Accepted taxonomic relationships have been confirmed via calculations based upon the analysis of hydrocarbons, terpenes, carotenoids, total fatty acids, unsaturated fatty acids, wax acids, wax alcohols, phospholipids and cyanolipids. Other classes of lipids may yet find similar use. In the past, lipids have been considered poor characters for taxonomic use and have been generally avoided by taxonomists, but in the future, analytical data for lipids will find increased use as an adjunct to taxonomy.

Because the metabolic pathways and hence, metabolic products in wide ranges of plants, animals and microorganisms are similar, it is quite likely that patterns of lipids in widely divergent taxa may be similar. Because the body fats of the Lepidoptera and milk fats of Mammalia have similar patterns of fatty acid composition does not make the cow an insect or the butterfly a kangaroo. The use of these chemical aids to taxonomy would seem to have their greatest use within limited taxa, to help decide whether varieties differ greatly enough to be considered species, or which members of a genus are metabolically most closely related. With the credence of the approach and the method confirmed, quantitative chemical taxonomy based upon composition of lipids can now be addressed to taxonomic problems.

Acknowledgements—This investigation was supported in part by National Institutes of Health Program Project Grant HL 08214, by a grant from the American Orchid Society Fund for Education and Research, and by The Hormel Foundation. The ingenuity and assistance of Dale Jarvis in development of the plotter programs for the pseudo three-dimensional matrix is gratefully acknowledged. During the preparation of this report, helpful suggestions were made by the following reviewers: Gunnar Bergström, R. L. Dressler, Holger Erdtman, Shoji Gotoh, Hiroshi Kaneko, Brian Livermore, Rebecca Northen, Peter Raven, R. R. Sokal, Leonard Thien, B. L. Turner, and E. von Rudloff.

REFERENCES

1. ADAMS, R. P. *Brittonia*, **27**, 305–316 (1975).
2. ADAMS, R. P. and HAGERMAN, A. *Biochem. Syst. Ecol.*, **4**, 75–79 (1976).
3. ADAMS, R. P. and POWELL, R. A. *Phytochemistry*, **15**, 509–510 (1976).
4. ALSTON, R. E. and TURNER, B. L. *Biochemical Systematics*, Prentice-Hall, Englewood Cliffs, New Jersey, 1963.

5. ANDERSON, W. H., HAWKINS, J. M., GELLERMAN, J. L. and SCHLENK, H. *J. Hattori Bot. Lab.*, **38**, 99–103 (1974).
6. BERTELSEN, O. *Ark. Kemi*, **32**, 17–26 (1970).
7. DEL CASTILLO, J. B., BROOKS, C. J. W., CAMBIE, R. C., EGLINTON, G., HAMILTON, J. R. and PELLITT, P. *Phytochemistry*, **6**, 391–398 (1967).
8. DOUGLAS, A. G. and EGLINTON, G., *Comparative Phytochemistry* (Ed. T. SWAIN) pp. 57–77, Academic Press, London and New York, 1966.
9. EGLINTON, G., GONZALEZ, A. G., HAMILTON, R. J. and RAPHAEL, R. A. *Phytochemistry*, **1**, 89–102 (1962).
10. EGLINTON, G., HAMILTON, R. J., RAPHAEL, R. A. and GONZALEZ, A. G. *Nature*, **193**, 739–742 (1962).
11. FLAKE, R. H., VON RUDLOFF, E. and TURNER, B. L. *Proc. Natl. Acad. Sci.*, **64**, 487–494 (1969).
12. FLAKE, R. H., VON RUDLOFF, E. and TURNER, B. L. *Recent Adv. Phytochem.*, **6**, 215–228 (1973).
13. GLASS, R. A. and JENNES, R. *Comp. Biochem. Physiol.*, **38B**, 353–359 (1971).
14. GLASS, R. L., TROOLIN, H. A. and JENNES, R. *Comp. Biochem. Physiol.*, **22**, 415–425 (1967).
15. HAGER, A. and STRANSKY, H. *Arch. Mikrobiol.*, **72**, 68–83 (1970).
16. HAGER, A. and STRANSKY, H. *Arch. Mikrobiol.*, **76**, 77–89 (1970).
17. HILDITCH, T. P. *The Chemical Constitution of Natural Fats*, John Wiley, New York (1940).
18. HOLMAN, R. T. *Handbook of Nutrition*, CRC Press, in press (1977).
19. HOLMAN, R. T. and HEIMERMANN, W. H. First Symposium on the Scientific Aspects of Orchids (Ed. H. H. SZMANT and J. WEMPLE) pp. 75–89, Department of Chemistry, University of Detroit (1976).
20. JACOB, J. and GLASER, A. *Biochem. Syst. Ecol.*, **2**, 215–220 (1975).
21. JACOB, J. and POLTZ, J. *J. Lipid Res.*, **15**, 243–248 (1974).
22. JANTZEN, E., BERGAN, T. and BOVRE, K. *Acta Pathol. Microbiol. Scand. Sect. B.*, **82**, 785–798 (1974).
23. KANEKO, H., HOSOHARA, M., TANAKA, M. and ITOH, T. *Lipids*, **11**, 837–844 (1976).
24. KARLSSON, H. and ODHAM, G. *Ark. Kemi*, **31**, 143–158 (1969).
25. LINNAEUS, C. *Systema Naturae*, J. W. deGroot, Leiden (1735).
26. LIVERMORE, B. The Comparative Lipid Compositions of Several Spirochetes, Ph.D. Thesis, Univ. of Minn., June (1974); see also Livermore, B. P. and Johnson, R. E. *J. Bacteriol.*, **120**, 1268–1273 (1974).
27. MABRY, T. J. *Pure Appl. Chem.*, **34**, 377–399 (1973).
28. MIKOLAJCZAK, K. L. *Progress in The Chemistry of Fats and Other Lipids* (Ed. R. T. HOLMAN) Pergamon Press, Oxford, in press, 1977.
29. MIKOLAJCZAK, K. L., MIWA, T. K., EARLE, F. R. and WOLFF, I. A. *J. Am. Oil Chem. Soc.*, **38**, 678–681 (1961).
30. ODHAM. G. *Ark. Kemi*, **21**, 379–392 (1963).
31. ODHAM, G. *Ark. Kemi*, **23**, 431–451 (1965).
32. ODHAM, G. *Ark. Kemi*, **25**, 543–554 (1966).
33. ODHAM, G. *Ark. Kemi*, **27**, 263–288 (1967).
34. ODHAM, G. *Ark. Kemi*, **27**, 289–294 (1967).
35. ODHAM, G. and STENHAGEN, E. *Acc. Chem. Res.*, **4**, 121–128 (1971).
36. POLTZ, J. and JACOB, J. *Biochim. Biophys. Acta*, **360**, 348–356 (1974).
37. POLTZ, J. and JACOB, J. *Biochem. Syst. Ecol.*, **3**, 57–62 (1975).
38. POWELL, R. A. and ADAMS, R. P. *Am. J. Bot.*, **60**, 1041–1050 (1973).
39. SNEATH, P. H. A. and SOKAL, R. R. *Numerical Taxonomy—The Principles and Practice of Numerical Classification*, W. H. Freeman, San Francisco, 1973, 573 pp.
40. STRANSKY, H. and HAGER, A. *Arch. Mikrobiol.*, **71**, 164–190 (1970).
41. STRANSKY, H. and HAGER, A. *Arch. Mikrobiol.*, **72**, 84–96 (1970).
42. TETENYI, P. In: *Chemistry in Botanical Classification*. Proc. 25th Nobel Symposium (Eds. G. BENDZ and J. SANTESSON) Nobel Foundation, Stockholm, 1973.
43. THIEN, L. B., HEIMERMANN, W. H. and HOLMAN, R. T. *Taxon*, **24**, 557–568 (1975).
44. VASEK, F. C. and SCORA, R. W. *Am. J. Bot.*, **54**, 781–789 (1967).
45. VON RUDLOFF, E. *Can. J. Bot.*, **45**, 891–901 (1967).
46. VON RUDLOFF, E. *Biochem. Syst. Ecol.*, **2**, 131–167 (1975).
47. VON RUDLOFF, E. *Phytochemistry*, **14**, 1695–1699 (1975).
48. WOLFF, I. A. and KWOLEK, W. F. *Chemotaxonomy of the Leguminosae* (Eds. L. B. HARBORNE, D. BOULTER and B. L. TURNER) pp. 231–255, Academic Press, New York, 1971.
49. YOUNG, R. G. Cornell Univ. Expt. Stat. Mem., no. 401, 185 (1967).

Prog. Chem. Fats other Lipids. Vol. 16, pp. 31–44. Pergamon Press, 1978. Printed in Great Britain

BIOLOGICAL AND CHEMICAL ASPECTS OF THE AQUATIC LIPID SURFACE MICROLAYER

Göran Odham,* Börje Norén,† Birgitta Norkrans,‡ Anders Södergren§ and Håkan Löfgren‖

CONTENTS

I. Introduction 31

II. Sampling Devices for Collection of the Lipid Microlayer and Associated Microorganisms 33

III. Physical and Chemical Aspects 35

IV. Studies with Model Systems 36

    A. Interaction between microorganisms and model surface layers 36
    B. Formation of surface films in continuous flow systems 40
    C. Sediment model ecosystems 41

V. Field Studies 42

VI. Discussion 43

Acknowledgements 44

References 44

I. INTRODUCTION

The air/sea interface has received much attention in recent years. This is not surprising because this interface plays an important part in the understanding of many environmental problems such as pollution. Natural waters—marine or limnic—are covered by a surface film which contains various types of lipids. These can form coherent films consisting of a mono-molecular layer or multifilms (slicks) of as many as ten layers in certain regions. It is obvious that many of the characteristic properties of the air/sea interface, e.g. the exchange of gas, liquid or solid, are greatly influenced by the lipid film. The damping effect of the film on the capillary waves influences profoundly the aerosol formation and hence, the transport mechanism from the sea to the atmosphere. The lipid film consists, under natural conditions, mainly of saturated and unsaturated fatty acids and glycerides produced in the aquatic environment.[16] The processes behind the formation are complicated and little understood. Multifilms appear to form as a result of saturation.[15] The amount of neutral lipids (glycerides) relative to fatty acids is always higher in the surface multifilms than in the subsurface water. The solubility of the free fatty acids, however, is not exceeded and they accumulate at the air/sea interface to reduce the free surface energy. Rising bubbles, particles, and diffusion all aid in the transport of material from the subsurface water to the surface.

Recently a fairly extensive literature has accumulated dealing with the chemical composition of the natural lipid film.[7,10,13–15] The film consists mainly of free fatty acids in the C_{12}–C_{18} range, alcohols, and glycerides, of which triglycerides are dominant.

*Laboratory of Ecological Chemistry, Ecology Building, University of Lund, Helgonavägen 5, 223 62 Lund, Sweden (to whom correspondence should be addressed).
†Department of Microbiological Ecology, Ecology Building, University of Lund, Sweden.
‡Department of Microbiology, Botaniska Trädgården, University of Gothenburg, Sweden.
§Institute of Limnology, University of Lund, Sweden.
‖Department of Medical Chemistry, University of Gothenburg, Sweden.

Figures 1(a, b) show the free fatty acid pattern of a surface film and the corresponding subsurface water from the Fjord of Gullmaren in Sweden in 1974.[16] The two chromatograms differ considerably, particularly with respect to the proportions between saturated and unsaturated material. The film contains more unsaturated fatty acids.

The natural lipid films are more or less contaminated with pollutants. Many of them, particularly the hydrophobic ones such as petroleum and chlorinated hydrocarbon residues, accumulate in the surface film. Since the lateral diffusion in a condensed surface film is extremely slow, the time for equilibrium after local changes, for example by an oil-spill, is long. The occurrence of organochlorine residues in the lipid film is of considerable interest. Compared to the levels of these components in seawater, a profound accumulation occurs in the surface film. Duce *et al.*[5] and Larsson *et al.*[16] report enrichment factors of 10^3–10^4 for PCB in samples taken from Narragansett Bay, Rhode Island and Jorefjorden, Sweden, respectively.

Accumulation of heavy metals in the lipid microlayer has been observed by several authors.[1,5,26] The accumulation as compared to the subsurface water can range from values very close to zero up to unity for the alkali metals and exceed 3 or 4 orders of magnitude for the transition metals.[26] Zn^{2+}, Cu^{2+}, Fe^{3+} and Ni^{2+} are often found in high concentrations in the microlayer. It is likely that these accumulations are a result of complex formation in the lipid microlayer due to the presence of organic acids, proteinaceous material and other polar organics.

Relative to the subsurface water, considerable amounts of particulate and dissolved organic carbon and nitrogen as well as dissolved inorganic nitrogen and phosphorus have been observed in the surface film of the sea.[20,22,34] In fresh water lakes, a similar enrichment pattern was found by Saijo *et al.*[28] who also noted that these substances were conspicuous in films of oligotrophic lakes and were not evident in eutrophic lakes.

The organisms inhabiting the surface microlayer are usually referred to as neuston,[21] which can be separated into different systematical levels. The smallest components are the bacteria, usually referred to as bacterioneuston and characterized by large bacterial numbers and great diversity. Early works on fresh water neuston mainly carried out in stagnant water and often concerned with medical entomology (i.e. malaria) seemed to indicate that neuston complex could develop only in calm ponds and puddles. Later data, however, have revealed that the surface films formed on fresh water lakes differ significantly from the subsurface water from a biological point of view. Not only is

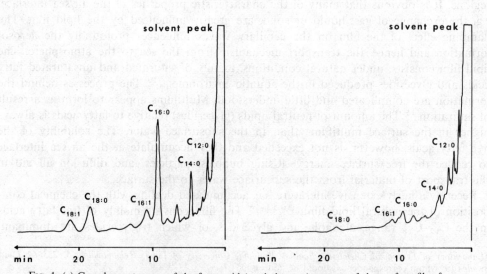

FIG. 1. (a) Gas chromatogram of the fatty acid (methyl esters) pattern of the surface film from the Fjord of Gullmaren; (b) gas chromatogram of the fatty acid (methyl ester) pattern of the corresponding subsurface water. 20% DEGS on Chromosorb W (AW–DMCS treated, 80–100 mesh) as stationary phase, *n*-hexane as solvent. Temperature 170°C.

the composition of the microflora different but the number of bacteria is also considerably greater in the microlayer, the enrichment factor ranging from 10 to 200.[12,24]

On the oceans, the density of bacteria in the surface film is greater than in the subsurface water, the contrast being especially evident in the open sea with its sparse bacterial populations.[33] The bacterial enrichment anticipated in early works (e.g.[35]) was clearly demonstrated when more adept sampling devices were developed and used.[6,10,29] There are a few reports indicating that the biochemical capacity of the bacterioneuston differs from that of the bacteria in the subsurface water. The former are reported to be specialized in the degradation of fats, carbohydrates and starch, whereas the latter evidently act more readily on proteins.[29,33]

The information on the effect of solar radiation of the bacterioneuston is somewhat contradictory. In the Pacific, the bacterioneuston has been reported to exist and flourish even under conditions of intense solar radiation, a fact which is partly ascribed to the large number, the great diversity, and the many pigmented bacteria present in the surface film. However, maximum solar radiation diminishes the bacterial number and harmful effects of the u.v.-rays of the solar radiation on bacteria and diatoms of the surface film have also been reported.[19,27]

A review of the literature reveals that very little is known about the biological interactions behind the formation of the lipid microlayer, its transformation due to biological and physical break-down, its dependence on physical parameters like temperature, light and wind, and its relation to the microbial life in the uppermost regions of the marine and limnic waters.

II. SAMPLING DEVICES FOR COLLECTION OF THE LIPID MICROLAYER AND ASSOCIATED MICROORGANISMS

The recent attention to aquatic surface microlayers is no doubt a result of the development of reasonably simple techniques for the collection. Garrett[6] introduced a 16-mesh window screen of Monel metal for the purpose of sampling the film. The surface of a liquid is removed more or less intact when the screen is placed either in horizontal contact with or drawn vertically through the liquid surface. The main mechanism for film removal is the entrapment between the wires. Only a minor adsorption occurs. Garrett reported a recovery of approximately 75% after the first surface contact. Using this technique, oleic and stearic acid and oleyl alcohol monolayers were removed in the yield stated. A 0.15 mm layer of water was collected by the 16-mesh screen made from 0.14 mm diameter Monel wire. The size was 75×60 cm and about 200–250 dippings were performed.

The screen samplers were reduced in size to fit a portable autoclave by Sieburth[29] who used the sampler in studying microorganisms (neuston) in the top 0.25 mm layer. Sieburth found stainless steel equally suitable as wire material.

Harvey[10] found the time necessary for processing large samples and difficulties arising from subsurface contamination inconvenient and developed a collecting device of a surface skimmer type. It consists of a rotating cylinder of stainless steel coated with a hydrophilic ceramic material. A neoprene blade is pressed tightly to the surface of the cylinder to remove continuously the film and water. During operation, the apparatus is pushed ahead by a boat at slow speed. The thickness of the layer collected by the skimmer varies with temperature. At 20°, it samples a layer of approximately 60 μm.

A bacterioneuston collector consisting of a two-necked bottle was constructed and used by Tsyban.[32] The device does not allow collecting thin layers.

With the demand of ease and simplicity in mind, Miget et al.[20] developed a sampler consisting of a 2 mm teflon disk which is attached to a 4 mm aluminium backing by means of bolts. For field use, a wooden pole is attached to the hinge so that the teflon face and water surface are parallel.

FIG. 2. Photograph illustrating the teflon plate used for collection of the lipid microlayer
(15 × 15 cm). The holes in the plate have conical openings downwards. The plate is attached
to a teflon holder.

Larsson *et al.*[16] designed a teflon plate densely perforated with conical holes (Fig.
2). Teflon was chosen because it is strongly hydrophobic and no contaminants leach
out during extraction. The yield of recovery is increased by the perforation. The geometry reduces the water/air contact area at the dipping. The plate was tested on known
monomolecular surface film phases. Solid-condensed phases of methyl stearate and of
behenic acid monomolecular films were prepared and the amounts of lipid recovered
after one dipping by the plate through the surface was measured. The recovery of the
surface film was 70–90%. In the case of oleic acid in a liquid condensed state, the
yields were 90–100%. The water droplets remaining on the collection plate correspond
to a water-layer thickness of 50–100 μm. It was also found that the teflon plate served
efficiently in the collection of microorganisms associated with the lipid microlayer.

By using a small germanium slide, it has been possible to examine the film *in situ*
by infrared spectroscopy (c.f. ref. 18).

Attempts to compare the different methods of collection have been made. Hatcher[12]
and Parker and Hatcher[27] compared collection and recovery efficiencies of the metal
screen,[6] the rotating drum,[10] a tray,[25] and a glass plate.[11] They found that the
plate and drum collect appreciably larger amounts of the surface films than the screen
or tray. Furthermore, the plate collects considerably less water than the drum. However,
as Garrett points out,[8] the experimental techniques used by Hatcher and Parker do
not reproduce natural films of monomolecular nature but rather systems with a 200-fold
excess of film. It is likely that the lipids under such conditions will be found in clusters
on the surface. Obviously, these deviations from natural conditions may lead to
erroneous conclusions as to the relative merits of the sampling devices.

Ledet and Laseter[17] compared the wire screen and the plain teflon disk with respect
to efficiency of retrieving hydrocarbons, aromatics, and free fatty acids. The results
suggest that each device is selective with respect to the ability to recover organic compounds. In general, the teflon disk appears superior to the screen in recovering fatty
acids, especially the saturated members.

III. PHYSICAL AND CHEMICAL ASPECTS

Blanchard[3] studied the dynamics of small bubbles produced or stranded at some depth. The bubbles may be produced by living organisms or by sea turbulence. The bubble ascends only slowly and model experiments have shown that organic molecules and particulate matter adsorb to the surface of the bubble, causing the microlayer to be enriched in organic matter. At the surface, the bubble eventually bursts to form aerosol droplets that are ejected at very high speed into the air. This formation of droplets has been described as a surface microtome, as the ejected droplets carry away material only from the bubble surface and the top micrometers of the sea. It has been estimated that the combined effects of the surface enlargement and the improved aerosol formation caused by the bubbles are sufficient to explain the rates of turn-over in the surface layer. One very important consequence of the bubble phenomenon is the coupling it causes between transport of salt into the air and the presence of organic substances in the surface microlayer. Some of these organic molecules should have a dramatic effect on the efficiency of the bubble process.

The lipids in the microlayer are almost insoluble in water and easily form two-dimensional supramolecular aggregates on the surface by ionic and van der Waal interaction. These substances are amphiphilic, because they contain both hydrophilic and hydrophobic groups. Therefore, each molecule has a preferred orientation with respect to the water surface—the hydrophobic parts extending into the air. As the density of the film varies greatly from one place to another, the physical forms of the film and effects thereof on the surface behavior will be dealt with in some detail.

At low densities, the film behaves as a two-dimensional gas (Fig. 3A–C). The molecules move almost independently of each other. Such a film is known to interact with larger molecules like polymers and proteins and with microorganisms. The interaction is usually very weak but involves sometimes complex formation. It follows that a large amount of organic matter can be accommodated within such an open structure.

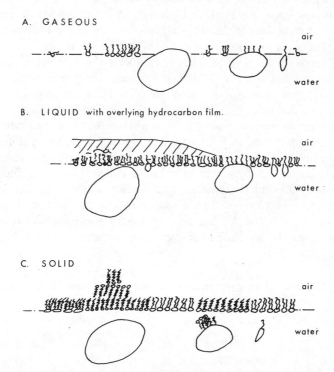

Fig. 3. Aggregation states of a lipid microlayer. The wavy lines are hydrophobic chains and the open structures polar groups: (A) Microlayer of low density behaving as a two-dimensional gas; (B) Liquid surface monolayer. It aids in forming extended sheets of oily films; (C) Solid monolayer. It usually contains small crystalline regions. Bulky molecules are excluded. Crystalline chains are represented as ⌇. Large bodies represent particles.

If not gaseous, the film may be liquid or solid. The solid state is either a monomolecular film or consists of a stratified multilayered film. With increased condensation, the ability to form complexes with larger entities decreases but instead a continuous hydrocarbon phase is formed. This hydrocarbon phase makes possible the solubilization of nonpolar hydrocarbons of varying chain-length. It may become fairly thick before it is broken up into islands or drops. If the film consists mainly of long-chain polar hydrocarbons, it is easily crystallized and the film is transformed into particulate matter anchored at the sea surface by the buoyancy effect. Apart from the van der Waal interaction between hydrocarbon chains, there exist strong ionic forces because the polar groups usually are charged at sea water pH. From surface balance experiments, it is known that ionic bonding between molecules strongly influences the rheological characteristics of the film. Further, the ionized groups facilitate interaction with other constituents of sea water. The polar groups are often strong dipoles as well. Their ordered structure in the surface, therefore, dramatically changes the surface potential to further upset the normal ion-balance of the surface region, e.g. the pH might be very different.

The monomolecular film seems to be important in connection with transport of material to and from the surface. By decreasing the surface tension of newly formed bubbles at the surface or in deeper strata, the growth of the bubbles is improved and, by micellation at living organisms, the formation of three-dimensional aggregates of water insoluble matter is made possible. Thus, organic matter and microorganisms are helped to the surface.

IV. STUDIES WITH MODEL SYSTEMS

A. *Interaction between Microorganisms and Model Surface Layers*

An accumulation of microorganisms at the lipid film of a surface microlayer on aquatic systems has been observed.[2,4] However, it is not known if bacteria differ in their tendency to accumulate at the surface microlayer or if the bacterial accumulations differ with the chemical composition of the lipid film. It is also an open question if the amount of lipids in the surface microlayer affects the number of bacteria accumulated. Furthermore, what is the relationship between the transformation of the surface lipids and the associated microorganisms? Studies in laboratory model systems may contribute to the understanding of the phenomenon of bacterial accumulation.[23]

The film forming substances—oleic acid, a mixture of oleic and palmitic acids, and triolein—were each dissolved in light petroleum: ethanol (9 : 1 v/v). The solutions were spread dropwise on the surface by a micrometer syringe in amounts calculated to cover the surface by a mono- or decalayer in a liquid condensed state. The teflon sampler was dipped in the film and the bacteria removed by washing in 2% Tween 80 in an appropriate sodium chloride solution. Plate counts were performed on this solution and on the subsurface water. All bacteria used were gram-negative and isolated from marine environments. In all experiments, an accumulation of bacteria in the surface

TABLE 1. Enrichment Factor (E) for Some Bacteria Isolated from Marine Environment in Model Systems with Lipid Film of Oleic Acid, Monolayer

	Number of bacteria/ml $\times 10^6$ in samples from		
Testorganisms	Surface film	Subsurface water	E
Aeromonas dourgesi ATCC 23211	32	3.1	10
Pseudomonas fluorescens ATCC 13525	46	5.9	8
Pseudomonas halocrenaea ATCC 19712	49	3.4	14
Serratia marinorubra ATCC 19279	390	4.0	97

TABLE 2. Enrichment Factor (E) for *Serratia marinorubra* ATCC 19279 in Model Systems with Different Types of Lipid Film

Lipid films	Number of bacteria/ml $\times 10^6$ in samples from		E
	Surface film	Subsurface water	
Oleic acid, monolayer	370	13.4	28
Oleic acid, decalayer	800	17.2	47
Oleic–palmitic acids— triolein (equimolecular amounts) monolayer	370	13.8	27

microlayer did occur, although at varying degrees. Table 1 gives the enrichment factor (E) for the bacteria, i.e. the number of bacteria in the surface microlayer relative to that in the subsurface water. The enrichment factor was about ten times higher for *Serratia marinorubra* than for the three other bacterial strains used. Some gram-positive bacteria tested gave still lower enrichment factors. The accumulation of *S. marinorubra* was of the same magnitude, whether the monolayer film was formed of oleic acid solely or of a mixture of oleic and palmitic acids plus triolein (Table 2). However, the accumulation in a decalayer of oleic acid was approximately twice that in the monolayer.

The variation of the enrichment factor for *S. marinorubra* may be a result of the difference in cell density in the subsurface water being 10^7 cells in the latter experiments and 4×10^6 in the former. This explanation was supported in experiments in which the cell density of the subsurface water was varied in the range from about 2×10^6 to 50×10^6 cells/ml resulting in enrichment factors from about 100 to 7, respectively (Table 3). The results may indicate a saturation of the film at these high concentrations of subsurface bacteria.

Using the surface balance technique, it has been possible to measure the change in the pressure-area isotherm of a phospholipid monolayer in the presence of bacterial extracts.[27] A lipopolysaccharide preparation from the envelope of the gram-negative bacterium *Salmonella typhimurium* was used. We thought that similar interactions should occur between intact microorganisms and a lipid monolayer. Our model systems,

TABLE 3. Enrichment Factor (E) for *Serratia marinorubra* ATCC 19279 in Model Experiments with Lipid Film of Oleic Acid Monolayer at Different Densities of Cells in the Bulk

Sample No	Number of bacteria/ml $\times 10^6$ in samples from		E
	Surface film	Subsurface water	
1	170	1.9	90
2	190	1.9	98
3	35	3.4	10
4	370	3.8	97
5	340	4.3	79
6	410	4.3	97
7	950	5.7	167
8	140	7.0	20
9	300	10.8	28
10	213	11.4	19
11	329	12.7	26
12	320	13.2	24
13	420	13.5	31
14	337	16.0	21
15	568	29.4	19
16	490	30.3	16
17	198	35.0	5.7
18	372	53.2	6.7

although modified in certain respects, were therefore subjected to studies with a surface
balance of Wilhelmy model.[15] Washed bacterial cells were mixed into a solution in the
trough to a density of 10^6 cells/ml. Monolayer studies were performed on oleic acid,
a mixture of oleic and palmitic acids, triolein solely or with dipalmitoyl-phosphatidylcholine. The lipids were spread to a gaseous monolayer on the surface and the system
was equilibrated for 2 hr before compression of the film started. The graph of the
surface pressure, π, (=decrease in surface tension) was recorded as a function of the
alloted molecular area. For *Acholeplasma laidlawii* and *S. marinorubra*, the experimental
isotherms (Fig. 4a, b) clearly show the influence of bacteria on the film of oleic acid
and dipalmitoyl-phosphatidylcholine. The effect was less pronounced or not observable
on other films or for other bacteria tested. The differences in enrichment factors for
the various bacteria could be expected. During the experimental time (2 hr), there was
no removal of lipid molecules from the surface by the microorganisms; the only effect
observed is the different abilities among the species to obstruct the formation of a
condensed monolayer. This effect was pronounced for *A. laidlawii* and very small for
Pseudomonas fluorescens (Table 4). In the presence of cells from *P. fluorescens*, the film
is more viscous although the isotherms are unchanged. Therefore, we suggest that cells
from *A. laidlawii* and *S. marinorubra* enter the film and can be expelled, whereas cells
from *P. fluorescens* form a second layer directly underneath the film.

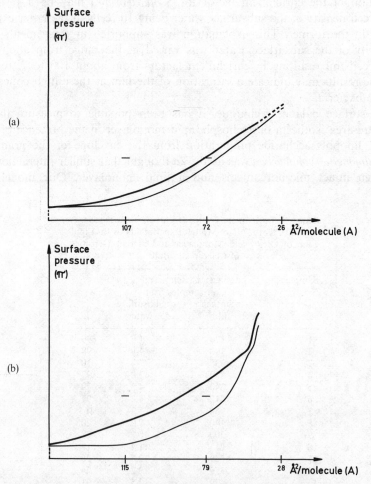

FIG. 4. (a) Surface pressure, π, as a function of allotted molecular area for a monolayer of
oleic acid on saline solution with (——) and without (——) cells of *Serratia marinorubra*. The
two horizontal lines are reference marks for calculating the molecular area at a given moment;
(b) Surface pressure, π, as a function of allotted molecular area for a monolayer of dipalmitoyl–
phosphatidyl choline on saline solution with and without cells of *Acholeplasma laidlawii*. Same
legends as in (a).

TABLE 4. The Degree of Interaction between Various Bacteria and Monolayered Lipid Films on a Subsurface Medium of Saline Solution (pH 10.5) or Artificial Sea Water (pH 8.7), Based on Calculations for the Increase of Molecular Area at a Given Value of π

Bacteria	Subsurface medium	Lipid monolayer	Interaction
G−			
Serratia marinorubra	"sea water"	oleic acid	−
	NaCl	oleic acid	+
Pseudomonas fluorescens	"sea water"	oleic acid	−
	NaCl	oleic acid	(+)
G+			
Bacillus subtilis	NaCl	oleic acid	−
Cell-wall less:			
Acholeplasma laidlawii	NaCl	oleic acid	+ +
	NaCl	dipalmitoyl-lecithin	+ + +

In order to study the transformation of the surface microlayer, microorganisms from the surface of the marine environment of Öresund, a channel between Sweden and Denmark, were collected with the teflon plate. Two strains of gram-negative bacteria (probably *Pseudomonas*) able to grow on tri-isotridecanoin (TIT), were used in the experiments. The test system consisted of a decalayer of TIT at 15° on synthetic sea water inoculated with 10^3 bacteria per ml of each type. After collecting the microlayer at different times of incubation, the teflon plate was extracted with chloroform–methanol mixtures. Methyl esters of the free fatty acids were prepared by means of diazomethane and of the bound fatty acid esters by transesterification with methanol–hydrochloric acid.

Figure 5 shows a gas chromatogram of the methyl esters of the total fatty acids after two days of incubation. The iso-acid originally bound to glycerol had reached levels of the same order as that of the straight chain C_{16} and C_{18} acids formed.

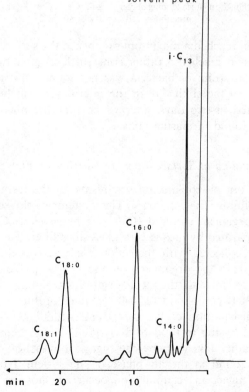

FIG. 5. Gas chromatogram of the methyl esters of the total fatty acids present in the film after two days of incubation. Experimental conditions same as in Fig. 1.

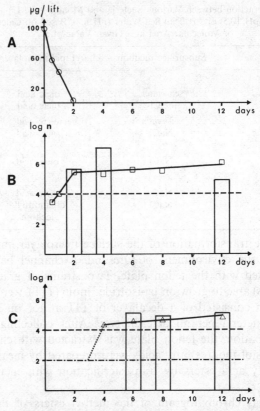

Fig. 6. (A) Tri-isotridecanoin (TIT) present in the film as a function of time. ○ Amount of TIT; (B) Presence of strain 1 bacteria as a function of time. n = number of bacteria per ml, □ = strain 1 in subsurface water, columns = strain 1 in surface microlayer; (C) Presence of strain 2 bacteria as a function of time. n = number of bacteria per ml, △ = strain 2 in subsurface water, columns = strain 2 in surface microlayer. ‑‑‑‑ = inoculated cell density of strain 1, —·—·— inoculated cell density of strain 2.

One strain of bacteria reached a maximum number in the surface after 4 days whereas, in the subsurface water, it reached a rather constant level after two days of incubation (Fig. 6A–C). The second strain of bacteria was not found until the fourth day in the subsurface water or until the sixth day in the microlayer. Furthermore, the lipid film (TIT) added disappeared in two days, after which time the microlayer was composed of the lipids normally found in marine films.

B. *Formation of Surface Films in Continuous Flow Systems*

The contributions from phytoplankton organisms on the formation of surface films were studied in continuous flow systems. The system employed was a modification by Södergren.[30] The organism cultured (*Chlorella pyrenoidosa Chick*) and the surface film produced were separated by means of a glass fiber filter. The arrangement allowed the surface film to be collected with the teflon plate and at the same time excluded possible interferences from the organisms on the sampling. To follow the formation of the surface film and to elucidate the mechanism of accumulation of organic pollutants in the film, a single PCB compound (2,′,4,4′,5-pentachlorobiphenyl) was added to the culture system. The distribution of the compound within the various components of the system was studied by the method of Södergren.[31] Upon entering the culture system, the PCB was readily taken up by the algae. It was not detected in the water and only traces were found in the air phase of the system. However, in the surface film produced by the algae, an accumulation occurred, showing a transport from the algae to the surface film. The content of the substance in the film, collected by one dipping with the plate (15 × 15 cm) ranged from 10 ng to 40 ng and was related to the amount added to the system and accumulated by the algae. In similar studies with DDT and a PCB mixture (Clophen A 50), a corresponding pattern of distribution between the water, algae, and surface film was observed.

The lipids simultaneously recovered from the surface film of the culture were dominated by fatty acids. Among the total fatty acids, palmitic acid constituted a major portion (28 μg), followed by stearic acid 8 μg, myristic acid 7 μg, and lauric acid 2 μg. A similar pattern of total fatty acids was found in cells of *C. pyrenoidosa*. In the surface film, the free fatty acids constituted about one-half of the total fatty acids whereas in the algae only one-third. In addition, a difference in the fatty acid pattern was observed. In one experiment, the flow was stopped and the cells of *C. pyrenoidosa* were left to die of nutrient depletion. In about one week, the amount of the fatty acids detected in the surface film had decreased by one order of magnitude. However, palmitic acid was still the dominating fatty acid.

C. *Sediment Model Ecosystems*

It is generally believed that the chemical components of the lipid microlayer are produced by the organisms living there and in the water below. Little attention has, however, been paid to the influence from the bottom sediment on the surface film. A simple model system was therefore designed to demonstrate the differences in the microlayer produced by bottom sediments, rich (mud) and poor (sandy) in organic matter. The model comprised a fresh water system kept under conditions preventing influence from external contaminations. A rich flora of bacteria soon developed in the surface film. In the mud system, the enrichment factors in the film were 4–8 whereas, in the sandy system, they were about 800. In the latter, the bacterial concentration in the subsurface water was less than one tenth of that in the former. Thus, even in this complex system undergoing growth, the enrichment appears to depend on the concentration of the subsurface bacteria. In addition, fluorescence-microscopy showed that a rather heterogeneous flora had developed in the mud system film whereas, in the sandy system, it was more homogeneous and was dominated by tetrade-forming coccoides and slime forming rods (Fig. 7).

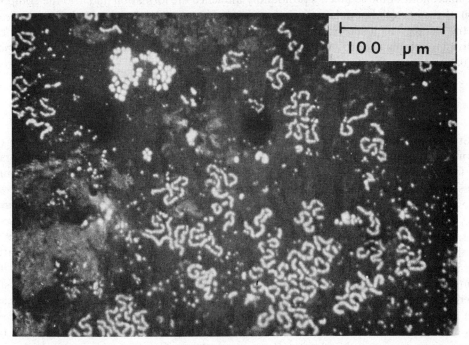

FIG. 7. The lipid microlayer containing bacterioneuston from the sandy model system, stained with 0.01% acridine-orange and photographed in incident light in a fluorescence-microscope.

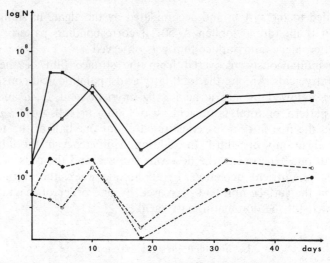

Fɪɢ. 8. The number of heterotrophic bacteria as a function of time in the surface microlayer and subsurface water in sediment model system. The number of heterotrophic bacteria was counted after 7 days of incubation at 20°C on peptone–yeast-extract agar, ■ surface microlayer at 20°C, □ surface microlayer at 5°C, ● subsurface water at 20°C, ○ subsurface water at 5°C.

To illustrate the distribution and transport from the sediment to the surface film, PCB-compounds were added to some sediments. None of the compounds were traced into the lipid film, indicating that no transport occurred.

In another study with a muddy sediment, designed to demonstrate the formation of film as a function of time and temperature, the concentrations of heterotrophic bacteria in the film and subsurface water reached their maximum after 3 days at 20° and 10 days at 5° (Fig. 8).

In both cases, a marked accumulation of bacteria occurred in the surface layer, the maximum enrichment factor being about 1000 at 20° and 700 at 5°. From the 10th day of incubation, the development in both the surface film and the subsurface water appeared to be rather parallel at the two temperatures used. The results indicate that the microflora of the sediments profoundly influence the formation and also the composition of the bacterial populations in the surface film. At 20°, the observed rapid increase of the number of heterotrophic bacteria in the film seems, however, to indicate that active growth may contribute to the enrichment more profoundly than to the transport of bacteria from the bottom and subsurface water to the film.

V. FIELD STUDIES

Studies in the model systems were carried out to investigate specific details in the formation, structure and properties of surface films, whereas field studies were made to follow seasonal variations in composition of the microlayer in marine and limnic ecosystems. Both polluted and nonpolluted environments were considered. Öresund was chosen as a representative of the former and the inner part of the fjord of Gullmaren was selected as a nonpolluted area. The two sampling sites differed also with respect to salinity, Öresund having about 10 ‰ salt and Gullmaren 28 ‰. The bacterial number in the subsurface water in Öresund was determined by plate count, but in Gullmaren, because of the low bacterial density, the membrane filter technique was used. In the nonpolluted water, there was an enrichment of bacteria in the surface microlayers and there seems to be a positive correlation between the number of viable bacteria and the amount of lipid (Table 5). In polluted water, however, there was no evidence of enrichment of bacteria in the microlayer. Interestingly, there seemed to be a larger proportion of terrestrial and limnic bacteria in the surface film than in the subsurface water.

TABLE 5. Enrichment Factors (E) for Bacteria in a Nonpolluted Area (Gullmarsfjorden) at Four Different Dates of the Year

| Date | Number of bacteria ($\bar{x}$)*/ml in samples from | | | Total amount (μg) of fatty acids in 10 samples from surface |
	Surface film	Subsurface water	E	
1974 Aug 23	12,600	165	76	24
Oct 24	2200	200	11	11
Nov 28	5200	340	15	13
1975 Feb 6	50,000	5000	10	16

†Average value from generally six samplings.

The model system experiments indicate that the enrichment of bacteria in the film decreases when the number of bacteria in the subsurface water reaches a concentration of approximately 10^7/ml. This may explain the absence of enrichment in the surface film in Öresund where the total number of bacteria in the subsurface water is estimated to 10^5–10^8/ml. It is not yet clear if the large proportion of terrestrial and limnic bacteria in the film results from a fallout of atmospheric bacteria or from an enrichment of bacteria transported to the sea by rivers, etc.

VI. DISCUSSION

Recent works dealing with the air/sea interface primarily report field studies and efforts to evaluate different kinds of sampling devices. Thus far, the results obtained are rather sketchy and do not add much to the understanding of the chemistry and ecology behind the formation of the surface microlayer. With respect to sampling techniques, each method seems to be selective regarding collection of both the surface microlayer and the organisms associated.

To study biological and chemical aspects of surface films and to deduce the mechanisms behind their formation, laboratory model systems seem to offer great advantages. In such systems, environmental factors may be kept constant and individual variables can be studied under controlled conditions. When comparing results from polluted and nonpolluted areas (Öresund and Gullmaren), it was observed that the enrichment of components in films on polluted waters is insignificant. This might be a result of the higher density of microorganisms in the subsurface water. Using *S. marinorubra* to study the enrichment factor as function of bacterial concentration, the result reached in the field study was verified. The laboratory experiments thus indicate a saturation of the film at high concentrations of subsurface bacteria. A competition for available binding groups in the film may occur, or a minimum of free space between bacteria may be necessary to eliminate an effect of electrostatic repulsion.

The surface balance technique has proved to be useful in the study of interactions between microorganisms and the lipid microlayer. Such experiments indicate that the forces binding microorganisms to the sea surface are not strong enough to scatter a condensed monolayer. They increase the strength of loosely packed monolayers and the lipids do not seem to become attached to the microorganisms.

The continuous flow technique was used to study various species of phytoplankton organisms and their ability to produce surface films. By incorporating consumer organisms into the system, the combined effects of producer and consumer organisms in the surface film can be followed. By this approach, a better knowledge of the mechanism of formation of the surface film may be gained, and their properties observed.

To transfer certain components of an aquatic ecosystem into the laboratory is another approach to study specific problems related to the lipid microlayer. Sediment model ecosystems have already produced interesting results. The system can also be made more complex by introducing, for example, molluscs, fish and plants. Such systems constitute a bridge between simple models systems and the complicated conditions in nature.

Acknowledgements—Much of the experimental work has been carried out by Messrs Ingemar Andersson, N. Håkansson, Staffan Kjelleberg, Bengt Stehn and Fred Sörensson to whom the authors are greatly indebted. The work has, to a large extent, been financed by the Swedish Natural Research Council.

REFERENCES

1. BARKER, D. R. and ZEITHIN, H. *J. Geophys. Res.*, **77**, 5076–86 (1972).
2. BEZDEK, H. F. and CARLUCCI, A. F. *Limnol. Oceanogr.*, **17**, 566–69 (1972).
3. BLANCHARD, D. C. *Prog. Oceanogr.*, **1**, 71 (1963).
4. BLANCHARD, D. C. and SYZDEK, L. *Science*, **170**, 626–28 (1970).
5. DUCE, R. A., QUINN, J. G., OLNEY, C. E., PIOTROWICZ, S. R., RAY, B. J. and WADE, T. L. *Science*, **176**, 161–63 (1972).
6. GARRETT, W. D. *Limnol. Oceanogr.*, **10**, 602–605 (1965).
7. GARRETT, W. D. *Deep Sea Res.*, **14**, 221–27 (1967).
8. GARRETT, W. D. *Limnol. Oceanogr.*, **19**, 166–67 (1974).
9. GOERING, J. J. and MEUZEL, D. W. *Deep Sea Res.*, **12**, 839–843 (1965).
10. HARVEY, G. W. *Limnol. Oceanogr.*, **11**, 608–13 (1966).
11. HARVEY, G. W. *Limnol. Oceanogr.*, **17**, 156–57 (1972).
12. HATCHER, R. 1974 Thesis, Virginia Polytechnic Institute and State University, Blacksburg, Virginia, U.S.A.
13. JARVIS, N. L. *Limnol. Oceanogr.*, **12**, 213–22 (1967).
14. JARVIS, N. L., GARRETT, W. D., SCHIEMAN, M. A. and TIMMONS, C. O. *Limnol. Oceanogr.*, **12**, 88–90 (1967).
15. KJELLEBERG, S., NORKRANS, B., LÖFGREN, H. and LARSSON, K. *Appl. Environmental Microbiol.* **31**, 609–611 (1976).
16. LARSSON, K., ODHAM, G. and SÖDERGREN, A. *Mar. Chem.*, **2**, 49–57 (1974).
17. LEDET, E. J. and LASETER, J. L. *Anal. Lett.* **7**, 553–62 (1974).
18. MacINTYRE, F. *Sci. Am.*, **230**, 621–67 (1974).
19. MARUMO, R., TAGA, N. and NAKAI, T. *Bull. Plankton Soc. Japan*, **18**, 36–41 (1971).
20. MIGET, R., KATOR, H., OPPENHEIMER, C., LASETER, J. L. and LEDET, E. J. *Anal. Chem.*, **46**, 1154–57 (1974).
21. NAUMANN, E. *Biol. Zentralbl.* **37**, No. 2 (1917).
22. NISHIZAWA, S. *Bull. Plankton, Soc. Japan*, **18**, 42–44 (1971).
23. NORKRANS, B. and SÖRENSSON, F. *Botanica Marina* **20**, 7 (1977).
24. PARKER, B. C. and HATCHER, R. F. *J. Phycol.*, **10**, 185–89 (1974).
25. PARKER, B. C. and WODEHOUSE, E. B. In: *Water for Texas*, Water Res. Inst. 15th Ann. Conf., Texas A and M Univ., U.S.A., 1974.
26. PIOTROWICZ, S. R., RAY, B. J., HOFFMAN, G. L. and DUCE, R. A. *J. Geophys. Res.* **27**, 5243–54 (1972).
27. ROTHFIELD, L. and ROMEO, D. *Bacteriol. Rev.*, **35**, 14–38 (1971).
28. SAIJO, Y., MITAMURA, O. and OGIYAMA, K. *Japan. J. Limnol.*, **35**, 110–116 (1974).
29. SIEBURTH, J. McN., *Ocean Science and Ocean Engineering*, Trans. Joint Conf. MTS and ASLO, Washington, D.C., 1064–68 (1965).
30. SÖDERGREN, A. *Oikos*, **19**, 126–138 (1968).
31. SÖDERGREN, A. *Oikos*, **24**, 30–41 (1973).
32. TSYBAN, A. V. *Gidrobiol. Zh.*, **3**, 84 (1967).
33. TSYBAN, A. V. *J. Oceanogr. Soc. Japan*, **27**, 56–66 (1971).
34. WILLIAMS, R. M. *Deep Sea Res.*, **14**, 791–800 (1967).
35. ZOBELL, C. E. In *Marine Microbiology*, Chronica Botanica, Waltham, U.S.A., 240 pp., 1946.

Prog. Chem. Fats other Lipids. Vol. 16, pp. 45–58. Pergamon Press, 1978. Printed in Great Britain

OCCURRENCE, SYNTHESIS AND BIOLOGICAL EFFECTS OF SUBSTITUTED GLYCEROL ETHERS

Bo Hallgren, Gunnel Ställberg

Research Laboratories, AB Astra, Mölndal, Sweden

and

Bernt Boeryd

Dept. of Pathology I, University of Göteborg, Sweden

CONTENTS

I. Introduction   45

II. Isolation and Characterization of Substituted Glycerol Ethers   45
  A. Methoxy-substituted saturated and monounsaturated glycerol ethers   45
  B. A methoxy-substituted polyunsaturated glycerol ether   51
  C. Hydroxy-substituted saturated and monounsaturated glycerol ethers   52

III. The Occurrence and Composition of Methoxy-Substituted Glycerol Ethers in Lipids from Man and Animals   52

IV. Syntheses of Methoxy- and Hydroxy-Substituted Glycerol Ethers   54
  A. 2-Methoxy-substituted, saturated and monounsaturated glycerol ethers   54
  B. 2-Hydroxy-substituted, saturated and monounsaturated glycerol ethers   57

V. Biological Effects of Substituted Glycerol Ethers   57

Acknowledgements   58

References   58

I. INTRODUCTION

Glycerolipids with ether-linked aliphatic moieties have attracted increasing interest and have during recent years been the object of extensive chemical and metabolic studies. Two major groups of ether-linked lipids, *viz.* alkyl ether lipids and alk-1-enyl ether lipids, have been shown to be present in the living cells of most animal tissues. A high content of ether lipids has been found in malignant tumors (cf ref. 13).

Glycerolipids containing methoxy-substituted *O*-alkyl groups were first isolated from Greenland shark liver oil.[9] This article will mainly deal with the isolation, identification and synthesis of methoxy-substituted glycerol ethers, but will also present some studies on their biological effects. The identification of glycerol ethers with a hydroxyl group in the 2-position of the long alkyl chain will also be described.

II. ISOLATION AND CHARACTERIZATION OF SUBSTITUTED GLYCEROL ETHERS

A. *Methoxy-substituted Saturated and Monounsaturated Glycerol Ethers*

When determining the content of glycerol ethers in the unsaponifiable fraction of liver oil from Greenland shark (*Somniosus microcephalus*) by chromatography on silicic acid columns, elution patterns of the type demonstrated in Fig. 1 were obtained.[9] Peak I represents a mixture of glycerol ethers with chimyl, batyl and selachyl alcohol (hexadecyl, octadecyl and octadecenyl glycerol) as principal components, whereas the material in peak II had characteristics differing from those of the ordinary glycerol ethers. Thin layer chromatography (TLC) (Figs. 2a and b) showed that the material in peak II had

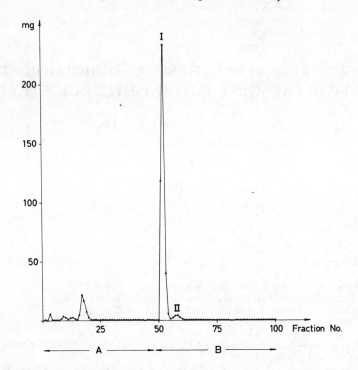

FIG. 1. Chromatography of nonsaponifiable material (467 mg) from Greenland shark liver oil on a silicic acid column (35 g). Eluting solvents: 5% diethyl ether in light petroleum, b.p. 60–80°C (A) and diethyl ether (B). Fraction volume: 20 ml. (Figs. 1–5 reproduced from Hallgren and Ställberg[9] by permission of *Acta Chem. Scand.*).

a somewhat lower R_f value than the ordinary glycerol ethers in peak I. The main part of the peak II material consisted of unsaturated components (Fig. 2b). The infrared (IR) spectra of the materials from peaks I and II were practically identical with a strong absorption band at $\sim 1100 \, \text{cm}^{-1}$, where the absorption of the peak II material was even somewhat stronger. The IR spectrum thus indicated that also the unidentified peak II material consisted of glycerol ethers. The nuclear magnetic resonance (NMR) spectrum showed a signal at $\delta = 3.3$ ppm indicating a methoxy substituent. A very slight band in the IR spectrum at $2830 \, \text{cm}^{-1}$ supported this assumption. As for the ordinary glycerol ethers, treatment of the peak II material with acetone in acid solution gave isopropylidene derivatives, showing that the attached long alkyl chain was bound to the glycerol in the α-position.

Gas liquid chromatography (GLC) of the isopropylidene derivatives of the substituted glycerol ethers before and after hydrogenation indicated two homologous series, one of which was saturated, the other unsaturated. The relative retention times of the four main components of peak II on a polar and a nonpolar stationary phase are summarized in Table 1.

The mass spectra of the dominating components 1, 2 and 4 of Table 1 are shown in Figs. 3–5. The fragmentation patterns could be explained by structures with a methoxy group attached to the 2-position of the long carbon chains, calculated from the glycerol ether bond. Mass spectrometry (MS) also indicated that the molecular weights of components 1–4 before hydrogenation were 384, 386, 398 and 412 and after hydrogenation 386, 386, 400 and 414.

The positions of the double bonds in the two dominating unsaturated components were determined by oxidative splitting of the double bonds and gas chromatographic and mass spectrometric analysis of the esterified acids formed. The presence of methyl dodecanoate and methyl tetradecanoate showed the double bond positions to be between carbon atoms nos. 12 and 13 and between nos. 14 and 15, respectively, calculated from the free ends of the long hydrocarbon chains.

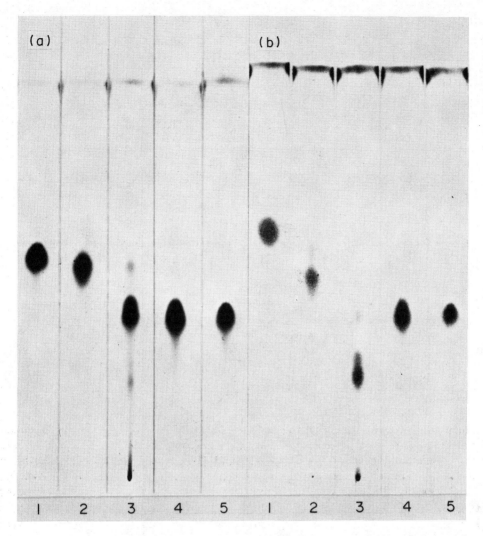

FIG. 2. (a) Thin layer chromatograms on Silica Gel G and (b) on silver nitrate impregnated silica gel G with trimethylpentane–ethyl acetate—methanol, 50:40:7, as solvent. (1) Batyl alcohol. (2) Ordinary glycerol ethers isolated by silicic acid column chromatography (peak I, Fig. 1). (3) Glycerol ethers eluted after the ordinary ones during the silicic acid column chromatography (peak II, Fig. 1). (4) *Ditto* after hydrogenation. (5) Synthetic 1-*O*-(2-methoxyhexadecyl)glycerol.

TABLE 1. Retention Times of the Isopropylidene Derivatives of the Glycerol Ethers from the Peak II Material of the Silicic Acid Chromatogram (Fig. 1) Relative to the Isopropylidene Derivative of Octadecylglycerol

Glycerol ethers (as isopropylidene derivatives)	Retention at 218°C on Apiezon L (1%)		Retention at 194°C on polyethylene glycol succinate (15%)	
	Before hydrogenation	After hydrogenation	Before hydrogenation	After hydrogenation
Octadecylglycerol	1.00	1.00	1.00	1.00
Component 1	0.59	0.68	1.30	1.21
Component 2	0.68	0.68	1.21	1.21
Component 3	0.84	0.97	1.67	1.55
Component 4	1.19	1.38	2.17	2.02

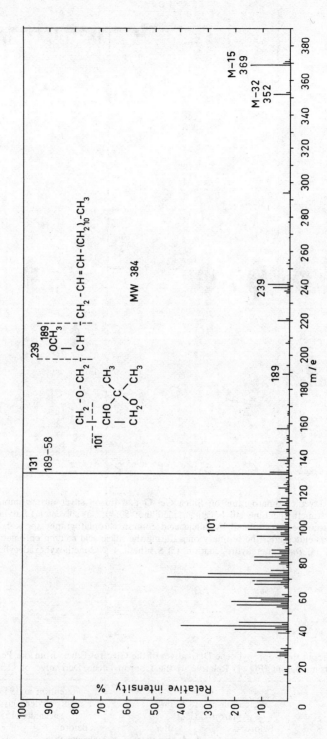

Fig. 3. Mass spectrum of 2,3-O-isopropylidene-1-O-(2-methoxy-4-hexadecenyl)glycerol from Greenland shark liver oil.

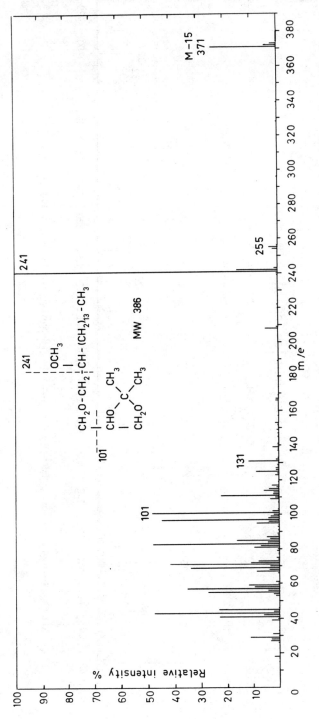

Fig. 4. Mass spectrum of 2,3-O-isopropylidene-1-O-(2-methoxyhexadecyl)glycerol from Greenland shark liver oil.

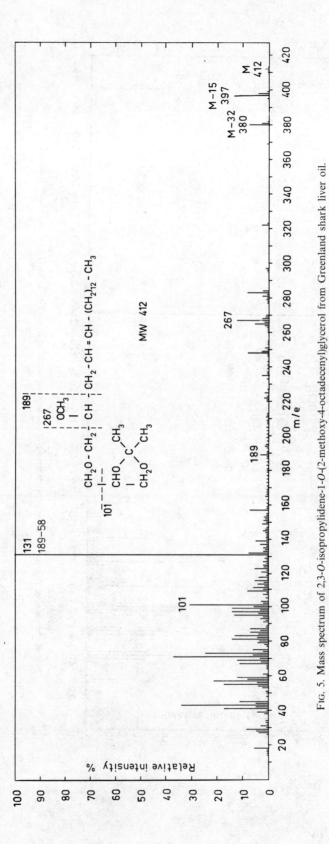

FIG. 5. Mass spectrum of 2,3-*O*-isopropylidene-1-*O*-(2-methoxy-4-octadecenyl)glycerol from Greenland shark liver oil.

The IR absorptions of the mixture of mainly unsaturated glycerol ethers at 1410 and 1655 cm^{-1} indicated a *cis*-form of the double bond. Thus, the compounds 1–4 of Table 1 would be the isopropylidene derivatives of:

1-*O*-(2-methoxy-4-*cis*-hexadecenyl)glycerol;
1-*O*-(2-methoxyhexadecyl)glycerol;
1-*O*-(2-methoxy-4-*cis*-heptadecenyl)glycerol, and
1-*O*-(2-methoxy-4-*cis*-octadecenyl)glycerol.

To confirm the structures, 1-*O*-(2-methoxyhexadecyl)glycerol and 1-*O*-(2-methoxy-4-*cis*-hexadecenyl)glycerol were synthesized. (See Section IV A). The synthetic compounds and the corresponding components from shark liver oil were identical with respect to TLC, MS, IR, NMR and to GLC on several different phases.

In addition to the four components mentioned, small amounts of other homologs were present in the shark liver oil. Table 2 gives the composition of the mixture of the methoxy-substituted as well as the unsubstituted glycerol ethers, determined by GLC and MS. The glycerol ether with a 4-hexadecenyl chain is predominant in the mixture of methoxy-substituted compounds, whereas a glycerol ether with a 9-octadecenyl chain constitutes the largest component of the unsubstituted glycerol ethers. A polyunsaturated component (See Section II B) was only found in the methoxy-substituted glycerol ethers. The methoxy-substituted glycerol ethers constituted about 4% of the total glycerol ether content of the shark liver oil.

B. *A Methoxy-substituted Polyunsaturated Glycerol Ether*

The mixture of methoxy-substituted glycerol ethers described above contained a small amount of a compound which was slightly more polar than the main part.[11] This compound was enriched to about 85% by repeated chromatography of the material (as isopropylidene derivatives) on silicic acid columns.

Analysis by combined GLC–MS of the compound before and after hydrogenation gave the molecular weights 458 and 470, respectively, consistent with the existence of six double bonds. A weak band at 1640 cm^{-1} and a strong band at 700 cm^{-1} in the IR spectrum indicated *cis* double bonds. That the compound was a 2-methoxy-substituted glycerol ether was shown by the MS fragmentation pattern of the hydrogenated isopropylidene derivative. The pattern was the same as for 1-*O*-(2-methoxyhexadecyl)-2,3-*O*-isopropylideneglycerol and 1-*O*-(2-methoxyoctadecyl)-2,3-*O*-isopropylideneglycerol.

TABLE 2. Composition of Unsubstituted and Methoxy-substituted Glycerol Ethers from Greenland Shark Liver Oil

Alkyl chain*	Methoxy-substituted glycerol ethers %	Ordinary glycerol ethers %
14:0	0.5	2.8
15:0	tr	0.5
16:0	14.6	10.6
16:1	53.7	12.1
17:0	1.1	0.5
17:1	3.3	1.9
18:0	1.7	5.8
18:1	21.0	62.0
19:0	0.1	tr
19:1	0.6	0.3
20:0	0.3	0.1
20:1	tr	2.8
22:0	tr	tr
22:1	tr	0.6
22:6	3.1	

*The first figure denotes the number of carbon atoms in the long carbon chain and the figure after the colon the number of double bonds.

The NMR spectrum of the polyunsaturated compound had multiplets at about $\delta = 2.1$, 2.8 and 5.3 ppm, which could originate from CH_2-groups adjacent to one doubly bonded carbon atom, CH_2-groups between two doubly bonded carbon atoms and $CH{=}CH$ groups adjacent to CH_2-groups, respectively.

To settle the positions of the double bonds, the polyunsaturated compound was converted to a polymethoxy compound, which was analyzed by combined GLC–MS.[12] The conversion was carried out by cis-addition of osmium tetroxide to the six double bonds giving six osmate ester groups, which by reductive cleavage were split to six vic-diol groups. Methylation of these twelve hydroxy groups and the two of the glycerol part of the molecule should give a compound with fifteen methoxy groups and a molecular weight of 818. The mass spectrum did not give any molecular ion peak but showed fragment ions originating from both ends of the molecule. The mass spectrum was in accordance with the structure

$$
\begin{array}{c}
\quad\quad\quad OCH_3 \quad\quad\quad OCH_3\ OCH_3 \\
CH_2O{-}CH_2{-}CH{-}(CH_2{-}CH{-}CH)_6{-}CH_2{-}CH_3 \\
CHOCH_3 \\
CH_2OCH_3
\end{array}
$$

which for the polyunsaturated compound implies the structure

$$
\begin{array}{c}
\quad\quad\quad OCH_3 \\
CH_2O{-}CH_2{-}CH{-}(CH_2{-}CH{=}CH)_6{-}CH_2{-}CH_3 \\
CHOH \\
CH_2OH
\end{array}
$$

C. Hydroxy-substituted Saturated and Monounsaturated Glycerol Ethers

Trace quantities of a mixture of compounds more polar than the methoxy-substituted glycerol ethers were isolated from the unsaponifiable fraction of shark liver oil. These compounds were identified as 2-hydroxy-substituted glycerol ethers[10] by the following data. By TLC, the material had the same R_f-values as synthetically prepared 1-O-(2-hydroxy-4-hexadecenyl)glycerol and 1-O-(2-hydroxyhexadecyl)glycerol.[15] Combined GLC–MS analyses of the isopropylidene derivatives of the components in the mixture gave, for the predominating compound, a mass spectrum which was practically identical with that of synthetic 1-O-(2-hydroxy-4-cis-hexadecenyl)-2,3-O-isopropylideneglycerol (Fig. 6). The retention times at GLC were also identical. Hydroxy-substituted tetradecyl, tetradecenyl, hexadecyl and octadecenyl glycerol ethers were also identified by GLC–MS.

III. THE OCCURRENCE AND COMPOSITION OF METHOXY-SUBSTITUTED GLYCEROL ETHERS IN LIPIDS FROM MAN AND ANIMALS

It was considered of interest to find out whether the methoxy-substituted glycerol ethers, which had been found in Greenland shark liver oil, are of more common occurrence. Various biological materials from other marine animals[8] and from mammals,[7] principally man, were therefore analyzed for their content of methoxy-substituted glycerol ethers. The contents of unsubstituted glycerol ethers were determined in the same materials. The materials from marine animals were herring and Baltic herring fillets, mackerel fillets, the edible parts of marine crayfish, shrimps and sea mussels, a commercial sample of cod liver oil and liver oil from cod caught in the Baltic sea. Fresh-water crayfish was also studied. As it had been shown that human milk had a comparatively

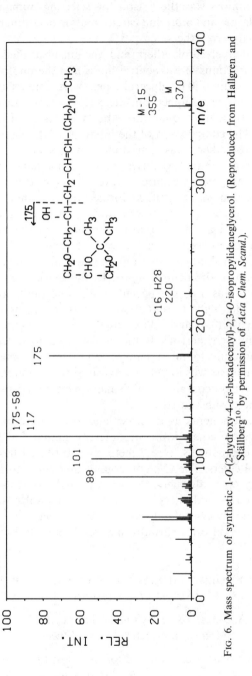

FIG. 6. Mass spectrum of synthetic 1-O-(2-hydroxy-4-cis-hexadecenyl)-2,3-O-isopropylideneglycerol. (Reproduced from Hallgren and Ställberg[10] by permission of Acta Chem. Scand.).

high content of glycerol ethers, methoxy-substituted glycerol ethers were sought in milk from different mammals. Human milk from different periods of lactation, cow's milk and sheep's milk were chosen for the study. The high content of glycerol ethers in red bone marrow and tumors was the reason for selecting human red bone marrow, red blood cells, blood plasma and a uterine carcinoma for determination of their content of methoxy-substituted glycerol ethers.

The methoxy-substituted glycerol ethers and the unsubstituted ones were isolated from the unsaponifiable fractions of the neutral lipids and the phospholipids by chromatography on silicic acid columns. The glycerol ethers were converted to their isopropylidene derivatives and rechromatographed on silicic acid columns or on thin-layer plates. When present in more than trace quantities, the two classes of glycerol ethers were estimated by weighing. The composition of the mixtures of the methoxy-substituted and the unsubstituted glycerol ethers was determined by gas chromatography and mass spectrometry. In a few cases, ethanol was used instead of methanol in the saponification, extraction and chromatographic procedures to exclude the possibility that the methoxy-substituted glycerol ethers might be artifacts formed by methanol treatment.

2-Methoxy-substituted glycerol ethers were found together with the unsubstituted ones in the neutral lipids as well as in the phospholipids of all the materials studied. In the different milk samples, as well as in the various materials from man, only trace quantities of methoxy glycerol ethers were found. The content was judged to be larger in the phospholipids than in the neutral lipids. In the marine animals, the percentage of methoxy glycerol ethers was higher, especially in the phospholipids, than in the mammalian tissues (Fig. 7). The highest content was found in the phospholipids from sea mussels and marine crayfish (0.47 and 0.35%, respectively). The content of unsubstituted glycerol ethers in the lipids of herring, Baltic herring and mackerel fillets is more or less comparable to the amounts found in mammalian tissues. A somewhat higher content of unsubstituted glycerol ethers was found in crayfish, shrimps and mussels. As compared to shark liver oil, cod liver oil contained only small quantities of glycerol ethers, unsubstituted as well as methoxy-substituted ones.

The same principal components as in Greenland shark liver oil were found in the methoxy-substituted glycerol ethers from the different animal and human tissues studied.[7,8] The dominating components were 2-methoxy-substituted hexadecyl, hexadecenyl and octadecenyl glycerol ethers. The C_{16} compounds amounted to 50–90% of the different mixtures of methoxy glycerol ethers. Particularly high contents of C_{16} compounds were found in human, cow's and sheep's milk. A methoxy-substituted docosahexaenyl glycerol ether, first found in Greenland shark liver oil, was also isolated from other sources, viz. human red blood cells, shrimps, mackerel and cod liver oil.

IV. SYNTHESES OF METHOXY- AND HYDROXY-SUBSTITUTED GLYCEROL ETHERS

A. 2-Methoxy-substituted, Saturated and Monounsaturated Glycerol Ethers

For confirmation of the structures proposed for the substituted glycerol ethers isolated from shark liver oil and also to get material for biological tests, 1-O-(2-methoxyalkyl)glycerols and 1-O-(2-methoxy-4-cis-alkenyl)glycerol were synthesized.[15] The routes, exemplified by the 16:0 and 16:1 compounds, are outlined in Fig. 8, in which the yields in the different steps are also given.

As starting materials, 2-methoxy-substituted esters were needed. The saturated methoxy ester was easily prepared from the bromo ester and sodium methoxide. The syntheses of the corresponding acetylenic esters[14] seemed more time-consuming as the different routes first tried were multi-step malonic ester syntheses. The overall yields were only about 25%. However, it was found that a reaction between an alkyl alkoxyacetate and a substituted propargyl halide, as exemplified in step 1 in route II (Fig. 8), directly

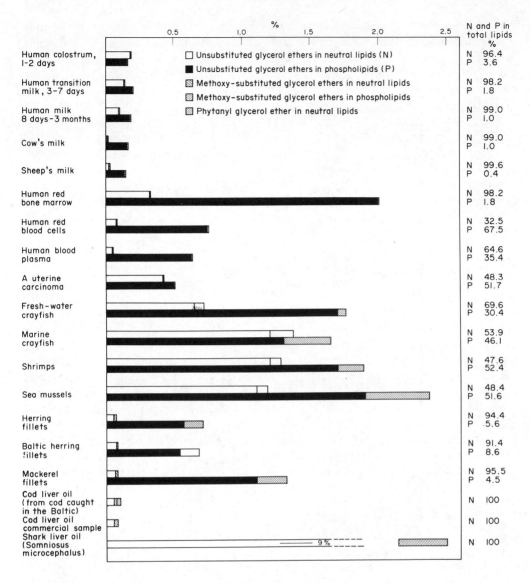

Fig. 7. Unsubstituted and methoxy-substituted glycerol ethers in neutral lipids and phospholipids from various human and animal sources. The contents of neutral lipids and phospholipids in the total lipids are given in the right column.

gave the desired ester of 2-alkoxy-4-alkynoic acid in a good yield. The methylene group of the alkyl alkoxyacetate was in fact reactive enough to be alkylated not only by substituted propargyl halides but also by substituted allyl halides and saturated halides, although in the latter cases to a lesser extent. The alkylations were carried out in tetrahydrofuran or 1,2-dimethoxyethane, with potassium, sodium or sodium hydride as condensing agents and with the alkyl alkoxyacetate in fairly large excess.

The reductions of the esters to alcohols were performed with lithium aluminium hydride in ether solution or with Red Al[5] in hexane (Red Al obtained as a 70% solution of sodium bis-(2-methoxyethoxy)aluminium hydride in benzene).

The alkylations[1] of isopropylideneglycerol were performed in heptane with either the p-toluenesulfonate or the methanesulfonate of the appropriate substituted alcohol. In most cases, potassium hydroxide was used for the salt formation of isopropylidene glycerol; the water formed was removed by azeotropic distillation. The ketal protecting group on the glycerol moiety of the condensation product was removed by treatment with a mixture of ether and concentrated hydrochloric acid at room temperature.

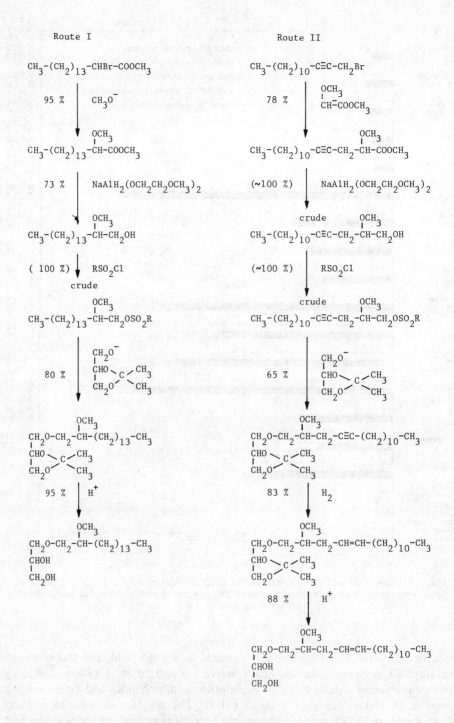

FIG. 8. Flow sheets for the syntheses of 1-*O*-(2-methoxyhexadecyl)glycerol, route I, and 1-*O*-(2-methoxy-4-*cis*-hexadecenyl)glycerol, route II.

The hydrogenation of the acetylenic glycerol ethers (or their isopropylidene derivatives) to the corresponding *cis*-olefinic compounds was achieved using palladium on barium sulfate as catalyst and pyridine as solvent.[6]

The four possible stereoisomers of the methoxy-substituted glycerol ethers have not yet been prepared. They would be of great interest for biological testing and for determination of the configuration of the naturally occurring compounds.

B. 2-Hydroxy-substituted, Saturated and Monounsaturated Glycerol Ethers

1-O-(2-Hydroxyhexadecyl)glycerol[15] was obtained by hydrogenolysis of 1-O-2-benzyl-oxyhexadecyl)-2,3-O-isopropylideneglycerol (95% ethanol, 10% palladium on charcoal as catalyst) and removal of the protecting ketal group by acid hydrolysis.

The monounsaturated compound, 1-O-(2-hydroxy-4-cis-hexadecenyl)glycerol,[10] was obtained by hydrogenation of 1-O-(2-benzyloxy-4-hexadecynyl)-2,3-O-isopropylidene-glycerol in pyridine solution and with 5% palladium on barium sulfate as catalyst.[6] This gave a mixture of benzyloxy- and hydroxy-substituted hexadecenyl compounds, which were separated by chromatography on a silicic acid column. Acid hydrolysis of the appropiate isopropylidene derivative then gave 1-O-(2-hydroxy-4-cis-hexadecenyl)-glycerol.

The benzyloxy compounds[14,15] needed for the hydrogenations were synthesized in the same way as outlined by routes I and II, Fig. 8, for the methoxy-substituted hexade-cyl and hexadecynyl glycerol ethers.

V. BIOLOGICAL EFFECTS OF SUBSTITUTED GLYCEROL ETHERS

In experimental studies, the methoxy-substituted glycerol ethers have demonstrated different biological activities, such as antibiotic effects against bacteria, fungistatic activity against dermatophytes, inhibition of tumor growth and of metastasis formation and stimulation of the immune reactivity.

The mixture of methoxy-substituted glycerol ethers from Greenland shark liver oil as well as synthetic 2-methoxyhexadecyl glycerol (mixture of stereoisomers) had an antibiotic effect in vitro against several types of bacteria, especially Corynebacterium hofmannii, Diplococcus pneumoniae, Staphylococcus pyogenes (A) and Staphylococcus pyogenes (H Oxford), Streptococcus pyogenes and Streptococcus viridans. The antibiotic activity was about as strong as that of nitrofurantoin.[3]

The methoxy-substituted glycerol ethers showed a fungistatic and fungicidal activity in vitro. The dermatophytes used in these studies were monosporically selected strains of Epidermophyton floccosum, Microsporum canis, Trichophyton mentagrophytes and Trichophyton rubrum. In the fungistatic tests, the mixture from Greenland shark liver oil at a concentration of 100 μg/ml inhibited dermatophyte growth in the range of 25–60%. The synthetic compounds, 2-methoxyhexadecyl and 2-methoxyhexadecenyl glycerol, at the same concentration demonstrated an inhibition of 5–15%. In the fungicidal tests, the methoxy glycerol ethers in high concentrations, 1000 μg/ml, inhibited the growth of T. rubrum and E. floccosum.

The substituted glycerol ethers inhibited tumor growth both in vitro and in vivo. In the in vitro tests, two cell lines were studied, a methylcholanthrene-induced murine sarcoma (MCGl-SS) and a juvenile osteogenic sarcoma (2T). A marked growth inhibition was noted for the mixture of 2-methoxy glycerol ethers from Greenland shark liver oil, different single components from Greenland shark liver oil (2-methoxyhexadecenyl glycerol before and after hydrogenation and 2-methoxyoctadecenyl glycerol) and various synthetic glycerol ethers (2-methoxyhexadecyl, 2-ethoxyhexadecyl, 2-methoxyhexade-cenyl, 2-methoxyhexadecynyl, 2-hydroxydodecyl and 3-methoxyhexadecyl glycerol).

In the in vivo tests, the effects on the growth of several solid tumors, ascites tumors, leukemias and lymphomas were studied. The glycerol ethers were incorporated into the feed. Growth inhibition was noted on Melanoma B16, and on a methylcholanthrene-induced sarcoma (MCG101) in C57BL/6J mice and on lymphoma LAA in A/Sn mice by synthetic 2-methoxyhexadecyl glycerol and on a spontaneous mammary carcinoma in C3H mice by methoxy-substituted glycerol ethers from Greenland shark liver oil. Spontaneous metastasis formation from a methylcholanthrene-induced sarcoma (MCG1-SS) in CBA mice was inhibited in lymph nodes and lungs by methoxy-substi-tuted glycerol ethers from Greenland shark liver oil and by synthetic 2-methoxyhexadecyl glycerol. Metastases induced in the liver by injection of MCG1-SS cells into a tail vein or into the portal vein were inhibited by 0.5% of methoxy-substituted glycerol ethers from Greenland shark liver oil in the feed.[3,4]

The methoxy-substituted glycerol ethers also stimulated the immune reactivity against sheep red blood cells (SRBC) and the cellular immune reactivity tested by graft-versus-host reaction. The number of plaque-forming cells (PFC) was determined mostly in CBA mice and in some studies also in C57BL/6J mice and DBA/2J mice. The mixture of methoxy-substituted glycerol ethers from Greenland shark liver oil or synthetic 2-methoxyhexadecyl glycerol given in a concentration of 0.1, 0.25 or 0.5% of the diet for 4 or 14 days before SRBC significantly increased the number of PFC. The effect on cellular immune reactivity was investigated by means of graft-versus-host reaction by determining spleen indices

$$\text{(spleen index = spleen weight} \div \text{body weight of treated mice/spleen weight} \div \text{body weight}$$
$$\text{of controls)}$$

in F_1 hybrid mice receiving spleen cells from the parental strain. Synthetic 2-methoxyhexadecyl glycerol given in a concentration of 0.1% of the feed for 44 days, significantly increased the spleen indices.[2]

In toxicological studies in rats and dogs, it was found that synthetic 2-methoxyhexadecyl glycerol (mixture of stereoisomers) in high doses is toxic, especially to the lymphatic system and to epithelial cells. The high dose levels were 1 g/kg body weight twice a day for 4 weeks to the rats and 350 mg/kg body weight twice a day for the same time period to the dogs. The substance was administered orally by means of a gastric tube to the rats and in gelatine capsules to the dogs. Degenerative changes (vacuolization and some reduction of lymphocytes) were noted in the spleen and in the lymph nodes. The thymus displayed an involution mostly involving the lymphocytic component. Degenerative changes were observed in the tubular renal epithelium and in the epithelium of the urinary bladder. A degeneration of the testis and the ovary and an atrophy of the prostate and uterus were also found. In spite of these widespread pathological findings, the bone marrow was normal. Thus, the methoxy-substituted glycerol ethers in lower doses stimulate the immune reactivity but in higher doses they are toxic to the lymphatic system.

Acknowledgements—The authors are grateful to Drs. Ng. Dinh-Nguyen and J. Vincent for performing the studies on dermatophytes and to Drs. N. O. Bodin, G. Magnusson, T. Malmfors and J. A. Nyberg for the toxicological investigations.

REFERENCES

1. BAUMANN, W. J. and MANGOLD, H. K. *J. Org. Chem.* **29**, 3055–3057 (1964).
2. BOERYD, B., HALLGREN, B., LANGE, S., LINDHOLM, L. and STÄLLBERG, G. Abstracts from the XVIIIth International Conference on the Biochemistry of Lipids, Graz, Austria, 1975.
3. BOERYD, B., HALLGREN, B. and STÄLLBERG, G. *Br. J. Exp. Pathol.* **LII**, 221–230 (1971).
4. BOERYD, B., HALLGREN, B. and STÄLLBERG, G. Abstracts from the XIth International Cancer Congress, Florence, Italy, Panels 2, p. 139, 1974.
5. CERNY, M., MALEK, J., CAPKA, M. and CHVALOVSKY, V. *Collect. Czech. Chem. Commun.* **34**, 1025–1032 (1969).
6. FIESER, L. F. and FIESER, M. *Reagents for Organic Synthesis* p. 567, Wiley, New York, 1967.
7. HALLGREN, B., NIKLASSON, A., STÄLLBERG, G. and THORIN, H. *Acta Chem. Scand.* **B28**, 1029–1034 (1974).
8. HALLGREN, B., NIKLASSON, A., STÄLLBERG, G. and THORIN, H. *Acta Chem. Scand.* **B28**, 1035–1040 (1974).
9. HALLGREN, B. and STÄLLBERG, G. *Acta Chem. Scand.* **21**, 1519–1529 (1967).
10. HALLGREN, B. and STÄLLBERG, G. *Acta Chem. Scand.* **B28**, 1074–1076 (1974).
11. HALLGREN, B., STÄLLBERG, G. and THORIN, H. *Acta Chem. Scand.* **25**, 3781–3784 (1971).
12. NIEHAUS, W. G., JR. and RYHAGE, R. *Anal. Chem.* **40**, 1840–1847 (1968).
13. SNYDER, F. (Ed.) *Ether Lipids: Chemistry and Biology*, Academic Press, New York and London, 1972.
14. STÄLLBERG, G. *Chem. Scr.* **7**, 117–124 (1973).
15. STÄLLBERG, G. *Chem. Scr.* **7**, 31–41 (1975).

Prog. Chem. Fats other Lipids. Vol. 16, pp. 59–99. Pergamon Press, 1978. Printed in Great Britain

TREHALOSE-CONTAINING GLYCOLIPIDS*

C. Asselineau and J. Asselineau

Centre de Recherche de Biochimie et de Génétique cellulaires, C.N.R.S., Toulouse, France

CONTENTS

I. Introduction
II. Distribution and chemical structure of natural trehalose-containing   59
    glycolipids   60
    A. General aspects   60
       1. Detection of trehalose-containing lipids   60
       2. Structure determination   62
       3. Synthesis   63
    B. Esters of α-D-trehalose with fatty acids of medium chain length   64
    C. Esters of α-D-trehalose with β-hydroxy α-branched long chain fatty acids   67
       1. Isolation and structure of cord factor   67
       2. Mycolic acids   68
       3. Purification and structure of cord factor isolated from various species
          of Mycobacteria   72
       4. Isolation of lower homologs of cord factor from bacteria other than
          Mycobacteria   74
          (a) Corynebacteria   74
          (b) Nocardia   76
          (c) Arthrobacter and miscellaneous species of bacteria   77
    D. Esters of α-D-trehalose with phleic acids   78
    E. Esters of α-D-trehalose with sulfuric and phthioceranic acids   80
III. Trehalose-containing glycolipids in the life of bacteria   82
    A. Subcellular localization   83
    B. Metabolism   83
       1. Mycolic acids   83
       2. Phleic acids   85
       3. Phthioceranic acids   85
    C. Possible roles   85
IV. Biological properties of trehalose-containing lipids   87
    A. Toxicity and action on mitochondria and microsomes   87
    B. Production of circulating antibodies   91
    C. Immunostimulation   92
    D. Role in the pathogenicity of bacteria   94
    References   95

I. INTRODUCTION

α-D-Trehalose is a nonreducing disaccharide (Scheme 1) occurring mainly in bacteria, fungi, algae, and insects where it seems to have a role as an energetic reserve compound. So far, only the α,α′-isomer has been found in nature, with the exception of 3-o-α-D-gluco-pyranosyl-α,β-trehalose isolated from *Streptococcus faecalis*.[74a] α,β- and β,β′-isomers (Schemes 2 and 3 respectively) have been prepared by chemical synthesis.[63a]

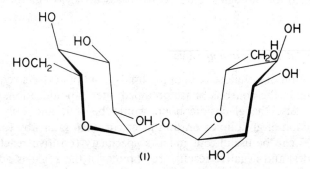

(1)

*This paper is dedicated to the memory of Einar and Stina Stenhagen to whom the authors are grateful for a part of their training during a one-year post-doctoral stay which was the beginning of a warm friendship.

The occurrence of trehalose in lipids was first noticed in 1933 by Anderson and Newman[7] who identified this disaccharide in the saponification products of fats extracted from tubercle bacilli. No definite compound could be isolated at that time, but it was concluded that "the acetone-soluble neutral fats of acid-fast bacteria are not glycerides, but esters of fatty acids with trehalose",[6] a statement that was not supported by subsequent studies.[27, 171]

Later (1956), 6,6'-dimycoloyl-α,α'-D-trehalose was isolated from the lipids of *Mycobacterium tuberculosis*, and afterwards several groups of trehalose-containing lipids have been isolated, mainly from Mycobacteria, Nocardia and Corynebacteria. These glycolipids contain fatty acid esterifying alcohol functions of the sugar, so that they can be considered as *acyl-sugars*, a kind of glycolipid having a much more restricted distribution than the glycosides of diglycerides. Table 1 shows the position of trehalose-containing lipids among the main kinds of natural glycolipids (in this table, glycosides of peptido-lipids have been omitted).

Most reviews on glycolipids give little or no room to glycolipids of Mycobacteria and akin bacteria,[83,105,197–199] which are the best sources of trehalose-containing glycolipids. Only review articles devoted to mycobacterial components[88, 139, 141] provide detailed information on esters of fatty acids and carbohydrates.

II. DISTRIBUTION AND CHEMICAL STRUCTURE OF NATURAL TREHALOSE-CONTAINING GLYCOLIPIDS

A. *General Aspects*

The chemical study of natural trehalose-containing glycolipids includes their detection, isolation and determination of structure. As the acyl-trehaloses so far known are structurally rather different, no general procedure for their isolation can be given. In order to confirm the results of structure determination, and also to obtain pure glycolipids (natural or modified) for biological studies, syntheses of acyl-trehaloses have been performed.

1. *Detection of Trehalose-containing Lipids*

The characterization of trehalose-containing lipids in a lipid extract requires the identification of trehalose in the aqueous phase obtained after saponification. Acid hydrolysis is not suitable because the glycosidic bond would be split and only glucose would be obtained. Detection of glycolipids on thin layer chromatograms by the periodic acid–Schiff procedure[196] can be useful, but it lacks specificity (positive reaction is obtained with monoglycerides) and negative results are obtained if the sugar is adequately esterified. Therefore, it is important to be able to identify trehalose on small samples of lipids. The most simple way is to desalt the aqueous phase obtained after saponification

TABLE 1. Different Kinds of Linkages between Lipid Components and Sugars*

Mode of linkage sugar-lipid	Kind of lipid	Lipid moiety	Sugar moiety	Distribution	References
Ester bond	acylated sugar	fatty acids	D-glucose	mycoplasma, bacteria, yeast, fungi	53, 126, 201, 227
			D-arabinose	bacteria	51, 222
			myoinositol monomannoside	bacteria	1, 32, 132
			β-D-mannopyranosyl-D-erythritol	fungus	183, 200
			α-D-trehalose	bacteria	75
Glycosidic bond	diglycerides	diglycerides	D-glucose, D-galactose, D-mannose	bacteria, plants, animals	193, 197, 198
	glycosides	hydroxy-acids	uronic acids		104, 105
	glycosides	hydroxy-acids	D-sophorose	fungi	52, 103, 205, 221
	ustilagic ac.	hydroxy-acids	D-glucose	fungi	129
	muricatin	hydroxy-acid	D-cellobiose	fungus	145
	pharbic acid	hydroxy-acid	D-glucose	plant	125
	rhamnolipid	hydroxy-acid	D-glucose, L-rhamnose	plant	173
	mycosides A, B or G	hydroxy-acid "phenol-phthiocerol"	L-rhamnose	bacteria (Pseudomonas)	63, 231a
			deoxy-sugars, _o_-methyl-sugars	mycobacteria	18, 139
	ascarosides	long-chain polyols	3,6-_bisdeoxy_-sugar	nematode	76
	glycolipid	long-chain polyol	glucose, galactose	algae	56, 130
	cerebrosides, gangliosides	ceramides	D-glucose, D-galactose, L-fucose	animals	228
			sialic acids, amino-sugars		105
		phytoceramides	various sugars, _myoinositol_	plants, yeasts, fungi	104
Amide bond	lipid A	hydroxy-acids	D-glucosamine	gram negative bacteria	156
	glycolipid	_iso_-C$_{17}$ acid	D-glucosamine (inside an oligoside of diglyceride)	_Flavobacterium thermophilum_	175

*Phosphoglycolipids, containing the glycerophosphate skeleton, are not considered here; terpene derivatives are not included in the lipid compounds.

by ion exchange chromatography, and, after concentration of the solution, to perform paper or thin layer chromatography (TLC). Detection with alkaline silver nitrate[219] is more sensitive than the periodic acid-benzidine spray.[84] Vapor phase chromatography, as trimethylsilyl derivative for example, can also be used.[209]

The α,α'-linkage of the two glucose residues and their D series can be established by optical rotation measurements (α-D-trehalose: $[\alpha]_D + 197°C$ in water). D-glucose obtained after acid hydrolysis can be specifically characterized and estimated by D-glucoseoxidase. α-D-trehalose can be directly identified also by its specific splitting by trehalase.[97] Crystalline α-D-trehalose octaacetate, long prismatic needles m.p. 97–98°C, $[\alpha]_D + 161°$ (chloroform), has also been used for identification of α-D-trehalose.[170, 210]

2. Determination of Structure

When a pure glycolipid has been obtained and the presence of trehalose established, the molecular ratio between the fatty acids and the trehalose can be determined by estimation of the fatty acids (by vapor phase chromatography, for instance) and/or of the trehalose (by anthrone colorimetric determination, for example). By using stationary phases such as silicone SE-30 or SE-52, fatty acid methyl esters and pertrimethylsilyl-trehalose can be estimated at the same time by vapor phase chromatography, making easier determination of their molar ratio. Usually some information on the degree of substitution of the trehalose has been obtained during the isolation process from the migration rate on thin layer chromatograms. This point will be settled during the next step.

The location of the acyl groups on the trehalose molecule is a delicate problem that can be attacked in two different ways. The usual one consists of irreversibly blocking the free hydroxyl groups of the sugar moiety by methylation: methyl iodide and silver oxide in dimethylformamide is one of the best reagents.[128] It is very important to make sure that the methylation is complete (by infrared spectroscopy, for example). Then the permethylated glycolipid is hydrolyzed in acid medium, giving poly-o-methyl-glucose. The places where the fatty acids were linked to the trehalose moiety in the initial glycolipid are pointed out by the alcohol groups that remain free in the methylated sugar after this sequence of reactions. Thus, the problem of location of the acyl groups is transformed into a problem of identification of one or several poly-o-methylglucoses, which is not necessarily an easy one because of the scarcity of reference samples (for examples, see refs 86 and 170).

The alkaline medium required by the methylation step may promote acyl migrations. For example, the easy isomerization of 4-o-acyl-sugars into 6-o-acyl isomers is well known. [191] This problem was solved by avoiding alkaline medium before the protection of the free hydroxyl groups of the glycolipid. This prerequisite requires a pathway that is almost the reverse of the preceding one. The free hydroxyl groups are reversibly protected by acetalation (by methylvinyl ether[42] or by dihydropyranne[181] in the presence of p-toluene-sulfonic acid). Then the product is either deacylated in alkaline medium and methylation by Kuhn's method for example,[128] or successively treated with dimsyl sodium ($CH_3-SO-CH_2^- Na^+$) and methyl iodide.[181] Acid hydrolysis frees the hydroxyl groups that had been protected by acetalation, and splits the glucosidic bond. Now places where fatty acids were linked to the trehalose moiety in the initial glycolipid are labelled by the o-methyl groups. Moreover, this second method usually gives only mono- or di-o-methylglucose which is easier to identify than poly-o-methylglucose. This method, checked with four different methyl palmitoyl-α-D-glucopyranosides, gave very neatly only the expected mono-o-methylglucose.[181]

Protonmagnetic resonance (PMR) spectroscopy does not seem to be of a great help in the structure determination because of the great complexity of the spectra of the glycolipids; however, some information can be obtained in the simpler case of acylated glucoses.[220] So far, mass spectrometry distinguishes only between symmetrical and dissymmetrical diesters of trehalose.[223]

TABLE 2. Synthetic Esters of α-D-Trehalose with Fatty Acids of Medium Chain Length

Acyl-α-D-trehalose	Formula	M.P. °C	$[\alpha]_D$ °C	Method used	References
6-palmitoyl-	$C_{28}H_{52}O_{12}$	117–118	+95 (dioxanne)	transesterification	218
		92–96	+114 (methanol)	splitting of pertrimethyl-silyl-derivative	181
6,6'-dipalmitoyl-	$C_{44}H_{82}O_{13}$	88–90	+75 (dioxanne)	transesterification	218
		154–155	+68 (chlorof.)	potassium salt and diiodo-derivative of trehalose	212
2,3,4,2',3',4'-hexa-stearoyl-	$C_{120}H_{226}O_{17}$	61	+82.2 (chlorof.)	acid chloride and ditrityl-derivative of trehalose	229
2,3,4,6,2',3',4',6'-octastearoyl-	$C_{156}H_{298}O_{19}$	65	+81.8 (chlorof.)	direct esterification	229

3. Synthesis

The first esters of trehalose were prepared by Willstaedt and Borggård in 1946[229] by direct esterification of trehalose or of 6,6'-ditrityl-trehalose by stearoyl chloride in pyridine and the octa- and hexastearoyl-α-D-trehaloses were obtained (Table 2). Later, taking advantage of the higher reactivity of the primary alcohol groups in positions 6 and 6' of trehalose, Gendre and Lederer[82] prepared 6,6'-diacetomycoloyl-α-D-trehalose by using 2 moles of the chloride of acetylated mycolic acid and 1 mole of trehalose in pyridine. Chromatography of the reaction product gave 6-monoester, 6,6'-diester and 2,6,6'-triester fractions. The best yield of diesters (5–10% calculated from the amount of acetylated mycolic acid) was obtained by keeping the reaction mixture at room temperature for 2 or 3 weeks. The location of the ester groups in positions 6 and 6' was not directly established, but was assumed because of the higher reactivity of these alcohol groups previously observed in the case of partial esterification of methyl α-D-glucopyranoside[16] and the similarity of the infrared spectrum with that of a reference compound and the biological properties of this diester.[82]

Because of the low yield of the reaction, the necessity to protect the hydroxyl group of hydroxy acids and also some uncertainty about the homogeneity of the diester fraction, another synthetic approach was devised. Hydrogenolysis of 6,6'-ditrityl-2,3,4,2',3',4'-hexaacetyl-α-D-trehalose gave a hexaacetate which was transformed into the known 6,6'-ditosyl-2,3,4,2',3',4'-hexaacetyl-α-D-trehalose. The attacking of this ditosyl derivative by the anion of the chosen acid in dimethylformamide resulted in the peracetylated 6,6'-diester of the fatty acid being obtained in good yield. However, it was very difficult to remove the protective acetyl groups without losing the fatty acyl residues[177] and for this reason, the preparation of 6,6'-ditosyl-α-D-trehalose by direct partial esterification of trehalose with tosyl chloride was proposed.[55] Later it was shown that the ditosyl-trehalose fraction can be separated by column chromatography into two isomers, identified by nuclear magnetic resonance (NMR) spectrography and mass spectrometry with 6,6'- and 2,6-ditosyl-α-D-trehalose.[217]

A general procedure for the synthesis of 6,6'-diesters of trehalose, giving a high yield of a pure product, was recently described.[212] This procedure is based on the specific reaction of the crowded reagent triphenylphosphine-N-iodosuccinimide with the primary alcohol groups of trehalose. The sequence of reactions is illustrated in Schemes 4–7. A similar synthesis was simultaneously published.[216]

Transesterification between methyl palmitate and trehalose (catalyzed by potassium carbonate) gave a mixture of mono- and dipalmitates, from which 6,6'-dipalmitoyl-α-D-trehalose could be isolated.[218] 6-Palmitoyl-α-D-trehalose was synthesized by an adaptation of the procedure used by Tulloch and Hill[220] to prepare partially acylated β-D-glucopyranosides. Trehalose was transformed into its pertrimethylsilyl derivative (Scheme 8) and one ether group was selectively cleaved by potassium carbonate in methanol at low temperature, giving 2,3,4,2',3',4',6'-hepta(trimethylsilyl)-α-D-trehalose

(Scheme 9). By esterification and hydrolysis by aqueous methanol of the protective groups, the expected monopalmitate (Scheme 11) was obtained.[181]

B. *Esters of α-D-trehalose with Fatty Acids of Medium Chain Length*

From the lipids of *Mycobacterium fortuitum* (*M. minetti*), a glycolipid fraction was isolated because of its low solubility in ether, but still contaminated by some free saturated fatty acids.[224] After treatment with diazomethane, the fatty acid methyl esters were easily removed by chromatography and a pure glycolipid could be obtained. This glycolipid, m.p. 77–81°C, $[\alpha]_D + 45°$ (chloroform), contained 37% carbohydrates. By saponification, it gave a mixture of palmitic and stearic acids, 65 and 25%, respectively, of the total fatty acids, along with small amounts of saturated and unsaturated C_{19} and C_{20} acids, and a hydrosoluble component identified as α-D-trehalose by the melting point and the optical rotation of its octaacetate.

The location of the fatty acyl groups on the trehalose molecule was studied by complete methylation of the glycolipid and after acid hydrolysis, only one spot of methylated glucose was observed by paper chromatography. This derivative was identified as 2,3,4-tri-o-methylglucose by comparing behavior in paper chromatography with that of the 2,3,6-, 2,4,6- and 3,4,6-tri-o-methylglucose. Moreover, it rapidly used 1 mole of periodic acid, as does 2,3,4-tri-o-methylglucose. From these results, it was concluded that the glycolipid isolated from *M. fortuitum* is 6,6'-dipalmitoyl-α-D-trehalose (Scheme 12), contaminated by some homologous material.[224]

(12)

Later, from the lipids of the same strain of *M. fortuitum*, another glycolipid, giving trehalose after saponification, was obtained by using the same isolation procedure as above. Apparently it was obtained in the place of the 6,6'-dipalmitoyl derivative (Scheme 12) because no separation between them was described.[223] This new glycolipid preparation was purified by TLC and peracetylated by a mixture of acetic anhydride and pyridine. The mass spectrum of the peracetylated glycolipid is shown in Fig. 1. The molecular peak at m/e 1150 corresponds to the formula $C_{62}H_{102}O_{19}$. Thus the glycolipid contained two fatty acid residues per mole of trehalose and the fatty acids were $C_{19}H_{36}O_2$, i.e. odd-numbered unsaturated acids. Cyclopropane fatty acids that could have fitted this requirement in the C_{15}–C_{19} range are present only in trace amounts in the lipids of Mycobacteria and their presence in this glycolipid was considered unlikely. Splitting of the glucosidic bond by the electron beam gave an oxonium ion at m/e 802, containing the two fatty acid residues and two acetate groups (Scheme 13) along with an oxonium ion at m/e 331 containing 4 acetate groups (Scheme 14). The peak at m/e 279, due to the ion $C_{18}H_{35}$—$C{\equiv}O^+$, proved that the two fatty acids were C_{19} acids and not a mixture of C_{18} and C_{20} acids. From these mass spectrometric data, it was concluded that this new glycolipid was unsymmetrical diester of trehalose, having the two acyl groups on the same glucose moiety.[223]

(13)   (14)

A sample of this glycolipid was permethylated and hydrolyzed as above. Three spots of methylated glucose were observed by paper chromatography, corresponding to tetra-, tri- and di-o-methylglucose. They were identified with 2,3,4,6-tetra-o-methylglucose, 2,3,4-tri-o-methylglucose and 2,3-di-o-methylglucose, respectively.

All these results showed that the unsymmetrical glycolipid of *M. fortuitum* is a 4,6-diacyl-α-D-trehalose (Scheme 15), probably contaminated with a small amount of a 6,6'-diester because of the characterization of 2,3,4-tri-o-methylglucose.[223] A 6,6'-diester of mycolic acids and trehalose ("cord factor", see below) has also been characterized in the lipids of *M. fortuitum*.[31]

From the lipids produced by *Micromonospora* (strain Sp F3), a glycolipid (about 5% of the total free lipids) was isolated, the saponification of which gave trehalose and a mixture of fatty acids (iso-C_{15}, iso-C_{16}, branched-chain C_{17} and C_{18}). The mass spectrum of the hexaacetyl derivative showed that the main component of the glycolipid fraction was a symmetrical diester of trehalose, possibly contaminated by some unsymmetrical diester.[210] The occurrence of acylated trehalose in the lipids of *Propionibacter-*

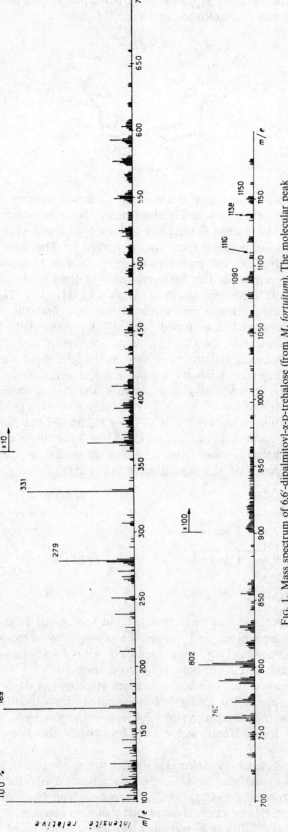

FIG. 1. Mass spectrum of 6,6'-dipalmitoyl-α-D-trehalose (from *M. fortuitum*). The molecular peak is at *m/e* 1150; peaks at *m/e* 331 and 802, respectively, correspond to oxonium ions (Schemes 14 and 13). Reproduced with the permission of the authors and *Chem. Phys. Lipids*.

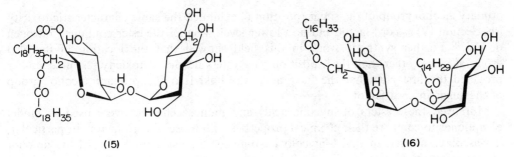

(15)   (16)

ium shermanii is made very likely by the characterization of trehalose in the saponification products. No information on their nature was obtained.[183] Acylated trehalose has been detected in the lipids of the fungi *Pullularia pullulans*,[150] *Agaricus bisporus*[57] and *Claviceps purpurea*;[59] but no structural study was performed.

Perhaps some connection can be made between the diacyl-trehalose found in several bacterial species and diesters of α-D-mannopyranosyl-2-*myo*inositol that were characterized in the lipids of Propionibacteria (Scheme 16).[183, 200] In both cases, they are lipophilic derivatives of a non-reducing disaccharide or analog of disaccharide. The position of the acyl groups on the sugar moiety (mannose) is 6, similar to the 6 position of a glucose moiety of trehalose, but the 1 position of the inositol part is rather different from that found in trehalose derivatives.

Table 2 gives a few properties of synthetic esters of α-D-trehalose with palmitic or stearic acids.

C. Esters of α-D-Trehalose with β-hydroxy α-branched Long-Chain Fatty Acids

1. Isolation and Structure of "Cord Factor"

The cultures of virulent strains of the tubercle bacillus can exhibit parallel arrangements of cells on the surface of the broth resembling curved cords, whereas cultures of avirulent strains have a disorganized aspect.[155] Assuming that the cord formation arose from the division of the bacterial cells inside a lipid matrix, Bloch[43] washed cultures of young virulent tubercle bacilli with petroleum ether, the cords were disrupted and a complex mixture of lipid components was recovered from the organic solvent, apparently containing the factor responsible for the formation of cords.

Subsequent studies on this material[168, 202] showed the presence of a low amount of a *toxic* compound that retained the name "cord factor" because it was first extracted from cord-forming cultures of virulent tubercle bacilli.

By using the toxicity as an index (see Section IV), wax fractions obtained by extraction of whole cells of the tubercle bacillus according to the procedure of Anderson[6] as modified by Aebi *et al.*[3] appeared to be best starting material for the isolation of the cord factor.[19] So far, cord factor has been isolated from every species of mycobacteria studied for this purpose.[18]

The purification of cord factor was performed by means of adsorption chromatography on magnesium trisilicate and silicic acid. Whereas cord factor is strongly adsorbed on silicic acid (Mallinckrodt type), it is not retained by a column of silica gel (Davison type).[25, 169]

As soon as some purification was achieved, it became obvious that cord factor was a derivative of the specific high molecular weight fatty acids of mycobacteria, *mycolic acids* (see below). Saponification of this partly purified cord factor gave free mycolic acids,[25,169] and a glycoside derivative which could be cleaved into D-glucose plus an unidentified compound by acid hydrolysis.[169] Because elementary analyses had given some positive response for nitrogen the presence of nitrogen in cord factor was assumed, and syntheses of mycolic acid esters of D-glucosamine derivatives and methyl α-D-glucopyranoside were undertaken.[20] Esters containing a mycolic acid residue esterifying the primary alcohol group of the sugar (position 6) exhibited the same characteristic toxicity (see Section IV) as cord factor, but at a lower level, whether the ester contained nitrogen or not.[20] Further synthetic work in this field showed that the location of the acyl group at places other than the 6 position gave esters devoid of toxicity.[24] These studies provided the first clue that the fatty acid was linked to the primary alcohol group of the sugar in cord factor.

Moreover, these esters of mycolic acids and monosaccharides were used as model compounds to compare their chemical properties with those of cord factor. In particular, it was observed that methyl 6-mycoloyl-α-D-glucopyranoside used 67% of the amount of periodic acid (corresponding to 2 moles of oxidant per mole of ester), and cord factor used 60% under the same conditions. These low yields are probably due to the organic medium required by the solubilization of mycolic acid derivatives and to steric hindrance. These results favored the presence of 2 α-glycol groups per glucose residue in the cord factor molecule and were only compatible with the binding of the fatty acid with the 6-alcohol group and the presence of free secondary alcohol groups at positions 2, 3 and 4.[25]

Careful study of the water-soluble components obtained after saponification of cord factor led to the isolation of α-D-trehalose and its characterization as crystalline octaacetate. Conventional methylation of cord factor, followed by acid hydrolysis, gave only 2,3,4-tri-o-methyl-D-glucose. All these results indicate that cord factor is a 6,6'-dimycoloyl-α-D-trehalose (Scheme 17a).[170]

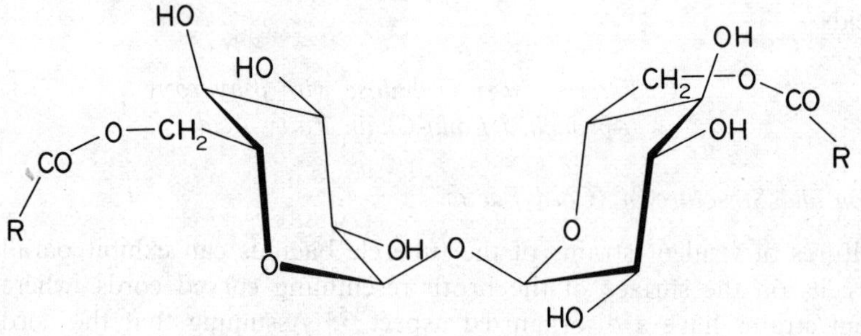

(17)   a : R—CO— = mycoloyl—
       b : R—CO— = corynomycoloyl—

2. Mycolic Acids

Mycolic acids were obtained for the first time by Stodola et al. in 1938[206] by saponification of wax fractions of the tubercle bacillus. Mycolic acids were shown to be high molecular weight methoxy hydroxy fatty acids, of about 88 carbon atoms, and could be split by pyrolysis into n-hexacosanoic acid and an unidentified material.

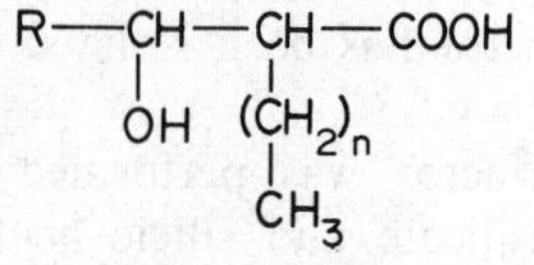

(18)   n=21 or 23

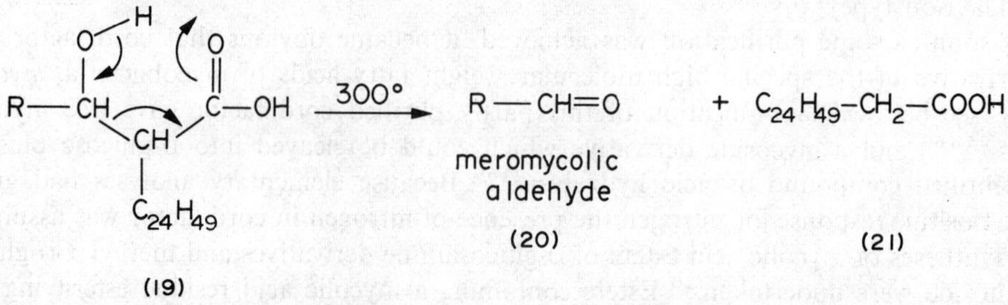

(19)   (20)   (21)

The heterogeneity of samples of mycolic acids isolated from a strain of Mycobacteria was demonstrated by adsorption chromatography, since every strain of Mycobacteria produces a complex mixture of such acids. Their general partial formula is shown in Scheme 18.[23] This structure explains the splitting of mycolic acids by pyrolysis with production of meromycolic aldehyde (Scheme 20) and of n-fatty acids (Scheme 21), the nature of which may vary slightly according to the origin of the mycolic acid. It was proposed that the name *mycolic acid* be the general name for "high molecular weight aliphatic β-hydroxy acids substituted at the α position with a long aliphatic chain".[23]

$$R_4-COOH + \underset{\underset{R_3}{|}}{CH_2}-COOH + \underset{\underset{R_2}{|}}{CH_2}-COOH + \underset{\underset{R_1}{|}}{CH_2}-COOH \longrightarrow R_4-\underset{\underset{R_3}{|}}{CH_2}-CH-\underset{\underset{R_2}{|}}{CH_2}-\underset{\underset{R_1}{|}}{CH}-\overset{\overset{OH}{|}}{CH}-CH-COOH$$

(22)   (23)

$$R_4-\underset{\underset{R_3}{|}}{CH_2}-CH-CH_2-(CH_2-CH_2)_n-\underset{\underset{R_2}{|}}{CH}-\overset{\overset{OH}{|}}{CH}-\underset{\underset{R_1}{|}}{CH}-COOH$$

(24)

In Scheme 18, the radical R may or may not contain another oxygenated group such as hydroxyl, methoxyl, keto, or carboxyl group. In this latter case, they are "dicarboxylic" mycolic acids never encountered in the lipids of *M. tuberculosis* or *M. bovis*.[18]

The first studies on the structure of the radical R, which still contains about 60 carbon atoms, were directed by a biogenetic hypothesis assuming that the mycolic acids having 88 carbon atoms resulted from the condensation of 4 fatty acid molecules (2×26 C + 2×18 C) by Claisen-like reactions (Schemes 22 and 23).[138] Some results giving support to this hypothesis were obtained.[17, 146] As studies of monomolecular films of mycolic acids had shown that "the cross section of the molecule at no part of its length can be larger than that occupied by three parallel hydrocarbon chains",[204] the structure (Scheme 24) was preferred as a hypothetic skeleton of mycolic acids.[18]

The complete elucidation of structure of the mycolic acid molecule had to wait for the application of vapor phase chromatography to long chain fatty acids and for the introduction of more elaborate techniques such as NMR spectrography and mass spectrometry (MS). The use of this latter technique in the lipid field was particularly developed by Ryhage and Stenhagen.[192]

Working on mycolic acids isolated from *M. smegmatis*, Etemadi et al.[72-74] showed in 1964-7 that the main components of the mycolic acid fraction had structures 25 and 26 (Table 3). This conclusion was drawn from the identification of the products obtained by ozonolysis and from NMR spectrography. In particular, a triplet signal near 3.3 τ (relative to tetramethylsilane with an apparatus working at 60 MHz) in the spectrum of the anhydro-acid (Scheme 27) was attributed to the proton on the β-carbon

$$\overset{\gamma}{R}-\overset{\beta}{CH_2}-\overset{}{CH}=\overset{\alpha}{\underset{\underset{C_{22}H_{45}}{|}}{C}}-COOH$$

(27)

because of the presence of 2 protons on the γ-carbon. This observation excluded the presence of a branching chain at position γ that would have occurred with structures 23 or 24.

At the same time, NMR spectrography allowed the detection of *cis*-cyclopropane rings in mycolic acids (mainly those isolated from *M. tuberculosis* and *M. bovis*) by signals at 9.4 and 10.3 τ.[80]

The localization of double bonds in an aliphatic chain can be done very easily by using splitting methods such as ozonolysis used by Etemadi et al. for the unsaturated mycolic acids of M. smegmatis, [72] but the location of a cyclopropane ring inside a long aliphatic chain requires the use of more sophisticated methods developed more recently.

Gastambide-Odier et al.[80] showed that cyclopropane rings are formed in Mycobacteria by the same process of addition of a C-1 unit (arising from S-adenosyl-methionine) on an unsaturated precursor, as in lactobacilli which contain lactobacillic acid (cis-11,12-methylene-octadecenoic acid) and cis-vaccenic acid (11,12-octadecenoic acid). Interpretation of the mass spectra of methyl mycolates was done to find specific peaks arising from the splitting of the polymethylene chain near the cyclopropane rings at m/e values that could be expected if the cyclopropane rings were located at the same places as the double bonds in the polyunsaturated mycolic acids isolated from the same strain of mycobacteria.[66] Etemadi et al. observed that, whereas no specific peaks occur in the mass spectrum of a long chain cyclopropane fatty acid methyl ester, the mass spectrum of the corresponding aldehyde exhibits peaks that were attributed to ions arising from the splitting of the molecule on sides of the cyclopropane ring.[74] In the mass spectrometer, mycolic acids are cleaved with formation of meromycolic aldehydes (Scheme 20), and their mass spectra were interpreted by giving credit to the peaks consistent with the presumed location of the cyclopropane rings. The difficulty of selecting the right peaks gave rise to several possible interpretations of the same mass spectrum, leading the authors to modify their conclusions.[64, 74, 157, 160]

As many fatty acids containing a cyclopropane ring were discovered, the location of such a ring in an aliphatic chain became a general problem. Among several proposed solutions, the introduction of a functional group just beside, or instead of, the ring appeared to be the best one in the case of mycolic acids.

Minnikin and Polgar[158] suggested opening the ring by reaction with BF_3 and methanol prior to the introduction of the sample into the mass spectrometer:

$$\underset{(28)}{\overset{\overset{\displaystyle CH_2}{\diagup\ \diagdown}}{R\text{-}CH\text{-}CH\text{-}R'}} \ \text{------→} \ \underset{(29)}{\overset{\displaystyle CH_2OCH_3}{\underset{\displaystyle |}{R\text{-}CH^+}}} \ + \ \underset{(30)}{\overset{\displaystyle CH_2OCH_3}{\underset{\displaystyle |}{R'\text{-}CH^+}}}$$

Chromic acid oxidation of cyclopropane derivatives easily gives a mixture of the two mono-α-keto-cyclopropane compounds:[179]

$$\underset{(31)}{\overset{\overset{\displaystyle CH_2}{\diagup\ \diagdown}}{R\text{-}CH_2\text{-}CH\text{-}CH\text{-}CH_2\text{-}R'}} \ \text{----→} \ \underset{(32)}{\overset{\overset{\displaystyle CH_2}{\diagup\ \diagdown}}{R\text{-}CO\text{-}CH\text{-}CH\text{-}CH_2\text{-}R'}}$$

$$+ \ \underset{(33)}{\overset{\overset{\displaystyle CH_2}{\diagup\ \diagdown}}{R\text{-}CH_2\text{-}CH\text{-}CH\text{-}CO\text{-}R'}}$$

The presence of an α-keto group induces a specific splitting of the molecule in the mass spectrometer, leading to intense peaks. Mass spectra performed on the mixture of the two possible keto derivatives are easily interpreted.[179] This method, applied to the acetyl derivative of methyl α-mycolate Canetti, gave the 4 monoketo derivatives and the mass spectrum of the mixture (Fig. 2) gave the structure (38) (Table 3).[12]

Moreover, Puzo and Promé[186] observed that a cyclopropane fatty aldehyde exhibits intense specific peaks in its mass spectrum only when $n = 4$ or > 10 (Scheme 34):

$$\underset{(34)}{\overset{\overset{\displaystyle CH_2}{\diagup\ \diagdown}}{CH_3\text{-}(CH_2)_m\text{-}CH\text{-}CH\text{-}(CH_2)_n\text{-}CHO}}$$

These values of n allow the easier interactions between the aldehyde group and the cyclopropane ring. These results stress the importance of using chemically modified derivatives of the cyclopropane acids when unknown aliphatic cyclopropane compounds have to be studied by mass spectrometry.

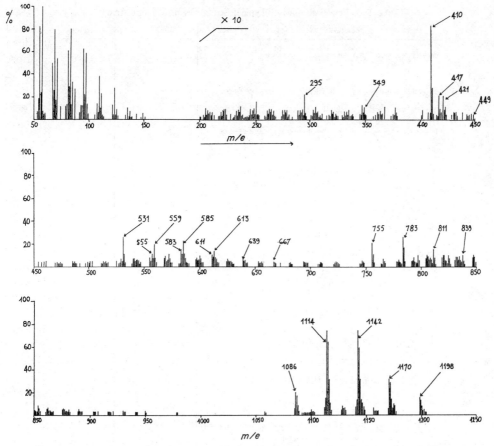

FIG. 2. Mass spectrum of the keto-esters obtained by oxidation of the acetyl derivative of methyl α-mycolate *Canetti*. The high mass part of the spectrum shows two series of peaks corresponding to ions (M-60-32)$^+$ and (M-60-32-28)$^+$ (the last ones resulting from a rearrangement of the α-keto-cyclopropyl derivatives into α-keto-unsaturated compounds with the loss of 28 daltons). Other peaks mainly result from the splitting of the molecules between the keto and the cyclopropane groups.[12]

We have mentioned above that a few strains of Mycobacteria contain "dicarboxylic" mycolic acids (Scheme 41) in their lipids, the molecular weight of which is always lower than that of monocarboxylic mycolic acids.[18] From the lipids of *M. paratuberculosis* (Johne's bacillus), Lanéelle and Lanéelle[135] isolated the keto-mycolic acid (Scheme 39) along with the monoester of dicarboxylic acid (Scheme 40, Table 3).

Formulae 35–41 (Table 3) show the structures of several types of mycolic acids. It must be noted that whereas mycolic acids isolated from *M. tuberculosis* or *M. bovis* have mainly a C_{24} α-branched chain (and give *n*-hexacosanoic acid by pyrolysis), mycolic acids isolated from *M. avium*, saprophytic mycobacteria (*M. phlei*, *M. smegmatis*,...) or atypical mycobacteria (*M. kansasii*, *M. fortuitum*,...) have a C_{22} α-branched chain (and give *n*-tetracosanoic acid by pyrolysis).

All the mycolic acids have the same partial structure, with two assymmetric centers at positions 2 and 3 having the R configuration (Scheme 42). The correlation was made

$$
\begin{array}{c}
CH_3 \\
|\\
(CH_2)_{23} \\
|\\
CH_3OOC\!-\!\overset{\displaystyle |}{C}\!-\!H \\
H\!-\!\overset{\displaystyle |}{C}\!-\!OH \\
|\\
R
\end{array}
$$

(42)

TABLE 3. A Few Examples of Typical Structures of Mycolic Acids*

Structure	Source	References
$CH_3-(CH_2)_m-CH=CH-(CH_2)_n-CH=CH-(CH_2)_p-CHOH-CH-COOH$ $\underline{25}$ $\quad C_{22}H_{45}$	M. smegmatis	72, 73
$CH_3-(CH_2)_m-CH=CH-(CH_2)_{n-1}-CH=CH-CH-(CH_2)_p-CHOH-CH-COOH$ $\underline{26}\quad CH_3 \qquad C_{22}H_{45}$	M. smegmatis	74
$CH_3-(CH_2)_m-CH-CH-(CH_2)_{n-1}-CH=CH-CH-(CH_2)_p-CHOH-CH-COOH$ $CH_2\quad\underline{35}\qquad CH_3\qquad C_{22}H_{45}$	M. smegmatis	74
$CH_3-(CH_2)_m-CH-CH-(CH_2)_n-CH-CH-(CH_2)_p-CHOH-CH-COOH$ $CH_2\qquad\underline{36}\quad CH_2\qquad C_{22}H_{45}$	M. kansasii	71
$\underline{25},\underline{26},\underline{35},\underline{36}:\quad m=15,17,19\quad n=14,16\quad p=15,17$		
$CH_3-(CH_2)_{19}-CH-CH-(CH_2)_n-CH-CH-(CH_2)_p-CHOH-CH-COOH$ $CH_2\qquad\quad CH_2\qquad C_{24}H_{49}$ $\underline{37}\quad n=14,16\quad p=9,11,13$	M. tuberculosis (strain Canetti)	12
$CH_3-(CH_2)_{17}-CH-CH(OCH_3)-(CH_2)_{14}-CH-CH-(CH_2)_{17}-CHOH-CH-COOH$ $CH_3\qquad\qquad\qquad CH_2\qquad\quad C_{24}H_{49}$ $\underline{38}$	M. tuberculosis (strain Test)	64, 70
$CH_3-(CH_2)_{17}-CH-CO-(C_nH_{2n-2})-CHOH-CH-COOH$ $CH_3\qquad\underline{39}\qquad C_{22}H_{45}$	M. paratuberculosis	135
$CH_3-(CH_2)_{17}-CH-O-CO-(C_nH_{2n-2})-CHOH-CH-COOH$ $CH_3\qquad\underline{40}\qquad C_{22}H_{45}$	M. paratuberculosis	135
$HOOC-(C_nH_{2n-2})--CHOH-CH-COOH$ $\qquad\qquad\qquad C_{22}H_{45}$ $\underline{41}$	M. paratuberculosis	135
$\underline{39},\underline{40},\underline{41}:\ n=34\ to\ 39$ (mixtures of molecules with or without cyclopropane ring)		

*This table is not exhaustive as regards structures or sources.

by transforming several mycolic acids into the corresponding β-diols and then isopropylidene derivatives.[15] Examination of the infrared spectra of mycolic acids and of the epimers obtained by alkaline treatment gave results that are in agreement with the *threo* structure (Scheme 42) of mycolic acids.[64, 159] Epimers of mycolic acids (pre-α-mycolic acids) are observed in small amounts in the samples of mycolic acids obtained after saponification.[65]

3. Purification and Structure of Cord Factor Isolated from Various Species of Mycobacteria

As we have seen, there are many kinds of mycolic acids according to the species or even to the strain of Mycobacteria. Cord factor is a 6,6′-diester of α-D-trehalose with mycolic acids (Scheme 17) and, as a consequence, there are many kinds of cord factor that differ one from the others by the nature of their mycolic acid moieties.

Beside the procedures of isolation of cord factor from wax fractions of the tubercle bacillus and purification described by Noll and Bloch[169] and Asselineau and Lederer,[25] based on adsorption chromatography on magnesium trisilicate and silicic acid or silica gel, chromatographic purification of cord factor was also performed by Nojima[166] and by Mishima *et al.*[162] Free mycolic acid is a tenacious contaminant of cord factor preparations (it was suggested that it might arise from some partial hydrolysis on the absorbent during chromatography[91]) and its complete removal is rather troublesome. Complete elimination of free mycolic acid requires either a brief treatment of a solution of cord factor with diazomethane followed by a rapid chromatography, or by a chromatography of the preparation on DEAE-cellulose (acetate form).[91] Purification by centrifugal microparticulate gel chromatography has also been used,[188] but this last method is mainly an analytical one.

TABLE 4. Strains of Mycobacteria from which Cord Factor was Isolated and Characterized

M. tuberculosis	M. bovis	M. avium	Saprophytic strains of Mycobacteria	Atypical strains of Mycobacteria
H-37 Rv (25)	Vallée (60)	n°802 (161)	M. phlei (29)	M. fortuitum (31)
Brévannes (1969)	Marmorek (60)		M. smegmatis (29)	M. marianum (152)
Aoyama B (30)	Peurois (2)		M. butyricum (2)	M. kansasii (36)
Nakano (172)	B.C.G. (46,28,170)		M. paratuberculosis*	n° 6 (29)
PN + DT + C (169)				n° 22 (29)
				P-16 (31)
				n° 1217 (136)

*Lanéelle, G., unpublished results.

Usually the purification of cord factor ends by a precipitation of its solution in ether by an excess of methanol while stirring the mixture. The main interest of this procedure, usable for many kinds of mycolic acid derivatives, is to obtain a fine white powder which is easier to handle than the waxy crude product. Cord factor has about 170 carbon atoms and all the preparations isolated from various species and strains of Mycobacteria (see Table 4) have about the same melting points (ca 44–45°C) and optical rotations. Reviews devoted to cord factor were published by Noll[167] and Goren.[90]

The purified preparations of cord factor are still mixtures of molecular species of 6,6'-dimycoloyl-α-D-trehalose, because their saponification gives mixtures of mycolic acids that can be separated by usual chromatographic techniques. So far, the chemical and biological properties of cord factor have been studied only on such purified mixtures.

However, it is possible to separate these different molecular species in groups. Saponification of a preparation of cord factor isolated from M. phlei gave two kinds of mycolic acids, a monocarboxylic acid and a dicarboxylic acid, and both are mixtures of homologs. The pertrimethylsilyl derivative of the cord factor fraction could be separated into 5 fractions by preparative TLC. Each fraction was hydrolyzed in very mild conditions to recover the free glycolipids. The five fractions thus obtained were the 6-monomycolates (Schemes 43 and 44), the two symmetrical 6,6'-dimycolates and the dissymmetrical

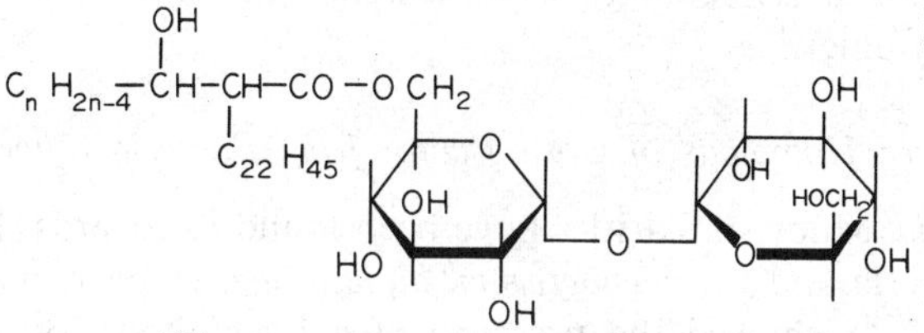

(43)

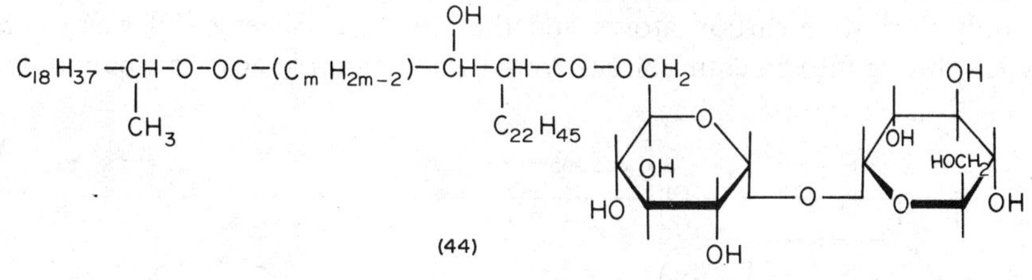

(44)

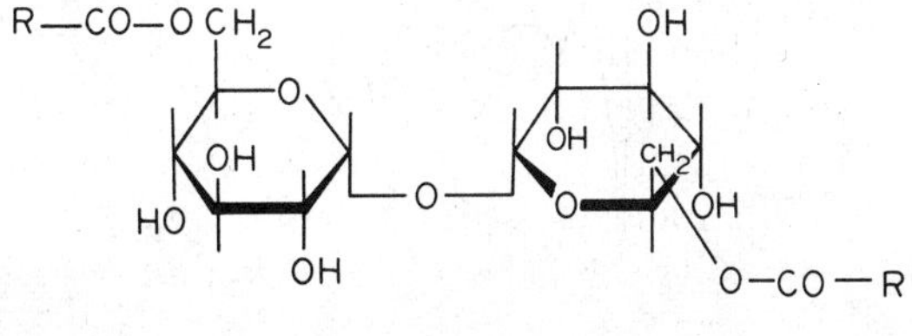

6,6'-dimycolate (Scheme 45). It must be noted that the ω-carboxyl group of the dicarboxylic mycolic esters is not free, but esterified by eicosanol-2.[181] In these glycolipids, the mycolic acids seemed to be distributed randomly.

$$R- = C_n H_{2n-4}-CHOH-CH- \quad (R_1)$$
$$\phantom{R- = C_n H_{2n-4}-CHOH-}\overset{|}{C_{22}H_{45}}$$

$$R- = C_{18}H_{37}-CH-O-OC-(C_m H_{2m-2})-CHOH-CH- \ (R_2)$$
$$\phantom{R- = C_{18}H_{37}-}\overset{|}{CH_3}\phantom{-O-OC-(C_m H_{2m-2})-CHOH-}\overset{|}{C_{22}H_{45}}$$

$$R- = \text{one mole of } R_1 \text{ and one mole of } R_2$$

A 6-monomycoloyl-α-D-trehalose was isolated from the human strain *Aoyama B* of *M. tuberculosis*.[121] The formation of such trehalose monoesters by partial saponification of cord factor during the chromatographic fractionation cannot be excluded.

The infrared spectrum of cord factor (Fig. 3) contains a strong bonded OH band at $3350 \, cm^{-1}$ and a strong CO ester band at $1715 \, cm^{-1}$. The presence of trehalose is associated with the occurrence of five bands at 1057, 1079, 1104, 1152 and $1175 \, cm^{-1}$ which give a characteristic pattern to the spectrum[167] and are a convenient guide to follow cord factor purification. A small band at $808 \, cm^{-1}$ is "found consistently in every trehalose derivative so far examined" (Goren and Brokl[91]). Because of the long polymethylene chains of the mycolic acid residues, the band at $720 \, cm^{-1}$ is unusually strong.

(aceto–mycoloyl)–O–CH₂

(46)

The mass spectrum of peracetylated cord factor was studied by Adam *et al.*[2] Because the molecular weight is higher than 3000, no molecular peak could be observed. The most useful peaks were those corresponding to the oxonium ions (Scheme 46) having different kinds of acetylated mycolic acid residues. No information on the location of the acyl groups on the sugar rings could be drawn from the spectra. Because of the too high molecular weight, and therefore the absence of the high mass part of the spectrum, it was not possible to observe whether the different molecular species of cord factor were symmetrical.[2]

4. *Isolation of Lower Homologs of Cord Factor from Bacteria other than Mycobacteria*

(a) *Corynebacteria*. Many similarities have been found as regards the chemical composition of Mycobacteria and *Corynebacterium diphtheriae*. In particular, as early as 1951, Lederer and Pudles[143] showed the presence of a lower homolog of mycolic acids in the lipids of *C. diphtheriae*. This α-branched chain β-hydroxy acid, called *corynomycolic acid*, had only thirty-two carbon atoms and the structure (Scheme 47) was established by pyrolytic splitting into an equimolecular mixture of palmitic acid and palmitaldehyde,

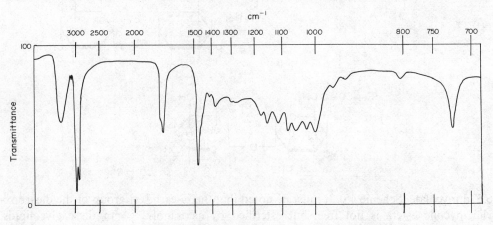

FIG. 3. Infrared spectrum of cord factor (isolated from BCG).

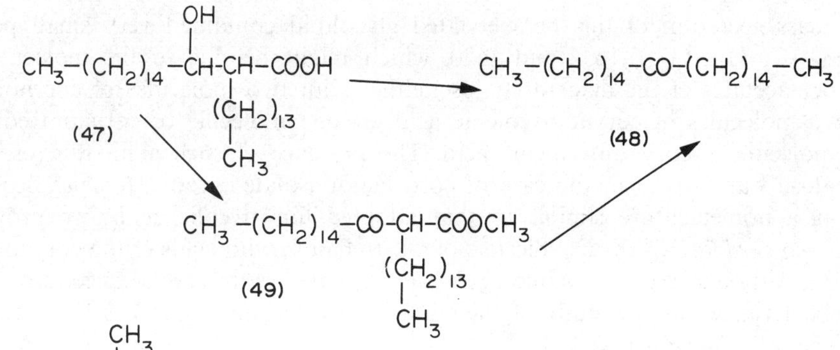

OH
|
CH₃–(CH₂)₁₄–CH–CH–COOH ⟶ CH₃–(CH₂)₁₄–CO–(CH₂)₁₄–CH₃
|
(CH₂)₁₃
|
CH₃

(47)   (48)

CH₃–(CH₂)₁₄–CO–CH–COOCH₃
|
(CH₂)₁₃
|
CH₃

(49)

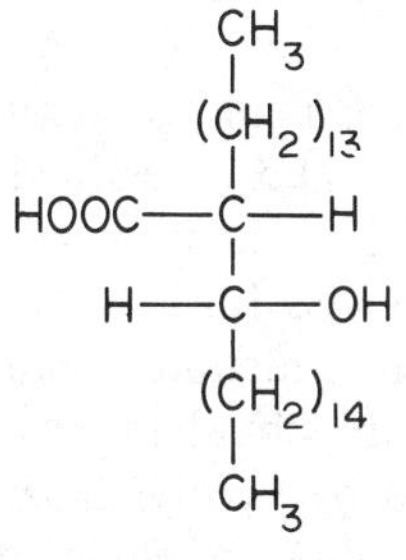

CH₃
|
(CH₂)₁₃
|
HOOC—C—H
|
H—C—OH
|
(CH₂)₁₄
|
CH₃

(50)

OH
|
CH₃–(CH₂)₅–CH=CH–(CH₂)₇–CH–CH-COOH
|
C₁₄H₂₉

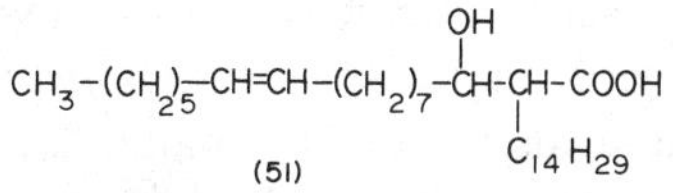

(51)

and by oxidative transformation into palmitone (Scheme 48).[143, 144] This structure was checked by synthesis of the mixture of diastereoisomers from two molecules of methyl palmitate and subsequent reduction of the methyl 2-tetradecyl-3-oxooctadecanoate (Scheme 49).[142] Later, the diastereoisomers were separated[178] into the two racemic (*threo* and *erythro*) compounds by chromatography. The stereochemistry of natural corynomycolic acid was established as 2R,3R –*threo* isomer) (Scheme 50). This conclusion was based on the transformation of corynomycolic acid into appropriate derivatives[9] and on the synthesis of corynomycolanediol from a paraconic acid of known configuration.[213]

In the lipids of *C. diphtheriae*, corynomycolic acid is accompanied by a C_{32}-monounsaturated acid, *corynomycolenic acid* (Scheme 51).[184, 185]

Although *C. diphtheriae* and *C. ovis*[61] produce both saturated and unsaturated C_{32}-corynomycolic acids almost devoid of homologs, other species of Corynebacteria can synthesize more or less complex mixtures of saturated and unsaturated corynomycolic acids.[68, 226] Yano and Saito[234] could separate (by vapor phase chromatography of the trimethylsilyl derivatives of the methyl esters), twelve major corynomycolic acids ranging from C_{20} to C_{32} from the lipids of *C. ulcerans*.

The presence of a cord factor-like compound in the lipids of *C. diphtheriae* was suggested by the work of Alimova,[4] who isolated esters of trehalose with unidentified fatty acids from the lipids of this bacteria. These trehalose-containing glycolipids were located at the surface of the bacterial cell, for they were removed by washing the cells with a 2% aqueous solution of sodium taurocholate.[5] Moreover, these glycolipids had the ability to induce toxic phenomena in the skin of guinea-pigs and rabbits by intradermal injection[5] (see Section IV).

The lipids of *C. diphtheriae* were separated according to their solubility in boiling acetone. By chromatography of the insoluble fraction on silicic acid and silica gel, Ioneda *et al.*[98] isolated a glycolipid as a colorless wax, m.p. 110–115°C, $[\alpha]_D + 64°$ (chloroform). Saponification of this glycolipid gave equimolecular amounts of α-D-trehalose, corynomycolic acid and corynomycolenic acid. The infrared spectrum supported the structure of a diester of trehalose with these hydroxy acids.[98] No information on the positions of the acyl groups on the glucose rings of the trehalose could be obtained from the mass spectrum of the peracetylated glycolipid. Hydrolysis of the permethylated derivative gave a tri-*o*-methylglucose which was identified with 2,3,4-tri-*o*-methylglucose by vapor phase chromatography. Thus, the glycolipid isolated from *C. diphtheriae* is 6,6′-dicorynomycoloyl-α-D-trehalose (Scheme 17*b*).[194]

The mass spectrum of this peracetylated glycolipid contained very small peaks at m/e 1634 ($C_{82}H_{162}O_{23}$), 1632 and 1630, which might be due to the molecular ions of the octaacetates of the diacyl-trehalose, either with two molecules of corynomycolic acid, two molecules of corynomycolenic acid, or one molecule of corynomycolic acid and 1 molecule of corynomycolenic acid. The presence of such a mixture of diesters of trehalose was proved in the case of cord factor isolated from *M. phlei*, see above. By using a nomenclature similar to that adopted for mycolic acids, we propose to call "*coryno-cord factor*" the 6,6'-diesters of α-D-trehalose with acids of the corynomycolic acid type. Although corynomycolic acids were isolated from several species of aerobic Corynebacteria, so far the study of their esters with trehalose appears to be restricted to the species *C. diphtheriae*.

(b) *Nocardia*. *Nocardia* is the genus closest to *Mycobacterium* and sometimes the classification of a strain in one of these two genera requires long and diversified studies (example: *N. rhodocrous*). In Nocardia, mycolic acids have 40–50 carbon atoms, a size intermediary between those of corynomycolic and mycolic acids. Initially they were called nocardic acids,[153] but the name *nocardomycolic acid*, which denotes relationship to mycolic acids, is preferred. Usually nocardomycolic acids are obtained as mixtures of saturated, monounsaturated and diunsaturated acids, cleaved by pyrolysis with liberation of normal chain C_{12}–C_{18} fatty acids. Unsaturations are most often localized in the main chain. Their general formula is (52).[153]

$$CH_3-[R]-CH-\underset{\underset{C_nH_{2n+1}}{|}}{\overset{\overset{OH}{|}}{CH}}-COOH$$

(52)

$$R = C_m H_{2m} \qquad m: \text{even number}$$
$$C_m H_{2m-2} \qquad 26 \leqslant m \leqslant 34$$
$$C_m H_{2m-4}$$

$$n = 10, 12, 14, 16$$

Detailed analyses of the nocardomycolic acid fraction isolated from the lipids of *N. corallina*[33] and *N. erythropolis*[235] by a combination of vapor phase chromatography and MS showed the presence of at least fifteen nocardomycolic acids without considering possible isomers arising from the positions of the double bonds. The carbon numbers of these acids varied from 32 to 48. The stereochemistry of the nocardomycolic acids is the same as that of corynomycolic acid and mycolic acids.[15]

An overlapping occurs between corynomycolic acids (C_{20}–C_{36}) and nocardomycolic acids (C_{32}–C_{56}). *C. rubrum* is unique because it produces corynomycolic acids with carbon numbers of 29–46.[66] Thus, the names of these acids are quite arbitrary and only indicate the bacterial source. The situation is rather different for mycolic acids, which have carbon numbers of 74–90 and sometimes structural details such as cyclopropane rings and/or one more oxygen-containing functional group. However, small amounts of C_{60}-mycolic acids were characterized in the lipids of *M. smegmatis*[127] and *N. kirovani* produced nocardomycolic acids $C_{58}H_{110}O_3$–$C_{66}H_{124}O_3$.[149] These latter acids contain three or four double bonds whereas mycolic acids have never more than two double bonds, and the two kinds of hydroxy acids differ by the length of the α-branch (10–16 carbon atoms in nocardomycolic acids instead of 22 or 24 carbon atoms in mycolic acids). This last property has been used for taxonomic studies of some Mycobacteria and Nocardia strains which are difficult to characterize by usual bacteriological methods.[134, 137]

By chromatography of the lipids insoluble in boiling acetone isolated from *N. asteroides*, Ioneda *et al.*[97] obtained a waxy compound, m.p. 76–78°C, $[\alpha]_D + 62°$ (chloroform), the saponification of which gave α-D-trehalose (identified by its splitting by trehalase).

The infrared spectrum of the glycolipid was identical with that of coryno-cord factor. The mass spectrum of its permethylated derivative, as well as the mass spectrum of the methyl esters of the acids obtained after saponification, showed that the trehalose core was esterified by two molecules of mono- or diunsaturated corynomycolic acids having 32–36 carbon atoms. The location of the acyl groups could not be determined because of the paucity of material, but was assumed to be 6,6'. Thus, this strain of N. asteroides is likely to contain coryno-cord factor.

Saponification of the "bound" lipids of the cell wall of the same bacteria gave a series of nocardomycolic acids, $C_{50}H_{100}O_3$–$C_{58}H_{114}O_3$, characterized by MS.[97] We have just seen that in the "free" lipids, corynomycolic acids are specifically esterified with trehalose. That would say that in the same cell, a segregation of the mycolic acids occurs according to their chain length. Previous studies on the branched-chain hydroxyacids of N. asteroides, obtained by direct saponification of the whole cells, gave only C_{46}–C_{58}-nocardomycolic acids.[69] This discrepancy might be due to different culture conditions.

The presence of diesters of α-D-trehalose with acids of the mycolic acid type in the lipids of N. asteroides was confirmed by Yano et al.[233] These authors increased the yield of trehalose-containing glycolipids by growing the bacteria in a medium in which glycerol was replaced by glucose. Two-dimensional TLC showed that some monoacyl-trehalose accompanied the diacyl-trehalose. These latter diesters had the same toxic properties as cord factor[233] (see Section IV).

By chromatography of the crude phospholipid fraction of N. rhodocrous on silicic acid and silica gel, a glycolipid (1.5% of the phospholipid fraction) was isolated as a wax m.p. 59–60°, $[\alpha]_D + 50°$ (chloroform). Its saponification gave α-D-trehalose and a mixture of nocardomycolic acids in the range $C_{38}H_{76}O_3$ to $C_{46}H_{90}O_3$. Its sugar content (18%) was in agreement with that of a diester of nocardomycolic acids and trehalose. The mass spectrum, performed on the permethylated derivative, confirmed the presence of an acyl group on each glucose moiety, but gave no information on their precise locations. As the infrared spectrum was practically identical with that of cord factor, it was concluded that this glycolipid is 6,6'-dinocardomycoloyl-α-D-trehalose.[97] In this case, it is a true "nocardo-cord factor". The same range of nocardo-mycolic acids was found in N. calcarea, another member of the rhodocrous group of bacteria.[157]

The presence of acyl-trehalose was detected in the lipids of N. polychromogenes and N. coelica grown in a medium containing glucose.[126] No information on the structure of these glycolipids was given. They were accompanied by glucose-containing compounds having the behavior of a triacyl-glucose in N. polychromogenes and of a diacyl-glucose in N. coelica.[126]

(c) Arthrobacter and miscellaneous species of bacteria. If Arthrobacter paraffineus is grown in a jar fermentor in a medium containing 10% n-paraffin (mainly C_{12}–C_{14} hydrocarbons) under agitation and aeration, an emulsion layer appears, and consists of bacterial cells, lipids and n-paraffin. The emulsion layer was collected by centrifugation and extracted. Adsorption chromatography of the extract gave a trehalose-containing lipid.[208] α-D-Trehalose, the only water-soluble component obtained after saponification, was identified by paper chromatography and by its degradation into D-glucose with glucoseoxidase.

The acid moiety of this glycolipid was obtained as a colorless compound, m.p. 68–70°C, consisting of hydroxy acids as shown by the formation of acetate (m.p. 20–21°C) and methyl ester (m.p. 51–52°C). GLC and MS showed that this fraction contained at least seven hydroxy acids, having the general formula R_1—CHOH—CH(R_2)—COOH, where R_1 was an alkyl chain C_{18}–C_{23} and R_2 an alkyl chain C_7–C_{12}.[208]

Because two molecules of hydroxy acids were obtained per mole of trehalose, the glycolipid fraction was a mixture of diesters of corynomycolic acid-like acids and α-D-trehalose. The positions of the acyl groups on the sugar were not determined; however, when the permethylated glycolipid was methanolized, GLC showed a single peak, the retention time of which was different from those of methyl 2,3,4-tri-o-methyl-α- or β-D-glucoside. Periodic acid oxidation of the glycolipid used only 1 mole of periodate, making likely a dissymmetrical structure. This result is difficult to reconcile with the detection of a single methylated glucose after permethylation and hydrolysis.[208]

As this glycolipid might have some role in the absorption of the paraffin by the bacterial cells, Suzuki *et al*[208] looked for the possible production of trehalose-containing lipids by other species of bacteria grown on paraffin. They obtained positive results in the case of *Corynebacterium pseudodiphtheriae*, *C. fasciens*, several strains of *Arthrobacter*, several strains of *Brevibacterium*, one strain of *Nocardia* and five unidentified bacterial strains.

B. thiogenitalis, an oleic acid-requiring mutant, releases in the culture medium containing oleic acid the 6-monoacyl-D-glucose (Scheme 53), which is about one-half of the molecules of diesters of trehalose produced by *Arthrobacter*.[174]

$$CH_3-(CH_2)_7-CH{=}CH-(CH_2)_7-CHOH-CH-(CH_2)_6-CH{=}CH-(CH_2)_7-CH_3$$

(53)

Brevibacterium vitarumen contains a 6,6′-dicorynomycoloyl-α-D-trehalose.[133]

D. *Esters of α-D-Trehalose with Phleic Acids*

A highly unsaturated lipid fraction was detected in a petroleum ether extract of *Mycobacterium phlei* cells. By saponification and chromatography of the fatty acid methyl esters on silicic acid and silver nitrate impregnated silicic acid, a series of polyunsaturated fatty esters could be isolated. Vapor phase chromatography of these methyl esters demonstrated the presence of five components, one of them being largely predominant. The name *phleic acids* was given to these acids because they were isolated for the first time from *M. phlei*. It was found later that phleic acids always occur as esters of α-D-trehalose.[13, 14]

The mass spectrum of the methyl esters of phleic acids showed a molecular peak at m/e 540 for the main component, and at m/e 486, 512, 566 and 594 for the minor ones, corresponding to the formulae $C_{37}H_{64}O_2$, $C_{33}H_{58}O_2$, $C_{35}H_{60}O_2$, $C_{39}H_{66}O_2$ and $C_{41}H_{70}O_2$, respectively (as methyl esters). Catalytic hydrogenation of the mixture of methyl esters gave saturated derivatives, the mass spectrum of which exhibited molecular peaks at m/e 550 (main component), 494, 522, 578 and 606. The products of molecular weight (m.w.) 540 (main component) and 512 had thus five double bonds and differed one from the other only by 2 CH_2 (28 daltons). Methyl phleates of m.w. 566 and 594 had six double bonds and also differed by 2 CH_2, whereas the methyl phleate m.w. 486 had only four double bonds.[13]

The mixture of perhydrogenated methyl esters was transformed into hydrocarbons through the alcohols, and hydrogenolysis of the tosylates. According to their infrared and NMR spectra and their vapor phase chromatographic behavior, these hydrocarbons were of normal chain. On the other hand, no conjugation of double bonds was observed in the U.V. spectrum of methyl phleates. Degradative oxidation of heptyl phleates gave palmitic and myristic acids, succinic acid and heptyl hydrogen succinate.[13] All these results demonstrated the general formula (Scheme 54) for the five phleic acids. The main component of the mixture is hexatriaconta-4,8,12,16,20-pentaenoic acid (Scheme 54, $m = 14$ and $n = 5$); it makes about 75% of the total phleic acids.[13]

$$CH_3-(CH_2)_m-(CH{=}CH-CH_2-CH_2)_n-COOH$$

(54) phleic acids

$$m=14 \quad n = 4,5 \text{ or } 6$$
$$m=12 \quad n = 5 \quad \text{ or } 6$$

TABLE 5. Main Peaks Observed in the NMR Spectrum of Methyl Phleates

Formula	Chemical shift τ	Signal	Proton number
CH_3	9.08	multiplet	2.9
$(CH_2)_{n-1}$	8.70	singulet (apparent)	26.1
CH_2	8.00	multiplet	18.0
CH	4.62		
CH			
CH_2	7.88		
CH_2			
CH	4.62	multiplet (symmetrical)	9.3
CH			
CH_2 $\}_{m-1}$	7.70	doublet (apparent)	4.3
CH_2			
CO			
O			
CH_3	6.35	singulet	3.0

A broad band of low intensity at 990 cm^{-1} in the infrared spectrum and the absence of a band near 960 cm^{-1} are in agreement with the *cis* geometry of the double bonds, as well as the presence of a band at 4760 cm^{-1} in the near infrared.[96] The NMR spectrum of the mixture of methyl phleates is in agreement with the structure (54). Its main features are collected in Table 5. The complete interpretation of this spectrum was performed by using data published on synthetic fatty acids containing some of the groups present in methyl phleates.[13]

The question of the form under which they occur in the bacterial cell arose from the isolation of phleic acids. By fractionation of the lipids of *M. phlei* on silver nitrate impregnated silicic acid, trehalose phleates were isolated, which appeared to be the main natural derivatives. These esters were obtained as mixtures of trehalose esters differing one from the others by the number of acyl groups carried by the sugar core. The isolation of a pure compound was difficult and could only be achieved for the main component of the mixture, which was also the stablest and least polar member of the series.[14]

This glycolipid which is liquid at room temperature and chromatographically homogeneous was saponified to give only phleic acids and α-D-trehalose. The quantitative determination of phleic acids as methyl esters and trehalose as the trimethylsilyl derivative by vapor phase chromatography and the absence of OH vibration in the infrared spectrum were in agreement with the structure of an octa-ester. NMR spectrography

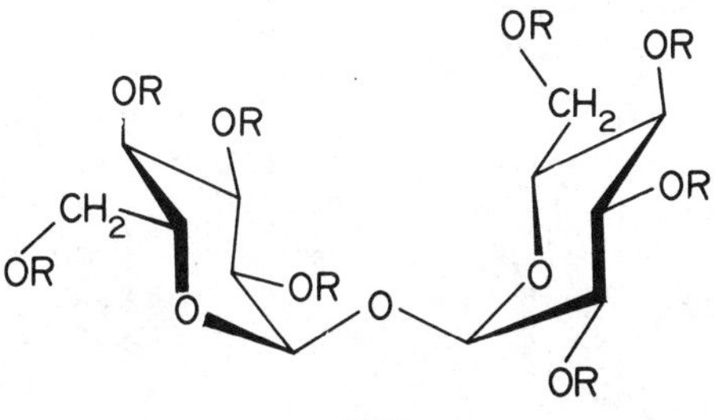

(55)

R = phleyl

did not give quantitative results precise enough to make the distinction between hepta-
or octa-esters. The behavior of this glycolipid on thin layer chromatograms was not
altered by treatment with diazomethane and BF_3. Therefore, this less polar ester was
considered to be α-D-trehalose octaphleate (Scheme 55),[14] a lipid of molecular weight
about 4400 and having *ca* 40 double bonds. It was probably a mixture of molecular
species because the phleic acid fraction obtained after saponification was heterogeneous.

Other trehalose polyphleates were observed in smaller amounts in the lipid extracts
and were found to have free hydroxyl groups as shown by their infrared spectra and
their behavior on thin layer chromatograms. No acid other than phleic acids could
be detected. Because of their low stability, possible deacylation and easy isomerization,
it was difficult to make a distinction between natural partial esters and artifacts.[14]

E. Esters of α-D-Trehalose with Sulfuric Acid
and Phthioceranic Acids

In 1948, Dubos and Middlebrook[62] observed that virulent tubercle bacilli fix neutral
red in their salt form from a slightly alkaline aqueous medium and are red colored,
whereas avirulent mycobacteria are yellow. Looking for the acidic material responsible
for the salt formation with neutral red, Middlebrook *et al.*[154] isolated a new lipid
fraction from virulent tubercle bacilli by extraction with hexane containing 0.05% of
decylamine. The extract was contaminated by some acidic phospholipids.

By chromatographic fractionation of the crude product, a sulfur-containing lipid was
isolated. Its molecular weight was estimated to be about 3000 from the determinations
of the amount of bound neutral red, the sulfur content and the acidity. The infrared
spectrum exhibited a prominent carboxylic acid ester band (1740 cm^{-1}). It was assumed
that "the material contained a peculiarly high proportion of methyl groups suggestive
of the methyl-branched chain fatty acids described by several previous investigators
in hydrolyzed preparations of the acetone soluble lipids of tubercle bacilli".[154]
Phthienoic type of fatty acid was excluded because of the lack of unsaturation. The
sulfur was considered to occur as sulfonic acid groups (IR absorption bands at
$1020–1060 \text{ cm}^{-1}$ and $1140–1200 \text{ cm}^{-1}$). From its chromatographic behavior, it was
deduced that this sulfur-containing material was probably "a group of closely related
sulfolipids, with slight differences in polarity".[154] From tubercle bacilli grown on a
medium containing $^{35}SO_4^{2-}$, ^{35}S-labelled sulfolipid could be prepared. The purification
of the sulfolipid fraction was later improved, mainly by using the solubility characteristics of the salt of the acidic lipid and octylamine, and chromatography on silicic
acid.[78, 99]

The isolation of several well-defined groups of sulfolipids was achieved in 1970 by
Goren using DEAE-cellulose column chromatography and monitoring of the effluent
by TLC.[85] In this paper and the following one,[86] Goren established the structure of
the main sulfolipid (called *sulfolipid-I*) as 2,3,6,6′-tetraacyl-2′-sulfate-α-D-trehalose
(Scheme 56). Thus, this compound is not a sulfonate derivative, but a sulfate ester
characterized by absorption bands at 1250 cm^{-1} (S=O) and 824 cm^{-1} (O—S) in its in-

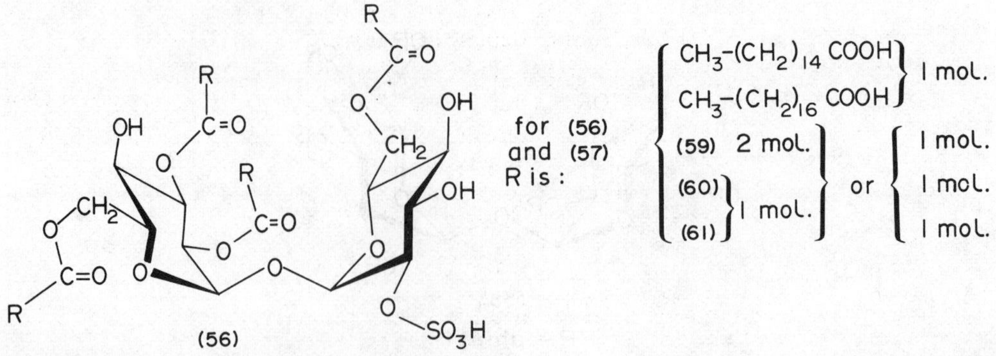

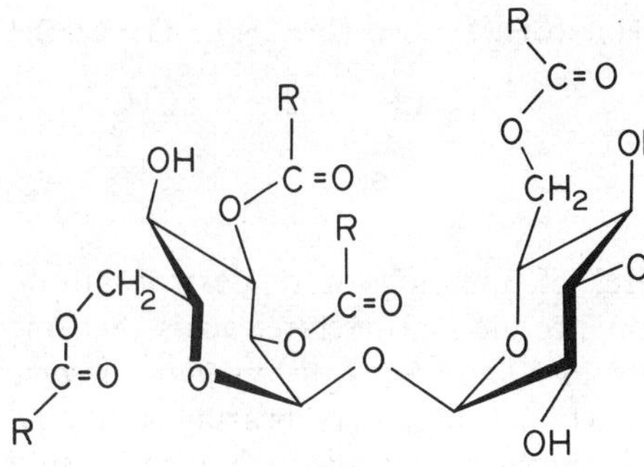

(57) 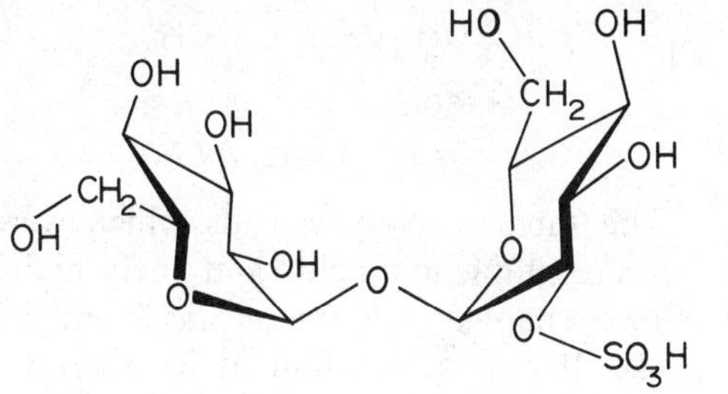   (58)

frared spectrum (see also ref. 94). Spontaneous elimination of the sulfate group from the ammonium salt of sulfolipid-I in anhydrous ether gave a nonionic sulfur-free lipid called *desulfolipid-I* (Scheme 57) and a water soluble product precipitable as barium sulfate. This desulfatation reaction was inhibited by small amounts of water, alcohols or pyridine and the mechanism of this reaction was studied.[87]

Sodium alkoxide-catalyzed ethanolysis of the sodium salt of sulfolipid-I afforded a water-soluble carbohydrate sulfate (Scheme 58). Acid-catalyzed hydrolysis of this carbo-hydrate sulfate gave α-D-trehalose, characterized as its octaacetate. Base-catalyzed per-methylation was performed on ^{35}S-labelled carbohydrate sulfate and the complete reten-tion of the radioactive sulfure was checked. Complete hydrolysis of the permethylated derivative gave equal amounts of 2,3,4,6-tetra-*o*-methyl-glucose and 3,4,6-tri-*o*-methyl-glucose, demonstrating that the sulfate group is located at a single 2 position of the trehalose core.[86] Localizations of the acyl groups were studied by methylation (by diazo-methane-BF$_3$) of desulfolipid-I, followed by saponification and acid hydrolysis. Equal amounts of 4-*o*-methyl-D-glucose and 2,3,4-tri-*o*-methyl-D-glucose were obtained. As pos-ition 2 had been demonstrated for the sulfate ester group, the structure (Scheme 56) was established.[85,86]

The lipid product recovered from alcoholysis of sulfolipid-I was fractionated by TLC, giving four classes of carboxylic esters. One of them was easily identified with methyl palmitate, mixed with some methyl stearate. Elucidation of the complete structures of the other esters required careful studies by NMR spectrography and mainly MS per-formed on each group of esters and the preparation of some derivatives. It was recog-nized that these esters, except for methyl palmitate, were closely related products, some of them bearing one hydroxyl group.[92]

The long suspected methyl branched structure was confirmed by NMR spectro-graphy which showed a large doublet at 9.1 and 9.2 τ, due to secondary methyl groups. Mass spectra of each group of esters bearing or not bearing a hydroxyl were in accord-ance with a multi-methyl branched structure and homology by 42 mass units (—CH$_2$—CH(CH$_3$)—) was prominent in both series. Peaks at m/e 88 and 101 showed the presence of a methyl branch in α position to the methyl carboxylate group. The secondary nature of the hydroxyl group was demonstrated by its transformation into a keto group.

Comparative examination of the NMR and mass spectra of the hydroxy- and keto-esters (methyl esters), the corresponding deuterated derivatives obtained by deuterium exchange, the corresponding ethyl esters obtained by transesterification, and the methoxy-esters obtained by direct permethylation of the sulfolipid allowed the localization of the methyl branches and the secondary alcohol group. From all these investigations, structures (Schemes 59–61) were suggested for these acids.[92]

$$CH_3-(CH_2)_{14}-CH_2-(CH-CH_2)_n-CH-COOH$$
$$\qquad\qquad\qquad\qquad CH_3 \qquad CH_3$$

(59)   n = 4 to 9

(average : 5 or 6)

$$CH_3-(CH_2)_{14}-\underset{\underset{OH}{|}}{CH}-(\underset{\underset{CH_3}{|}}{CH}-CH_2)_n-\underset{\underset{CH_3}{|}}{CH}-COOH \qquad CH_3-(CH_2)_{14}-\underset{\underset{OH}{|}}{CH}-(\underset{\underset{CH_3}{|}}{CH}-CH_2)_2-\underset{\underset{CH_3}{|}}{CH}-COOH$$

(60)   n = 4 to 9   (61)
(mainly 7)

The three carboxylic acids which were mixtures of homologs were dextrorotatory, as were phthienoic acids and their hydrogenation products. Therefore, acids (Scheme 59) were named *phthioceranic acids* and acids (Schemes 60 and 61), *hydroxy-phthioceranic acids*. It is assumed that at least most of the asymmetric centers bearing a methyl branch are related to the L series, as they are in phthienoic acids.[18] No conclusion was reached about the stereochemistry of the secondary alcohol group, but, according to the positive shift of rotation from acid (Scheme 59) to acid (Scheme 60), the D configuration is the most probable for this assymmetric center, provided that it has one methylene group on one side and a L-methyl-branched group on the other side.[9] Such a *threo* stereochemistry is the most favorable one to allow the observed dehydration of hydroxy acids (Scheme 60) without rearrangement of the carbon skeleton.

As mentioned above, palmitic acid was present as well as the phthioceranic acids, and the ratio of these four different classes of fatty acids was not equimolecular. As permethylation demonstrated that sulfolipid-I was a tetraacyl derivative, it was concluded that sulfolipid-I is a mixture of molecular species having 1 mole of palmitic (or stearic) acid, 1 mole of phthioceranic acid (59) and either 2 moles of the hydroxy acid (60) or 1 mole of (60) and 1 mole of (61).[92]

Sulfolipid-I was accompanied in the lipids of the tubercle bacillus by at least four minor products that could be separated by DEAE-cellulose column chromatography or by TLC. All of them appeared to be derivatives of α-D-trehalose-2-sulfate.[88] Sulfolipid-I', with a lower migration rate than the main product (sulfolipid-I) on thin-layer chromatograms, was said to be also 2,3,6,6'-tetraacyl-2'-sulfate-α-D-trehalose, containing the same acyl groups but in a different ratio. To reconcile its lower mobility in TLC with its higher content of fatty acids of low polarity, hydrogen bonding or steric hindrance might be taken into account. In sulfolipid-II, which had a higher migration rate than sulfolipid-I, phthioceranic acids (59) were not present and the content of hydroxy-phthioceranic became prominent, but the amount of this glycolipid was too small to make a study of the permethylation products. On the basis of the absence of acids (59), it was suggested that sulfolipid-II might be a triacylated trehalose-2'-sulfate,[92] but such a conclusion would not be in agreement with its faster migration rate in TLC. Moreover, sulfolipid-III, exhibiting a slower migration rate than sulfolipid-I, was characterized as the triacylated trehalose-2'-sulfate: 2-palmitoyl-3,6-*bis*-hydroxyphthioceranyl-2'-sulfate-α-D-trehalose (Scheme 62).[89] Whatever the number of acyl groups in sulfolipid-I' might be, it is likely that they were located at positions different from those bearing acyl groups in sulfolipid-I and it might be an artifact, as suggested by Goren.[86]

(62)   R is 60 or 61
R' = $C_{15}H_{31}$

III. TREHALOSE-CONTAINING GLYCOLIPIDS IN THE LIFE OF
BACTERIA

Although acyl-sugars are widely distributed among prokaryotes, only fragmentary information is available about their metabolism and their role in the bacterial physiology.[197] However, from a lot of non-conclusive but converging observations, it seems likely that acyl-trehalose compounds are located on the surface of the bacterial cell and play a role in relation with their environment. The very specific structures of the acyl moieties in most of the known acyl-trehaloses involve peculiar biosynthetic pathways and are probably useful for the role that the glycolipids have to play in the life of the bacteria. Subcellular localization of the trehalose-containing lipids, information on their biosynthesis and their implication in the bacterial metabolism will be considered here.

A. *Subcellular Localization*

The fact that Mycobacteria remained viable after extraction with petroleum ether suggested that cord factor is located on the outer cell surface,[47] a conclusion made plausible by the apparent normal function of the cytoplasmic membrane in the washed cells. Moreover, when the tubercle bacillus was grown in a medium containing large amounts of Tween 80, cord factor could be recovered from culture filtrates. It was suggested that cord factor, like a secretory product, was released into the surrounding medium by the detergent instead of accumulating on the cell surface as it was the case in a purely hydrophilic medium.[47]

Working on cell walls of the strain A-1 of *M. avium* kindly prepared by Portelance (University of Montréal) by means of a Ribi press,[176] Lanéelle[131] detected dimycoloyl-trehalose. Here again, this observation gives support to the localization of cord factor in the outer layers of the cell envelope, but is not a proof because during disruption of the cell and subsequent fractionation of the cellular organites, a glycolipid as insoluble in water as cord factor may be adsorbed on particles and sediment with them. Nevertheless, all these observations are in agreement with the localization of cord factor at the surface of the mycobacterial cell, and cord factor is considered as a component of the cell wall in recent reviews.[140, 141]

Sulfolipids seem also to be localized on the surface of the cells of *M. tuberculosis*. The bacteria fix the dye (neutral red) onto their surface, and a mild and brief extraction (with hexane containing 0.05% decylamine) is sufficient to remove sulfolipids from the cells (154). Likewise, trehalose phleates can be extracted from the cells of *M. phlei* by a simple washing of the bacterial cells with hexane that leaves the bacteria alive. When the subcellular components of the disrupted bacteria are separated by ultracentrifugation, trehalose phleates sediment in the same fraction as diaminopimelic acid (used as a marker of the cell wall), in spite of their lipid character for which they could have been considered as membrane components.[14]

B. *Metabolism*

Acyl-trehaloses contain a carbohydrate moiety and fatty acid groups. The metabolism of α-D-trehalose is well known, and information on the metabolism of the fatty acids is available, but very little is known of the metabolism of the esters themselves. In particular, nothing is known on the possible coupling of two monoacyl-D-glucose molecules to produce a diacyl-trehalose. α-D-trehalose arises from a condensation of UDP-glucose with glucose-6-phosphate, followed by the hydrolysis of the phosphate group. α-Glucosidases (trehalases) are known which are able to split trehalose into its two glucose units, and a trehalase was characterized in the tubercle bacillus.[49] It is not yet known if trehalases can use acyl-trehaloses as substrates. The main features of the metabolism of mycolic acids, phleic acids and phthioceranic acids are presented in the following.

1. *Mycolic Acids*

Several steps of the biosynthetic pathway of mycolic acids are now more or less elucidated, but their places in the overall metabolic process leading to a complex mycolic acid molecule are not yet known. Some hypotheses were proposed. Only the ones having experimental support are given here.

The main characteristic for defining all mycolic acids is the presence of an α-alkyl branch and a β-hydroxyl group (18). Such a structure can be the result of a Claisen-type condensation of two molecules of fatty acids, followed by the reduction of the β-keto-acid.[18, 26]

Actually the formation of 1,3-^{14}C-corynomycolic acid from two molecules of 1-^{14}C-palmitic acid by incubation with whole cells of *C. diphtheriae* was observed,[81] and a cell-free system, prepared from the same bacteria, produced the intermediary β-keto-ester.[225] Inhibition of this last system by avidin demonstrates a carboxylation step, making likely a tetradecylmalonate intermediate. Working on whole cells with labelled precursors, the β-keto-ester was isolated and identified as a β-keto-acyl-trehalose (monoester), $C_{44}H_{82}O_{13}$, by chemical degradation and field desorption MS. The implication of trehalose derivatives in the metabolism of mycolic acids is in agreement with the high turnover of lipophilic derivatives of trehalose observed by Winder *et al.*[230] In the case of *M. smegmatis*, the same kind of condensation is operating and allows the incorporation of labelled *n*-tetracosanoic acid into mycolic acids to form the C_{22}-α-branch (66). A fraction having docosanylmalonyl-CoA-ACP transacylase activity was isolated from a homogenate of this bacteria.[124] Similar enzymatic systems catalyze this condensation and the subsequent reduction during the synthesis of corynomycolic, nocardomycolic, and mycolic acids as the assymmetric centers 2 and 3 have the same configurations.[15]

Hypotheses have been proposed to explain the synthesis of the meromycolic acid chain containing about sixty carbon atoms.[66] The role of methionine in the formation of cyclopropane rings or methyl branches was demonstrated,[80, 102] but the step at which this addition occurs during the synthesis of mycolic acids is not yet known. The possible role of 5,6-monoethylenic long chain fatty acids in the synthesis of mycolic acids was discussed.[10]

Mycobacteric acids, long chain fatty acids containing about forty carbon atoms,[180] are likely to play a role in the metabolism of mycolic acids. Partial identity of structures between mycobacteric and mycolic acids isolated from the same strain of Mycobacteria gives support to this assumption, but it cannot be said if this role of intermediate is played during biosynthesis or catabolism. The inhibition of the biosynthesis of mycolic acids by isoniazid has been demonstrated.[211, 230]

One wonders whether the corynomycolic acids bound to a trehalose core were not intermediates to be elongated into nocardomycolic acids. Such an hypothesis would explain the isolation of trehalose dicorynomycolate from the free lipids of *N. asteroides* and nocardomycolic acids from the bound lipids of the cell wall of the same bacteria.[97] Observations supporting such a transformation were obtained by incubating labelled precursors with *N. asteroides* cells.[50]

An active synthesis of trehalose-containing lipids was demonstrated by incubating whole cells of *N. asteroides* with ^{14}C-glucose. An important peak was observed by radioactive scanning of the lipid extracts at the place of diacyl-trehaloses.[233]

Cord factor isolated from *M. phlei*, *M. avium* or *M. marianum* gives "dicarboxylic" mycolic acids (Scheme 41) by saponification. Etemadi and Gasche[67] suggested that dicarboxylic mycolic acids might arise from the oxidation by a biological analog of Baeyer–Villiger reaction of a keto-mycolic acid precursor. This hypothesis, supported by the isolation of octadecanol-2 and eicosanol-2,[6] was strengthened by the characterization in the same strain of Mycobacteria (*M. paratuberculosis*) of the keto-mycolic acid (Scheme 39) and the ester of mycolic acid (Scheme 40) with glycerol.[135] We have previously mentioned the occurrence of cord factor species containing the same kind of mycolic acid ester (Scheme 40) with trehalose, in *M. phlei*.[181]

No information on the catabolism of mycolic acids is available but the partial splitting of a mixture of the 4 stereoisomers of corynomycolic acid into palmitic acid in the mouse,[148] and the use of mycolic acids as growth-promoting factor for pathogenic leptospirae,[232] probably indicates activity of some degradation product.

2. Phleic Acids

Studies have been performed on the biosynthesis of phleic acids (Scheme 54) by *in vivo* incorporation of specifically labelled precursors, followed by isolation of the phleic acids and chemical degradation. Comparison of the incorporation of 1-^{14}C- or 2-^{14}C-acetate demonstrates that all the carbon atoms of phleic acids arise from acetate units by oriented condensations. Palmitic acid is the specific precursor of phleates (Scheme 54) with $m = 14$, whereas myristic acid is the specific precursor of phleates with $m = 12$. Longer chain fatty acids are incorporated into phleic acids only after partial degradation.

The hypothesis of formation of phleic acids by condensation of palmitic or myristic acids with a polyunsaturated compound is not in agreement with experimental data. Such a compound could not be detected, making unlikely its transitory accumulation. In the two pentaunsaturated phleates (Scheme 54, $n = 5$) the condensation of the hypothetical polyunsaturated compound would be much more effective with palmitic acid than with myristic acid, whereas it would be the reverse with an hexaunsaturated compound leading to hexaunsaturated phleates (Scheme 54, $n = 6$), more abundant in the case of the members derived from myristic acid.

An elongation process is the most likely one, provided that it undissociably concerns two acetate units, making a four-carbon elongation. Such a requirement would be in agreement with a precondensation of two molecules of acetate before their entrance into the elongation process. Thus, crotonate would be the most probable substrate for the elongation process.[11]

3. Phthioceranic Acids

Phthioceranic and hydroxy-phthioceranic acids (Schemes 59–61) contain a series of methyl branches located at even-numbered carbon atoms. Their structures are similar to those of mycocerosic acids which are synthesized by the tubercle bacillus by addition of propionate units to a normal chain fatty acid.[79] The presence of phthioceranic acids with 31, 34, 37, 40, 43 and 46 carbon atoms made likely the role of a C_3-unit in their biosynthesis and Goren[88] observed an incorporation of ^{14}C-propionate into these branched chain fatty acids.

The position attributed to the hydroxyl group in hydroxy-phthioceranic acids requires two parallel biosynthetic pathways for phthioceranic acids on one hand and hydroxy-phthioceranic acids on the other hand, whereas β-hydroxy-phthioceranic acids would have been direct precursors of phthioceranic acids.

C. Possible Roles

So far, no particular role has been demonstrated for most of the known acyl-trehaloses, so that many hypotheses have been proposed. The role of an energetic reserve compound has been ascribed to α-D-trehalose by analogy with that of sucrose, but extrapolation to acyl-trehalose is not evident, in particular, because sucrose usually is not acylated.

The turnover of trehalose in free trehalose and acyl-trehaloses has been studied in *M. smegmatis*.[231] The turnover of the acylated forms was about 13 times the rate of its net formation, while that of free trehalose was only 3 times. The high turnover of acyl-trehalose, either by itself or by comparison with free trehalose, in experiments in which the concentration of glucose remained high throughout, is an indication that its primary role is neither a storage role, nor solely a structural one.[231]

As shown above (Section II), *Arthrobacter paraffineus* grown on hydrocarbons synthesizes a mixture of corynomycolic acid esters with α-D-trehalose.[208] Although acyl-sugars are synthesized even when hydrocarbons are no longer the carbon source of the culture medium, the usefulness of compounds made of sugars and fatty acids for the assimilation of hydrocarbons is demonstrated by the work of Itoh and Suzuki[100] on the growth of a mutant of *Pseudomonas aeruginosa* deficient in hydrocarbon-utilizing ability. This mutant is able to grow normally on hydrocarbons when the medium is supplemented with the rhamnolipid (Scheme 63)[95] isolated from the wild strain. However, the structure of this rhamnolipid is quite different from those of trehalose dicorynomycolates (Scheme 17b).

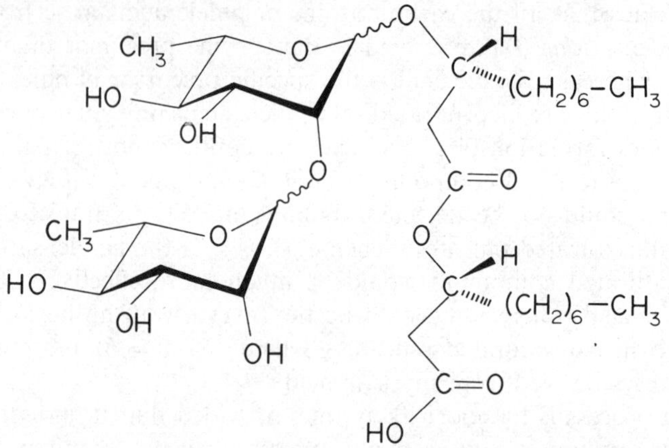

(63)

From this last observation, it might be deduced that trehalose-containing lipids might be replaced by other kinds of glycolipids for peculiar roles. Such an assumption would be in agreement with the fact that glycolipid-producing bacteria are able to incorporate different sugars in their glycolipids according to the carbohydrate source of the culture medium. For example, Corynebacteria and Mycobacteria grown in the presence of glucose contain large amounts of acyl-glucose in their lipids;[54] *Arthrobacter paraffineus* grown on sucrose synthesizes fatty acid esters of sucrose[207] and several species of bacteria, grown on fructose, produce mono- and diesters of fructose.[101] So far, the addition of trehalose to the culture medium does not seem to have been attempted.

It has been assumed that acyl-trehalose might be some form of transport of sugar across the cytoplasmic membrane. Specifically, trehalose-containing glycolipids might be used to transport trehalose to build more complex structural components such as polymers of 6,6'-dimannosyl-phosphate of α-D-trehalose which was described by Narumi and Tsumita.[165]

Phleates of trehalose cannot be considered as storage components since, unlike common fatty acids and their simple derivatives, they are neither used by *M. phlei* when no carbon source is supplied, nor accumulated when the main nitrogen source (asparagin) is lacking.[14]

An important decrease of the content of phleic acids is observed when a surfactant which prevents bacterial aggregation is added to the culture medium. In this case, it has been found that phleates of trehalose are not released into the surrounding medium. In shaken cultures of *M. phlei*, clumps of bacteria are formed, the size of which depends on the rate of shaking. A close correlation was observed between the size of the clumps and the content of trehalose phleates. The faster the shaking of the culture, the smaller the size of the clumps, and the lower was the content of trehalose phleates. These glycolipids seemed to play a role for the life of bacteria inside the clumps and they were not synthesized when the bacteria were grown in a dispersed state.[14] A lot of work has demonstrated the very active metabolism of α-D-trehalose after the aggregation stage of cellular differentiation in slime molds.[189] Acyl-trehaloses are surface components which seem to play a role in aggregation of bacteria.

Fatty acids such as mycolic, phthioceranic or phleic acids are probably easily catabolized only by a limited number of organisms. Therefore, acyl-trehaloses containing these acids might protect the bacteria against external aggression. It has been suggested for a long time that cord factor, the first known of these acyl-trehaloses, might play some role in the resistance of tubercle bacilli inside infected host cells.

IV. BIOLOGICAL PROPERTIES OF TREHALOSE-CONTAINING LIPIDS

A. *Toxicity and Action on Mitochondria and Microsomes*

As soon as crude cord factor preparations were obtained, Bloch[43] recognized their toxic properties that were used afterwards as a test in the course of the purification process of this glycolipid. Three to five intraperitoneal injections of 5 μg of cord factor (in paraffin oil) at 2–3 day intervals killed at least half of the mice within 3 weeks. The same result could be obtained by a single intravenous injection of 25 μg. A loss of weight of the animals was rapidly observed after the injections. Dead animals exhibited characteristic lung hemorrhages, due to some alteration of the capillary walls.[45] A single sublethal intraperitoneal injection of cord factor enhanced tuberculosis in infected animals.[48, 172]

Injection of cord factor to mice induced a decrease of the NAD content of the liver, probably because of a stimulation of NADase activity and, as a consequence, a marked decrease of the rate of oxygen uptake by incubation of liver homogenates with NADH-producing substrates.[8, 35] In tissues of cord factor treated animals, Kato[106, 107] observed a decrease of some enzymatic activities: succinate-oxidase system, succinate-cytochrome c reductase, and reduced coenzyme Q-cytochrome c reductase. No modification of these enzymatic activities was detected in heart tissue by contrast with other tissues (liver, lung, spleen, brain etc.). Similar results were later obtained by Murthy et al.[164] and Shankaran and Venkitasubramanian.[195]

The fact that these enzymatic activities were mainly located in mitochondria led Kato[108] to investigate how phosphorylative oxidation functioned in a liver homogenate prepared from cord factor-treated mice. An impairment of the electron transfer process was observed and it was suggested that it occurred at a site prior to cytochrome c (see Fig. 4). These results were confirmed by using mitochondria prepared from livers of mice that had been intraperitoneally injected with 0.1 mg of cord factor dissolved in 0.1 ml of paraffin oil 2 or 3 days before killing.[109] These functional alterations of mitochondria induced by cord factor were accompanied by morphological changes. Electron microscopy demonstrated swelling and partial disruption of mouse liver mitochondria *in vivo*.[119] The same changes could be obtained *in vitro* by incubation of liver mitochondria with a fine aqueous suspension of cord factor. This swelling was not dependent on either the electron transport or phosphorylation in mitochondria, for it was not prevented by inhibitors of oxidative phosphorylation. No alteration of mitochondria was observed when mycolic acids, or mixtures of mycolic acids and free trehalose, were used instead of cord factor.[110] Acetylated cord factor was unable to induce these morphological and functional alterations.[110]

More recently, Kato[111] found that preincubation of isolated mitochondria with a fine suspension of cord factor inhibited the phosphorylation coupled to the oxidation of either succinate or NADH-linked substrates. By using suitable electron acceptors and inhibitors of the mitochondrial electron transfer, Kato[111] localized the cord factor sensitive part of the phosphorylative oxidation system at the site II of phosphorylation (between cytochromes b and c_1; see Fig. 4).

The observed biochemical alterations of the mitochondrial phosphorylative oxidation system can be summarized as follows:[111] inhibition of respiration and phosphorylation in the succinate pathway as well as in the NADH pathway; loss of respiratory control; inhibition of the electron transport and of the coupling of phosphorylation to electron transfer at coupling site II, and stimulation of the mitochondrial adenosine-triphosphatase.

The dose of cord factor necessary to provoke these modifications on mitochondria *in vitro*, 25–50 μg of cord factor per milligram of mitochondrial protein was high compared with that required to induce an *in vivo* decrease of dehydrogenase activities *ca*

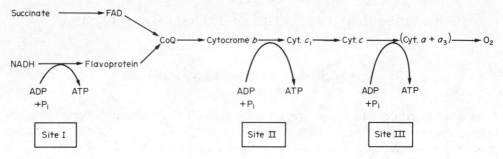

FIG. 4. Mitochondrial phosphorylative oxidation system (simplified scheme).

20 μg per animal. Such a discrepancy may be explained by the time of exposure to the toxin, 15 min for isolated mitochondria, versus 2 or 3 days for animals, and perhaps the physical state of water-insoluble cord factor in contact with the mitochondrial membrane, in vivo and in vitro. 6-Monomycoloyl-α-D-trehalose affected mitochondrial oxidative phosphorylation in a manner similar to that of cord factor, but to a lesser extent[121] (for toxicity, see Table 6).

As the structure of cord factor is relatively simple, it was possible to synthesize cord factor analogs and thus to study the relationships between in vivo toxicity, in vitro inhibition of mitochondrial oxidative phosphorylation, and chemical structure. So far, these studies have been mainly devoted to modifications of the sugar moiety of the glycolipid. 6,6'-Dimycoloyl-sucrose (Scheme 64) had qualitatively the same properties as cord factor, but quantitatively was less active. Lethal dose is given in Table 6. Higher amounts of dimycoloyl-sucrose were required to induce the same decrease of the P/O ratio in mitochondria (in the presence of either succinate or pyruvate + malate). Methyl 6-mycoloyl-α-D-glucopyranoside (Scheme 65) can be considered as one-half of the cord factor molecule (Scheme 17); the methyl glucoside group keeps the glucosidic bond in the α-configuration (as in cord factor). This glycolipid was only slightly less toxic than 6,6'-dimycoloyl-sucrose (Scheme 64) (Table 5). However, it failed to affect the transfer of electrons in the NADH pathway, in contrast to both diesters of disaccharides.[118] While the 6-mycolate of methyl glucoside is toxic, no toxicity was observed by injecting 1-mycoloyl-β-D-glucopyranose or methyl-2-mycoloyl-α-D-glucopyranoside,[24] demonstrating the importance of the location of the mycoloyl group on the sugar ring.

Methyl 6-mycoloyl-α-D-glucopyranoside (Scheme 65) which retained in vivo toxicity and in vitro decoupling properties at site II of phosphorylation was chosen as a reference compound to study the importance of the stereochemistry of the sugar moiety. The properties of methyl 6-mycoloyl-β-D-glucopyranoside (Scheme 66), -α-D-mannopyranoside (Scheme 67), -α-D-allopyranoside (Scheme 68) and -α-D-galactopyranoside (Scheme 69) were studied. The stereochemistry of the anomeric center (carbon 1) had little effect, at least in the case of a monoester of a monosaccharide, for α- and β-anomers had practically the same toxicity and the same inhibiting properties of the mitochondrial

TABLE 6. Toxicity of Cord Factor and Glycolipids Analogs in Mice

Compounds tested	LD$_{50}$ (μg)	References
Cord factor (6,6'-dimycoloyl-α-D-trehalose)	50	111
Coryno-cord factor (6,6'-dicorynomycoloyl-α-D-trehalose)	120	112
6,6'-dimycoloyl-sucrose	161	118
6,6'-dipalmitoyl-α-D-trehalose	200	123
methyl 6-mycoloyl-α-D-glucopyranoside	187	118
methyl 6-mycoloyl-β-D-glucopyranoside	199	21
6-mycoloyl-D-sorbitol	300	22
6-mycoloyl-α-D-trehalose	387	121
methyl 6-mycoloyl-α-D-mannopyranoside	500	21
methyl 6-mycoloyl-α-D-allopyranoside	>500	21
methyl 6-mycoloyl-α-D-galactopyranoside	>500	21

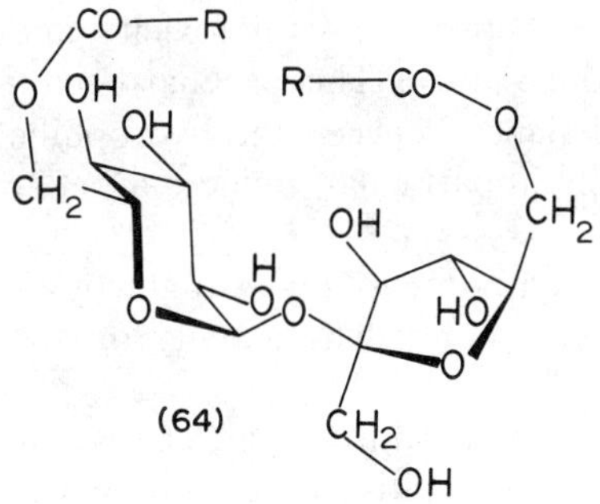

(64)

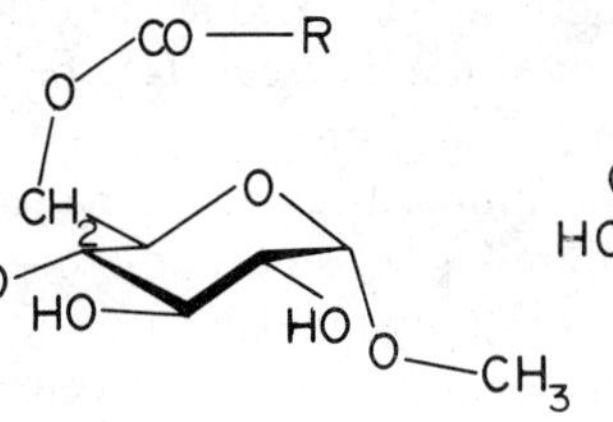

(65)

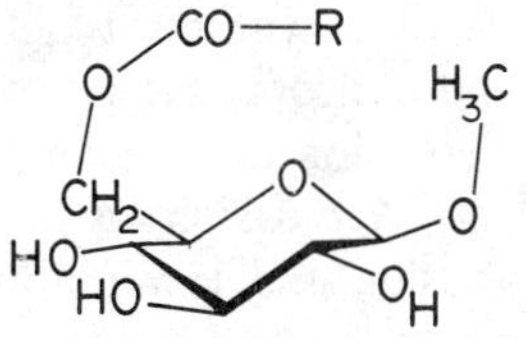

(66)

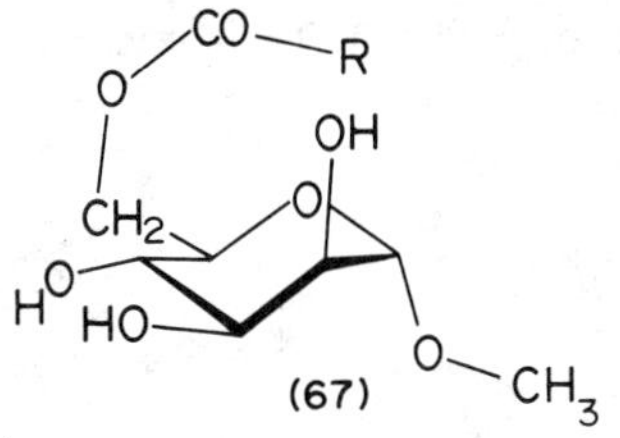

(67)

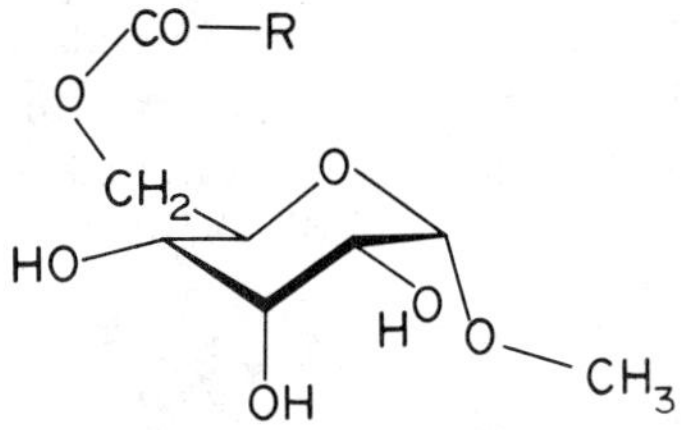

(68)

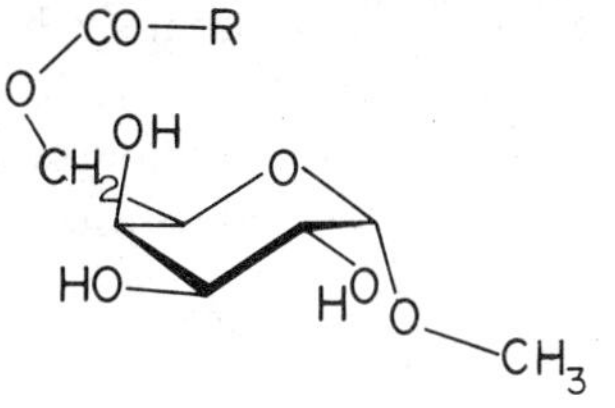

(69)

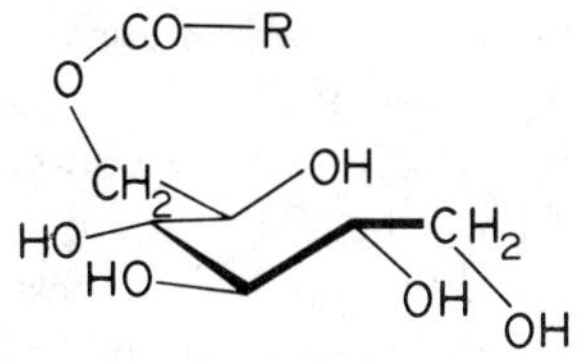

(70)

In schemes 64–70, R—CO$_2$—— is (ca. C$_{60}$) C$_{24}$H$_{49}$

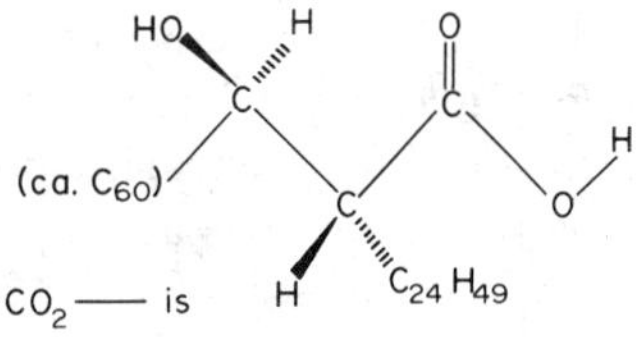

phosphorylative oxidation. The mannoside derivative was only slightly toxic and had weak inhibiting properties, and alloside and galactoside derivatives had no toxicity and no inhibitory properties with the amounts of material employed in these experiments.[21] Thus, the stereochemistry of the sugar is a very important factor and, from the four different monosaccharides used, only D-glucose led to toxic and inhibitory derivatives. 6-Mycoloyl-D-sorbitol (Scheme 70), the polar part of which retains the stereochemistry of D-glucose, was toxic and had inhibitory properties but to a lesser extent than the glucose derivatives, probably because of the higher number of possible unfavorable conformations of the molecule. It must be stressed that the conformation depicted in (Scheme 70), and which mimics the pyranose ring of glucose, is not the stablest one.[22]

From the examination of the properties of these various analogs, a rather good correlation was found between *in vivo* toxicity as expressed by LD$_{50}$ and *in vitro* inhibition of the mitochondrial phosphorylative oxidation system. Thus, it seems that the more toxic is a cord factor analog, the deeper is its ability to disorganize the mitochondrial phosphorylative oxidation system. The functional alterations of the mitochondria induced by these glycolipids seem to vary in a stepwise way.

Coryno-cord factor, a lower homolog of cord factor in which C$_{80}$-mycolic acids are replaced by C$_{32}$-corynomycolic acids, was a little less toxic than cord factor (Table 6).[112] Coryno-cord factor induced *in vitro* a swelling of mitochondria closely resembling that produced by cord factor. Like cord factor, coryno-cord factor inhibited the electron transport in the NADH-pathway in both phosphorylating and nonphosphorylating systems, but the succinate pathway was less affected. However, the aqueous suspensions of coryno-cord factor were able to directly affect the mitochondria whereas, in experiments with cord factor, a preincubation of the mitochondria with the glycolipid suspension was necessary to obtain inhibitory effects. Moreover, coryno-cord factor seemed to affect not only the coupling site II, but also site III.[113] Thus, coryno-cord factor had *in vitro* a stronger effect than cord factor, whereas *in vivo* its toxicity was a little lower. These observations disagree with the correlation between *in vivo* toxicity and *in vitro* phosphorylative oxidation inhibition just mentioned above. This peculiar behavior of coryno-cord factor might be due to a different balance between the hydrophilic and hydrophobic parts of the molecule (see below) which, in particular, induces a better dispersion in an aqueous medium.

In vivo toxicity is no more observed in the case of methyl 6-docosanoyl-α-D-glucopyranoside and lower homologs. *In vitro* these esters have a nonspecific activity on the mitochondrial phosphorylative oxidation.[122]

No simple relation seems to exist between the chemical structure of an acyl-sugar and its inhibiting properties towards mitochondrial phosphorylative oxidation. While 5-mycoloyl-L-arabinose had a low activity (as could be expected because its stereochemistry is rather different from that of glycopyranoside derivative), hydroxyethyl 5-mycoloyl-L-arabinofuranoside (Scheme 71) exhibited inhibiting activities about of the same order of magnitude as methyl 6-mycoloyl-glucopyranosides (65 or 66).[190]

$$R-CHOH-CH-CO-O-H_2C \cdots \quad H_2O-CH_2-CH_2OH$$
$$\underset{C_{22}H_{45}}{|} \qquad OH$$
(71)

These glycolipids, including cord factor, seem to act as highly specific detergents, with a bulky hydrophobic part and a polar part, the acetylation of which makes *in vivo* and *in vitro* activities disappear. The balance between these two parts of the molecules may have large variations, for cord factor (C_{80} fatty acids) and coryno-cord factor (C_{32} fatty acids) are both toxic *in vivo* and inhibitory of phosphorylative oxidation *in vitro*. The higher activity of diesters of trehalose compared with that of monoesters of methyl glucoside might be due to interactions with a broader region of the inner mitonchondrial membrane. Occurrence of an axial hydroxyl group in the mannoside, alloside or galactoside derivatives (67–69) hinders these interactions, eliminating their toxic properties.

It must be stressed that, in spite of the large use of rat mitochondria in studies of the biochemical properties of cord factor, this glycolipid is not toxic in rats. It is not yet known whether it is due to a lack of permeability of cell membrane of rat cells.[192b]

Sulfolipid-I (56), which is also a partially esterified trehalose derivative, has a peculiar behavior. This lipid, or its desulfated derivative (57), had no toxic properties *in vivo*, even at the high dosages of 100–500 μg in one single intraperitoneal injection, or several intraperitoneal injections of 200 μg at 2-day intervals, but mice that had received injections of sulfolipid-I were subsequently much more sensitive to cord factor injections. *In vivo* neither morphological changes nor functional alterations were detected in regard to mitochondria. However, *in vitro* incubation of mitochondria with an aqueous suspension of sulfolipid-I (5–20 μg per milligram of mitochondrial protein, for 15 min at 20°C) induced swelling and inhibition of oxidative phosphorylation, whether the substrates were succinate or NADH generators. Sulfolipid-I affected the phosphorylation at both coupling sites II and III. It was then a more powerful inhibitor of oxidative phosphorylation than cord factor. This effect was almost completely prevented by addition of bovine serum albumin to the incubation medium, which explains the *in vivo* absence of effect on mitochondria. This neutralizing action of serum albumin might be due to salt formation between the sulfate group of sulfolipid-I and basic groups of the protein.[120] Such a formation might be prevented either by elimination of the sulfate group (desulfolipid-I) or by esterification of the free acid group. It should be interesting to test the behavior of these neutral derivatives towards mitochondria without or with addition of bovine serum albumin.

When mice were intraperitoneally injected with cord factor *and* sulfolipid-I (or desulfolipid-I (Scheme 57)), the toxicity of cord factor was increased, up to nine-fold.[120] This enhancement of the toxicity of cord factor was observed even when both glycolipids were injected separately and by different routes, for example, intraperitoneal injection of sulfolipid-I and intravenous injection of cord factor. It has to be noted that this synergistic effect was not observed by using sulfolipid-I and simple analogs of cord factor such as toxic monomycolates of methyl glucosides. The simultaneous injection of sulfolipid-I and cord factor induced important morphological and functional alterations of mitochondria.[120] These results suggest that when sulfolipid-I is mixed with cord factor, the latter protects it against neutralization by serum albumin. It is remarkable that this phenomenon can be observed even when the two glycolipids are injected by different routes. Likewise, sulfolipid-III (Scheme 62), which has only three fatty acyl groups, was not toxic for mice and potentiated the activity of cord factor.

By working on liver homogenates prepared from cord factor-treated mice, Fukuyama et al.[77] also studied a fraction of *microsomes* corresponding to rough surfaced endoplasmic reticulum. This fraction, obtained by ultracentrifugation of a liver homogenate in a linear sucrose density gradient, decreased when mice had been intraperitoneally injected with 100 μg of cord factor 24 or 48 hr before. There was also an increase of the free ribosomes content.[77] These findings were in agreement with the electron micrographs of the liver of mice intoxicated with cord factor[119] and coryno-cord factor gave similar results.

By comparison of microsome preparations obtained from mice that have been intraperitoneally injected either with Bayol F (a paraffin oil fraction) or with 100 μg of cord factor dissolved in the same amount of Bayol F, a slight but significant decrease of the content of cytochrome P-450 and cytochrome b_5, and an important decrease of pyrazinamide deamidase and aminopyrine demethylase activities were observed.[215] On the other hand, a stimulation of NADase activity was found previously noted by Artman et al.[8] These observations were in agreement with the low pyrazinamide deamidase activity in the tissues of tuberculous animals or of cord factor-treated animals.[214] By contrast with mitochondria, *in vitro* incubation of microsomes with aqueous suspensions of cord factor did not affect microsomal enzymes,[215] at least under the conditions used.

B. *Production of Circulating Antibodies*

Specific antibodies cannot be detected in the blood of mice or rabbits after injections of aqueous suspensions of cord factor. To stimulate antibody formation, a solution of cord factor in propyleneglycol (10 mg/ml) was mixed with an equal volume of pig serum to form a suspension, mice were inoculated three times a day with such a suspension for several days, and the treatment repeated several times with intervals of 1 or 2 months. The serum of these vaccinated animals contained specific antibodies, the presence of which was demonstrated by Ouchterlony's agar diffusion technique by using an aqueous suspension of cord factor as an antigen.[172] The precise amount of cord factor received by each animal during these experiments is not clear, but it was rather astonishing to find them alive at the end of such a treatment. It may be that antibodies are rapidly produced, immunizing animals against subsequent inoculations with cord factor, for the same authors[172] observed a decrease of the toxicity of cord factor in vaccinated animals.

Much more detailed studies on the immunology of cord factor were recently performed by Kato[114] who used, instead of pig serum, methylated bovine serum albumin (MBSA), prepared from bovine serum albumin fraction V by methylation according to Mandell and Hershey.[147] The antigen was prepared by mixing an aqueous suspension of cord factor with an aqueous solution of MBSA and centrifugation to remove some insoluble material. A solution of this antigen containing 50 μg of cord factor was subcutaneously injected into mice twice a week for 3 weeks. Rabbits were also vaccinated with the same antigen in an amount equivalent to 1 mg of cord factor in the same manner. No toxic effects were observed by repeated vaccinations with this complex MBSA + cord factor. It must be noted that neither acetylated cord factor nor mycolic acids were able to give a complex with MBSA and thus to produce antibodies. By mixing antisera of the vaccinated animals with an aqueous suspension of cord factor, a precipitin reaction was observed; a large amount of cord factor was bound to the antibodies in the immune precipitate. From inhibition studies of the precipitin reaction, it was concluded that α-D-trehalose is the antigenic determinant of the antigen complex.[114]

Vaccinated mice showed a marked resistance to repeated injections of cord factor. Although the LD_{50} of cord factor in nonvaccinated mice is about 50 μg, it was higher than 1 mg in vaccinated animals. No morphological nor functional alterations of mitochondria were observed *in vivo* after injections of cord factor in vaccinated animals. In accordance with the role of trehalose as a determinant, vaccination with a complex of MBSA with 6,6'-dimycoloyl-sucrose or methyl 6-mycoloyl-α-D-glucopyranoside was unable to protect the animals against a challenge injection of cord factor.[114] The protective action of the two latter complexes against toxic effects induced by these cord factor analogs was apparently not studied.

Vaccination of mice with BCG did not seem to produce antibodies against cord factor, for the serum of vaccinated animals neither protected against cord factor toxicity, nor showed precipitin reaction with cord factor suspensions.[114] Sera of man and animals infected with tubercle bacilli did not give a precipitin reaction with cord factor.[115] The precipitin antibodies were found in the 19S macroglobulin fraction (immunoglobulin M) of rabbits vaccinated with the complex MBSA-cord factor.[115]

Tubercle bacillus multiplication in the organs of infected mice was prevented by vaccination with the cord factor-MBSA complex, or by the injection of an anti-cord factor serum, obtained from vaccinated mice. The protective role of these antibodies was specific and no protection was found against listeria, brucella or salmonella infections. This fact provides a good demonstration of the important role played by cord factor in the pathogenesis of tuberculosis.[116]

C. Immunostimulation

We have seen that cord factor may behave as an antigenic determinant when injected along with a protein component that plays the role of a carrier because antibodies appear, able to directly react with cord factor in a precipitin reaction.

Cord factor can also work as an adjuvant, increasing the production of antibodies. In animals injected in the footpad with an oil-in-water emulsion containing as little as 5 μg of cord factor, intense cellular reaction was observed at the site of inoculation (see below). After 20 days, the cord factor-treated animals were injected with a suspension of sheep red blood cells or with a solution of an antigen and the titers of antibodies in the sera of these animals were much higher than those observed in the sera of control animals.[41] No adjuvant effect was observed when cord factor and antigen were injected at the same time. A weaker reaction was obtained when cord factor was injected into one footpad and the antigen into another footpad, an observation that underlines the local character of the immunological response.[41]

Cord factor is a good adjuvant in mice and rats, but has little if any adjuvant effect in guinea pigs.[192a]

When 1–5 μg of cord factor was intravenously injected into mice, granuloma were found in the lungs 7 days later. Some qualitative differences were observed between granulomagenic activities of cord factor preparations from various origins. Cord factor from *M. kansasii* was the most active while cord factor from BCG was the weakest one. Differences between these compounds can only be due to differences of mycolic acids. Three intravenous injections of cord factor gave a more extensive response than did a single injection of the same total amount of the same cord factor.[34, 36] Mice having lung granuloma were protected against infection by the tubercle bacillus virulent strain H-37 Rv introduced by intravenous injection. Cord factor induces a marked inflammatory swelling at the sites of injection in mice, but not in rats and guinea pigs.[192a]

Injection of 10 μg of cord factor in an oil-in-water emulsion into the footpad of mice induced a granulomatous response at the site of injection and in draining lymph nodes. The nodules were composed of epithelioid cells, macrophages and lymphocytes, and their formation probably corresponded to defense mechanisms. Thus, cord factor seems to mobilize and activate macrophages as shown by an increase of acid phosphatase activity. Such injections provoked no cellular reaction in lungs, liver or spleen.[37]

Injection of an oil-in-water emulsion containing 20 μg of cord factor into mice infected with BCG bacilli induced a very extensive granulomatous response, much stronger than the one observed in normal mice, according to the number of granuloma and measurement of their surface sections.[40] Such a response was considered to express some form of hypersensitivity state that could only be detected by cord factor or by whole tubercle bacilli. Whereas, 6,6'-dimycoloyl-α-D-trehalose (cord factor), 6,6'-dimycoloyl-sucrose (Scheme 64) and 6,6'-dipalmitoyl-α-D-trehalose (Scheme 12) are all granulomagenic (see Table 7), only the mycolate derivatives, diesters of trehalose and sucrose, gave a hypersensitivity reaction by injection into BCG-treated mice.[40] Thus, in these experiments, the nature of the fatty acids is an important factor, while the structure of the sugar moiety is of minor importance. Various mycolate esters such as 1-mycoloyl-β-D-galacto-pyranose, are able to induce a hypersensitivity state in animals.[187]

The mode of introduction of cord factor into the animal body and the species of the animal appears to be of a great importance. Intravenous injection of an emulsion of cord factor into guinea-pigs did not induce a granulomatous response in their lungs, and injection into rabbits gave a minimal granulomatous response with a cellular composition different from that observed in mice.[40] These observations may explain the weak granulomatous response obtained in rabbits by Moore et al.[163] and its character of foreign body-type reaction.

The formation of granulomas induced by cord factor is quite similar to that induced by injections of living BCG and provokes a hyperplasia of the lymphoid tissue in the paracortical zone and macrophages accumulations, enabling the organism to resist the multiplication of bacteria and also of tumor cells. By three intravenous injections of an emulsion containing 10 μg of cord factor from M. kansasii at 7 or 5 day-intervals, a strong granulomatous response was maintained in the lungs of mice. Due to this stimulation of the defense mechanisms of the organism induced by granuloma, intraperitoneal injections of urethan were unable to provoke the formation of lung adenoma, which appeared in the lungs of control mice.[38]

Likewise, the growth of Ehrlich ascites cells was inhibited in animals pretreated with cord factor or some of its analogs and 3 days after intraperitoneal injections of cord factor (30–40 μg) or its analogs (40–100 μg) into mice, 2×10^6 Ehrlich ascites cells were intraperitoneally injected. On the 7th day after this latter injection, tumor cells were washed out of the animals and after centrifugation, the volume of the packed cells was measured. It is surprising to note that, while methyl 6-mycoloyl-α-D-glucopyranoside had a higher granulomagenic activity than 6,6'-dimycoloyl-sucrose (see Table 7), the

TABLE 7. Granulomagenic Activity of Cord Factor Analogs After Intravenous Injection of an Emulsion Oil-in-Water from Yarkoni et al.[236]

Glycolipid	Dose (μg)	Number of granulomas per 100 mm² of lung tissue				
		1	2	3	4	5
6,6'-dipalmitoyl-α-D-trehalose	100	246	*	336		
	16	73	217	194		
6-palmitoyl-α-D-trehalose	100	216	192	75	53	
methyl 6-mycoloyl-α-D-glucopyranoside	100	232	284	195	327	136
methyl 6-mycoloyl-β-D-glucopyranoside	100	158	109	79		
6,6'-dimycoloyl-sucrose	100	0	0	3	5	150

*Innumerable because of the presence of many confluencing granulomas.

former induced practically no protection against multiplication of the tumor cells. 6,6′-Dimycoloyl-sucrose and 6,6′-dipalmitoyl-α-D-trehalose had similar inhibiting properties, demonstrating that the inhibition of the multiplication of tumor cells induced by these glycolipids is specific neither to the sugar, nor to the fatty acids.[236] So far, too few glycolipids have been studied to draw conclusions regarding structural requirements. This inhibitory effect induced by cord factor was still evident 36 days after its administration. Cord factor and the tumor cells had to be injected at the same site[237] because of the local character of the response.

Three weekly injections of an emulsion containing 150 μg of each sulfolipid-III and cord factor were able to provoke a regression of established Line 10 tumors and metastases in about 65% of the infected guinea-pigs. Neither cord factor alone, nor sulfolipid-III alone, nor mixtures of sulfolipid-I or desulfolipid-III and cord factor had such tumor regression activity.[89]

The protective effect of delipidated preparations of cell wall of Mycobacteria, consisting almost exclusively of peptidoglycan, against several kinds of cancer tumors had been established (for example, 58). Better results can be obtained by treatment of the animals with a mixture of peptidoglycan preparations and cord factor, as shown by the works of Bekierkunst et al.[39] and of Azuma et al.[30] and Meyer et al.[151] Under the conditions used by Bekierkunst et al. established tumors in the skin and metastases in draining lymph nodes disappeared in 83% of the infected guinea-pigs. This rate of positive results was similar to that obtained by administration of lyophilized killed BCG cells.[39] Likewise, Azuma et al.[30] and Meyer et al.[151] could suppress the growth of hepatocarcinoma cells in guinea-pigs and observed a regression of established tumors by injections of a mixture of peptidoglycan and a glycolipid designated as glycolipid P3 by Azuma and Meyer, identical with cord factor.

D. Role in the Pathogenicity of Bacteria

Because the cord factor is known to be located at the surface of the bacterial cell, it is likely that its toxic properties may play a role in the multiplication of tubercle bacilli and their dissemination in the organs of infected animals.

Such a role could be demonstrated by the use of anti-cord factor antibodies induced by injection of a complex of cord factor and methylated bovine serum albumin.[114] Immunization of mice by this complex or by passive transfer of anti-cord factor serum obtained from immunized mice protected the animals against the toxic action of cord factor and the infection with virulent tubercle bacilli[116] (see above). The bacterial multiplication in the organs of the infected mice was prevented by the anti-cord factor antibodies, demonstrating the importance of cord factor in the pathogenesis of tuberculosis. The claims of the presence of another toxic lipid component different from cord factor in the tubercle bacillus, extractable with monochlorobenzene,[203] was not substantiated by further investigations.[46]

Gangadharam et al.[78] and Middlebrook et al.[154] have observed that the more virulent a strain of tubercle bacillus was, the higher the amount of sulfolipid material determined by the amount of ^{35}S found in the extracted lipids. No such correlation could be made between virulence and content of cord factor.[18, 19] Working on 40 strains of tubercle bacillus isolated from patients living in India or East Africa, Goren et al.[93] found a statistically significant correlation between the virulence for the guinea-pig and the content of "strongly acidic lipids" (SAL = phospholipids and sulfolipids). All the virulent strains had a high content of SAL and it was assumed that they could play a role in the pathogenesis of tuberculosis because of their activity against mitochondrial membranes. Moreover, a synergistic effect of sulfolipid-I and cord factor on mitochondria was demonstrated.[120] However, because a few attenuated strains of the tubercle bacillus also had a high content of strongly acidic lipids, it was concluded, in agreement with previous authors in this field, that the content of SAL was "evidently not a sufficient criterion for expression of virulence".[93]

Quite recently, it has been shown that sulfolipids are lysosomotropic and prevent the fusion of lysosome and phagosome membranes. These compounds would thus play an important role in the survival of virulent Mycobacteria inside macrophage cells.[93a]

REFERENCES

1. ACHARYA, N. V., SENN, M. and LEDERER, E. C. R. Acad. Sci. (Paris), Ser. C 264, 2173–2176 (1967).
2. ADAM, A., SENN, M., VILKAS, E. and LEDERER, E. Eur. J. Biochem. 2, 460–468 (1967).
3. AEBI, A., ASSELINEAU, J. and LEDERER, E. Bull. Soc. Chim. Biol. 35, 661–681 (1953).
4. ALIMOVA. E. K. Biokhimiya 20. 516–521 (1955).
5. ALIMOVA, E. K. Biokhimiya 24, 785–788 (1959).
6. ANDERSON, R. J. Fortschr. Chem. Org. Naturst. Vol. III. pp. 145–202. Springer Verlag. Wien and New York (1939).
7. ANDERSON, R. J. and NEWMAN, M. S. J. Biol. Chem. 101, 499–504 (1933).
8. ARTMAN, M., BEKIERKUNST, A. and GOLDENBERG, I. Arch. Biochem. Biophys. 105, 80–85 (1964).
9. ASSELINEAU, C. and ASSELINEAU, J. Bull. Soc. Chim. Fr. 1992–1997 (1966).
10. ASSELINEAU, C., LACAVE, C., MONTROZIER. H. and PROME, J. C. Eur. J. Biochem. 14, 406–410 (1970).
11. ASSELINEAU, C. and MONTROZIER, H. Eur. J. Biochem., 63, 509–518 (1976).
12. ASSELINEAU, C., MONTROZIER, H. and PROME, J. C. Bull. Soc. Chim. Fr. 592–596 (1969).
13. ASSELINEAU, C., MONTROZIER, H. and PROME, J. C. Eur. J. Biochem. 10, 580–584 (1969).
14. ASSELINEAU, C., MONTROZIER, H., PROME, J. C., SAVAGNAC, A. and WELBY, M. Eur. J. Biochem. 28, 102–109 (1972).
15. ASSELINEAU, C., TOCANNE, G. and TOCANNE, J. F. Bull. Soc. Chim. Fr. 1455–1459 (1970).
16. ASSELINEAU, J. Bull. Soc. Chim. Fr. 937–944 (1955).
17. ASSELINEAU, J. Bull. Soc. Chim. Fr. 135–141 (1960).
18. ASSELINEAU, J. The Bacterial Lipids, (Ed. HERMANN) Paris, and Holden-Day, San Francisco (1966).
19. ASSELINEAU, J., BLOCH, H. and LEDERER, E. Am. Rev. Tuberc. Pulm. Dis. 67, 853–858 (1953).
20. ASSELINEAU, J., BLOCH, H. and LEDERER, E. Biochim. Biophys. Acta 15. 136–137 (1954).
21. ASSELINEAU, J. and KATO, M. Biochimie 55, 559–568 (1973).
22. ASSELINEAU, J. and KATO, M. Jpn. J. Biol. Med. Sci. 28, 94–97 (1975).
23. ASSELINEAU, J. and LEDERER, E. Biochim. Biophys. Acta 7, 126–145 (1951).
24. ASSELINEAU, J. and LEDERER, E. Bull. Soc. Chim. Fr. 1232–1240 (1955).
25. ASSELINEAU, J. and LEDERER, E. Biochim. Biophys. Acta 17, 161–168 (1955).
26. ASSELINEAU, J. and LEDERER, E. In: Lipide Metabolism pp. 337–406 (Ed. BLOCH, K.) Academic Press, New York (1960).
27. ASSELINEAU, J. and MORON, J. Bull. Soc. Chim. Biol. 40, 899–911 (1958).
28. ASSELINEAU, J. and PORTELANCE, V. Recent Results Cancer Res. 47, 214–220 (1974).
29. AZUMA, I., NAGAZUKA, T. and YAMAMURA, Y. J. Biochem. (Tokyo) 52, 92–98 (1962).
30. AZUMA, I., RIBI, E. E., MEYER, T. J. and ZBAR, B. J. Natl. Cancer Inst. 52, 95–101 (1974).
31. AZUMA, I. and YAMAMURA, Y. J. Biochem. (Tokyo) 52, 82–91 (1962).
32. AZUMA, I. and YAMAMURA, Y. J. Biochem. (Tokyo) 53, 275–281 (1963).
33. BATT, R. D., HODGES, R. and ROBERTSON, J. G. Biochim. Biophys. Acta 239, 368–373 (1971).
34. BEKIERKUNST, A. J. Bacteriol. 96, 958–961 (1968).
35. BEKIERKUNST, A. and ARTMAN, M. Am. Rev. Resp. Dis. 86, 832–838 (1966).
36. BEKIERKUNST, A., LEVIJ, I. S., YARKONI, E., VILKAS, E., ADAM, A. and LEDERER, E. J. Bacteriol. 100, 95–102 (1969).
37. BEKIERKUNST, A., LEVIJ, I. S., YARKONI, E., VILKAS, E. and LEDERER, E. Infect. Immun. 4, 245–255 (1971).
38. BEKIERKUNST, A., LEVIJ, I. S., YARKONI, E., VILKAS, E. and LEDERER, E. Science 174, 1240–1242 (1971).
39. BEKIERKUNST, A., WANG, L., TOUBIANA, R. and LEDERER, E. Infect. Immun. 10, 1044–1050 (1974).
40. BEKIERKUNST, A. and YARKONI, E. Infect. Immun. 7, 631–638 (1973).
41. BEKIERKUNST, A., YARKONI, E., FLECHNER, I., MORECKI, S., VILKAS, E. and LEDERER, E. Infect. Immun. 4, 256–263 (1971).
42. BHATTACHARJEE, S. S., HASKINS, R. H. and GORIN, P. A. Carbohydr. Res. 13, 235–246 (1970).
43. BLOCH, H. J. Exp. Med. 91, 197–217 (1950).
44. BLOCH, H. Ann. Rev. Microbiol. 7, 19–46 (1953).
45. BLOCH, H. Experimental Tuberculosis, a Ciba Foundation Symposium, pp. 131–138 (Ed. WOLSTENHOLME, G. E. W. and CAMERON, C. M.) Churchill, London (1955).
46. BLOCH, H., DEFAYE, J., LEDERER, E. and NOLL, H. Biochim. Biophys. Acta 23, 312–321 (1957).
47. BLOCH, H. and NOLL, H. J. Exp. Med. 92, 1–16 (1953).
48. BLOCH, H. and NOLL, H. Br. J. Exp. Pathol. 36, 8–17 (1955).
49. BLOCH, H. and SÜLLMANN, H. Experientia 1, 94–95 (1945).
50. BORDET, C. and MICHEL, G. Bull. Soc. Chim. Biol. 51, 527–548 (1969).
51. BRENNAN, P. J., FLYNN, M. P. and GRIFFIN, P. F. S. FEBS-Lett. 8, 322–324 (1970).
52. BRENNAN, P. J., GRIFFIN, P. F. S., LÖSEL, D. M. and TYRELL, D. Progress in the Chemistry of Fats and Other Lipids, Vol. XIV, p. 51. Pergamon Press, Oxford (1974).
53. BRENNAN, P. J. and LEHANE, D. P. Biochim. Biophys. Acta 176, 675–677 (1969).
54. BRENNAN, P. J., LEHANE, D. P. and THOMAS, D. W. Eur. J. Biochem. 13, 117–123 (1970).
55. BROCHERE-FERREOL, G. and POLONSKY, J. Bull. Soc. Chim. Fr. 714–717 (1958).
56. BRYCE, T. A., WELTI, D., WALSBY, A. E. and NICHOLS, B. W. Phytochemistry 11, 295–302 (1972).
57. BYRNE, P. F. S. and BRENNAN, P. J. J. Gen. Microbiol. 89, 245–255 (1975).

58. CHEDID, L., LAMENSANS, A., PARANT, F., PARANT, M., ADAM, A., PETIT, J. F. and LEDERER, E. *Cancer Res.* **33**, 2187–2195 (1973).
59. COOKE, R. C. and MITCHELL, D. T. *Trans. Br. Mycol. Soc.* **52**, 365–372 (1969).
60. DEMARTEAU-GINSBURG, H. Thesis Dr.-Ing., University of Paris (1958).
61. DIARA, A. and PUDLES, J. *Bull. Soc. Chim. Biol.* **41**, 481–486 (1959).
62. DUBOS, R. J. and MIDDLEBROOK, G. *Am. Rev. Tuberc.* **58**, 698–699 (1948).
63. EDWARDS, J. R. and HAYASHI, J. A. *Arch. Biochem. Biophys.* **111**, 415–421 (1965).
63a. ELBEIN, A. D. *Adv. Carbohydr. Chem. Biochem.* **30**, 227–256 (1974).
64. ETEMADI, A. H. *J. Chem. Soc., Chem. Commun.* 1074–1075 (1967).
65. ETEMADI, A. H. *Chem. Phys. Lipids* **1**, 165–175 (1967).
66. ETEMADI, A. H. *Exposés annuels de Biochimie médicale*, Vol. 28, pp. 77–109. Paris, éd. Masson (1967).
67. ETEMADI, A. H. and GASCHE, J. *Bull. Soc. Chim. Biol.* **47**, 2095–2104 (1965).
68. ETEMADI, A. H., GASCHE, J. and SIFFERLEN, J. *Bull. Soc. Chim. Biol.* **47**, 631–638 (1965).
69. ETEMADI, A. H. and LEDERER, E. *Bull. Soc. Chim. Biol.* **46**, 107–113 (1964).
70. ETEMADI, A. H. and LEDERER, E. *Bull. Soc. Chim. Fr.* 2640–2645 (1965).
71. ETEMADI, A. H., MIQUEL, A. M., LEDERER, E. and BARBER, M. *Bull. Soc. Chim. Fr.* 3274–3276 (1964).
72. ETEMADI, A. H., OKUDA, R. and LEDERER, E. *Bull. Soc. Chim. Fr.* 868–870 (1964).
73. ETEMADI, A. H., PINTE, F. and MARKOVITS, J. *C. R. Acad. Sci. (Paris), Ser. C* **262**, 1343–1346 (1966).
74. ETEMADI, A. H., PINTE, F. and MARKOVITS, J. *Bull. Soc. Chim. Fr.* 195–199 (1967).
74a. FISCHER, W. and KRIEGLSTEIN, J. *Z. Physiol. Chem.* **348**, 1252–1255 (1967).
75. FLUHARTY, A. L. and O'BRIEN, J. S. *Biochemistry* **8**, 2627–2632 (1969).
76. FOUQUEY, C., POLONSKY, J. and LEDERER, E. *Bull. Soc. Chim. Biol.* **44**, 69–81 (1962).
77. FUKUYAMA, K., TANI, J. and KATO, M. *J. Biochem. (Tokyo)* **69**, 511–516 (1971).
78. GANGADHARAM, P. R. J., COHN, M. L. and MIDDLEBROOK, G. *Tubercle (London)* **44**, 452–455 (1963).
79. GASTAMBIDE-ODIER, M., DELAUMENY, J. M. and LEDERER, E. *Biochim. Biophys. Acta* **70**, 670–678 (1963).
80. GASTAMBIDE-ODIER, M., DELAUMENY, J. M. and LEDERER, E. *C. R. Acad. Sci. (Paris)* **259**, 3404–3407 (1964).
81. GASTAMBIDE-ODIER, M. and LEDERER, E. *Biochem. Z.* **333**, 285–295 (1960).
82. GENDRE, T. and LEDERER, E. *Bull. Soc. Chim. Fr.* 1478–1482 (1956).
83. GOLDFINE, H. *Adv. Microb. Physiol.* **8**, 1–58 (1972).
84. GORDON, H. T., THORNBURG, W. and WERUM, L. N. *Anal. Chem.* **28**, 849–855 (1956).
85. GOREN, M. B. *Biochim. Biophys. Acta* **210**, 116–126 (1970).
86. GOREN, M. B. *Biochim. Biophys. Acta* **210**, 127–138 (1970).
87. GOREN, M. B. *Lipids* **6**, 40–46 (1971).
88. GOREN, M. B. *Bacteriol. Revs.* **36**, 33–64 (1972).
89. GOREN, M. B. *Symposium intern. sur les immunostimulants bactériens* Paris, 1974, Résumé commun., p. 2.
90. GOREN, M. B. *Tubercle (London)* **56**, 65–71 (1975).
91. GOREN, M. B. and BROKL, O. *Recent Results Cancer Res.* **47**, 251–258 (1974).
92. GOREN, M. B., BROKL, O., DAS, B. C. and LEDERER, E. *Biochemistry* **10**, 72–81 (1971).
93. GOREN, M. B., BROKL, O. and SCHAEFER, W. *Infect. Immun.* **9**, 142–149 (1974).
93a. GOREN, M. B., D'ARCY HART, P., YOUNG, M. R. and ARMSTRONG, J. A. *The United States–Japan Cooperative Medical Science Program* pp. 228–241, 9th joint conference on Tuberculosis, San Francisco 1975; *Proc. Nat. Acad. Sci. USA*, **73**, 2510–2514 (1976).
94. HAINES, T. H. *Progress in the Chemistry of Fats and Other Lipids*, Vol. XI, p. 299, Pergamon Press, Oxford (1971).
95. HISATSUKA, K., NAKAHARA, T., SANO, N. and YAMADA, K. *Agr. Biol. Chem.* **35**, 686–692 (1971).
96. HOPKINS, C. Y. *Progress in the Chemistry of Fats and Other Lipids*, Vol. VIII, p. 213. Pergamon Press, Oxford (1965).
97. IONEDA, T., LEDERER, E. and ROZANIS, J. *Chem. Phys. Lipids* **4**, 375–392 (1970).
98. IONEDA, T., LENZ, M. and PUDLES, J. *Biochem. Biophys. Res. Commun.* **13**, 110–114 (1963).
99. ITO, F., COLEMAN, C. and MIDDLEBROOK, G. *Kekkaku (Tokyo)* **36**, 764–769 (1961).
100. ITOH, S. and SUZUKI, T. *Agr. Biol. Chem.* **36**, 2233–2235 (1972).
101. ITOH, S. and SUZUKI, T. *Agr. Biol. Chem.* **38**, 1443–1449 (1974).
102. JAUREGUIBERRY, G., LENFANT, M., DAS, B. C. and LEDERER, E. *Tetrahedron* suppl. 8, part I, 27–32 (1966).
103. JONES, D. F. *J. Chem. Soc., C* 479–484 (1967).
104. KATES, M. *Adv. Lipid Res.* **8**, 225–265 (1970).
105. KATES, M. and WASSEF, M. K. *Ann. Rev. Biochem.* **39**, 323–358 (1970).
106. KATO, M. *Am. Rev. Respir. Dis.* **93**, 411–420 (1966).
107. KATO, M. *Am. Rev. Respir. Dis.* **94**, 388–394 (1966).
108. KATO, M. *Am. Rev. Respir. Dis.* **96**, 998–1008 (1967).
109. KATO, M. *Am. Rev. Respir. Dis.* **98**, 260–269 (1968).
110. KATO, M. *Am. Rev. Respir. Dis.* **100**, 47–53 (1969).
111. KATO, M. *Arch. Biochem. Biophys.* **140**, 379–390 (1970).
112. KATO, M. *J. Bacteriol.* **101**, 709–716 (1970).
113. KATO, M. *J. Bacteriol.* **107**, 746–752 (1971).
114. KATO, M. *Infect. Immun.* **5**, 203–212 (1972).
115. KATO, M. *Infect. Immun.* **7**, 9–13 (1973).
116. KATO, M. *Infect. Immun.* **7**, 14–21 (1973).
117. KATO, M. *Kekkaku (Tokyo)* **49**, 229–238 (1974).
118. KATO, M. and ASSELINEAU, J. *Eur. J. Biochem.* **22**, 364–370 (1971).
119. KATO, M. and FUKUSHI, K. *Am. Rev. Respir. Dis.* **100**, 42–46 (1969).
120. KATO, M. and GOREN, M. B. *Infect. Immun.* **10**, 733–741 (1974).
121. KATO, M. and MAEDA, J. *Infect. Immun.* **9**, 8–14 (1974).

122. KATO, M., TAMURA, T., SILVE, G. and ASSELINEAU, J. *Eur. J. Biochem.* in press.
123. KATO, M., TOUBIANA, M. J. and TOUBIANA, R. unpublished results.
124. KERVABON, A., MASSON, P. and ETEMADI, A. H. *Biochimie* **57**, 811–824 (1975).
125. KHANNA, S. N. and GUPTA, P. C. *Phytochemistry* **6**, 735–739 (1967).
126. KHULLER, G. K. and BRENNAN, P. J. *J. Gen. Microbiol.* **73**, 409–412 (1972).
127. KREMBEL, J. and ETEMADI, A. H. *Tetrahedron* **22**, 1113–1119 (1966).
128. KUHN, R., TRISCHMANN, H. and LOW, I. *Angew. Chemie* **67**, 32–33 (1955).
129. LAINE, R. A., GRIFFIN, P. F. S., SWEELEY, C. C. and BRENNAN, P. J. *Biochemistry* **11**, 2267–2271 (1972).
130. LAMBEIN, F. and WOLK, C. P. *Biochemistry* **12**, 791–798 (1973).
131. LANEELLE, M.-A. Thesis Dr.Sci., University of Toulouse (1969).
132. LANEELLE, M.-A. and ASSELINEAU, J. *FEBS-Lett.* **7**, 64–67 (1970).
133. LANEELLE, M.-A. and ASSELINEAU, J. *Biochim. Biophys. Acta* **486**, 205–208 (1977).
134. LANEELLE, M.-A., ASSELINEAU, J. and CASTELNUOVO, G. *Ann. Inst. Pasteur* **108**, 69–82 (1969).
135. LANEELLE, M.-A. and LANEELLE, G. *Eur. J. Biochem.* **12**, 296–300 (1970).
136. LANEELLE, M.-A., LANEELLE, G. BENNET, P. and ASSELINEAU, J. *Bull. Soc. Chim. Biol.* **47**, 2047–2067 (1965).
137. LECHEVALIER, M. P., HORAN, A. and LECHEVALIER, H. *J. Bacteriol.* **105**, 313–318 (1971).
138. LEDERER, E. In: *Colloquium on the Chemotherapy of Tuberculosis* pp. 1–46 (Ed. by BARRY, V. C.) Trinity College Press, Dublin (1952).
139. LEDERER, E. *Chem. Phys. Lipids* **1**, 294–315 (1967).
140. LEDERER, E. *IUPAC 7th International Symposium on the Chemistry of Natural Products* pp. 135–165. Page Bros, Norwich (1971).
141. LEDERER, E., ADAM, A., CIORBARRU, R., PETIT, J.-F. and WIETZERBIN, J. *Mol. Cell. Biochem.* **7**, 87–104 (1975).
142. LEDERER, E., PORTELANCE, V. and SERCK-HANSSEN, K. *Bull. Soc. Chim. Fr.* 413–417 (1952).
143. LEDERER, E. and PUDLES, J. *Bull. Soc. Chim. Biol.* **33**, 1003–1011 (1951).
144. LEDERER, E., PUDLES, J., BARBEZAT, S. and TRILLAT, J. *J. Bull. Soc. Chim. Fr.* 93–95 (1952).
145. LEMIEUX, R. U., THORN, J. A. and BAUER, H. F. *Can. J. Chem.* **31**, 1054–1059 (1953).
146. MALANI, C. and POLGAB, N. *J. Chem. Soc.* 3092–3096 (1963).
147. MANDELL, J. D. and HERSHEY, A. D. *Anal. Biochem.* **1**, 66–77 (1960).
148. MARCEL, Y., DOUSTE-BLAZY, L. and ASSELINEAU, J. *Bull. Soc. Chim. Biol.* **48**, 693–704 (1966).
149. MAURICE, M. T., VACHERON, M. J. and MICHEL, G. *Chem. Phys. Lipids* **7**, 9–18 (1971).
150. MERDINGER, E., KOHN, P. and McCLAIN, R. C. *Can. J. Microbiol.* **14**, 1021–1027 (1968).
151. MEYER, T. J., RIBI, E. E., AZUMA, I. and ZBAR *J. Natl. Cancer Inst.* **52**, 103–111 (1974).
152. MICHEL, G. *Bull. Soc. Chim. Biol.* **41**, 1649–1669 (1959).
153. MICHEL, G., BORDET, C., and LEDERER, E. *C. R. Acad. Sci. (Paris)* **250**, 3518–3520 (1960).
154. MIDDLEBROOK, G., COLEMAN, C. and SCHAEFER, W. B. *Proc. Nat. Acad. Sci. U.S.* **45**, 1801–1804 (1959).
155. MIDDLEBROOK, G., DUBOS, R. J. and PIERCE, C. H. *J. Exp. Med.* **86**, 175–184 (1947).
156. MILNER, K. C., RUDBACH, J. A. and RIBI, E. E. In: *Microbial Toxins*, (Ed. WEINBAUM, G., KADIS, S. and AJL, S. J.) Vol, IV. pp. 1–65, Academic Press, New York (1971).
157. MINNIKIN, D. E., PATEL, P. V. and GOODFELLOW, M. *FEBS-Lett.* **39**, 322–324 (1974).
158. MINNIKIN, D. E. and POLGAR, N. *J. Chem. Soc. Chem. Commun.* 312–314 (1967).
159. MINNIKIN, D. E. and POLGAR, N. *J. Chem. Soc. Chem. Commun.* 648–649 (1966).
160. MINNIKIN, D. E. and POLGAR, N. *Tetrahedron Lett.* 2643–2647 (1966).
161. MIQUEL, A. M., GINSBURG, H. and ASSELINEAU, J. *Bull. Soc. Chim. Biol.* **45**, 715–730 (1963).
162. MISHIMA, T., OKADA, Y., TERAI, T., OGURA, K. and YAMAMURA, Y. *J. Biochem. (Tokyo)* **48**, 392–396 (1960).
163. MOORE, V. L., MYRVIK, Q. N. and KATO, M. *Infect. Immun.* **6**, 5–8 (1972).
164. MURTHY, H. S. R., KUMAR, K. S. and VENKITASUBRAMANIAN, T. A. *Biochim. Biophys. Acta* **146**, 584–586 (1967).
165. NARUMI, K. and TSUMITA, T. *J. Biol. Chem.* **242**, 2233–2239 (1967).
166. NOJIMA, S. *J. Biochem. (Tokyo)* **46**, 499–506 (1959).
167. NOLL, H. *Adv. Tuberc. Res.* **7**, 149–183 (1956).
168. NOLL, H. and BLOCH, H. *Am. Rev. Tuberc.* **67**, 828–852 (1953).
169. NOLL, H. and BLOCH, H. *J. Biol. Chem.* **214**, 251–265 (1955).
170. NOLL, H., BLOCH, H., ASSELINEAU, J. and LEDERER, E. *Biochim. Biophys. Acta* **20**, 299–309 (1956).
171. NOLL, H. and JACKIM, E. *J. Biol. Chem.* **232**, 903–917 (1958).
172. OHARA, T., SHIMMYO, Y., SEKIKAWA, I., MORIKAWA, K. and SUMIKAWA, E. *Jpn. J. Tuberc.* **5**, 128–143 (1957).
173. OKABE, H. and KAWASAKI, T. *Tetrahedron Lett.* 3123–3126 (1970).
174. OKAZAKI, H., SUGINO, H., KANZAKI, T. and FUKUDA, H. *Agr. Biol. Chem.* **33**, 764–770 (1969).
175. OSHIMA, M. and YAMAKAWA, T. *Biochemistry* **13**, 1140–1146 (1974).
176. PERRINE, T. D., RIBI, E. E., MAKI, W., MILLER, B. and OERTHI, E. *Appl. Microbiol.* **10**, 93–98 (1962).
177. POLONSKY, J., FERREOL, G., TOUBIANA, R. and LEDERER, E. *Bull. Soc. Chim. Fr.* 1471–1478 (1956).
178. POLONSKY, J. and LEDERER, E. *Bull. Soc. Chim. Fr.* 504–510 (1954).
179. PROME, J.-C. *Bull. Soc. Chim. Fr.* 655–660 (1968).
180. PROME, J.-C., ASSELINEAU, C. and ASSELINEAU, J. *C. R. Acad. Sci. (Paris), Ser. C* **263**, 448–451 (1966).
181. PROME, J.-C., LACAVE, C., AHIBO-COFFY, A. and SAVAGNAC, A. *Eur. J. Biochem.* **63**, 543–552 (1976).
182. PROME, J.-C., WALKER, R. W. and LACAVE, C. *C. R. Acad. Sci. (Paris) Ser. C*, **278**, 1065–1068 (1974).
183. PROTTEY, C. and BALLOU, C. E. *J. Biol. Chem.* **243**, 6196–6201 (1968).
184. PUDLES, J. and LEDERER, E. *Biochim. Biophys. Acta* **11**, 163–164 (1953).
185. PUDLES, J. and LEDERER, E. *Biochim. Biophys. Acta* **11**, 602–603 (1953).
186. PUZO, G. and PROME, J.-C. *Tetrahedron* **29**, 3619–3629 (1973).

187. RAFFEL, S., LEDERER, E. and ASSELINEAU, J. *Fed. Proc.* **13**, n° 1668 (1954).
188. RIBI, E. E., RIBI, K., GOODE, G., BROWN, W., NIWA, M. and SMITH, R. *J. Bacteriol.* **102**, 250–260 (1970).
189. ROSNESS, P. A. and WRIGHT, B. E. *Arch. Biochem. Biophys.* **164**, 60–72 (1974).
190. ROUANET, J. M. Thesis "Doctorat de spécialité", University of Toulouse (1973).
191. ROWELL, R. M. *Carbohyd. Res.* **23**, 417–424 (1972).
192. RYHAGE, R. and STENHAGEN, E. *J. Lipid Res.* **1**, 361–390 (1960).
192a. SAITO, R., TANAKA, A., SUGIYAMA, K., AZUMA, I., YAMAMURA, Y., KATO, M. and GOREN, M. B., *Infect. Immun.* **13**, 776–781 (1976).
192b. SAITO, R., TANAKA, A., SUGIYAMA, K. and KATO, M., *Am. Rev. Respir. Dis.* **112**, 578–580 (1975).
193. SASTRY, P. S. *Adv. Lipid Res.* **12**, 251–310 (1974).
194. SENN, M., IONEDA, T., PUDLES, J. and LEDERER, E. *Eur. J. Biochem.* **1**, 353–356 (1967).
195. SHANKARAN, R. and VENKITASUBRAMANIAN, T. A. *Am. Rev. Respir, Dis.* **101**, 401–407 (1970).
196. SHAW, N. *Biochim. Biophys. Acta* **164**, 435–436 (1968).
197. SHAW, N. *Bacteriol. Revs.* **34**, 365–377 (1970).
198. SHAW, N. *Adv. Appl. Microbiol.* **17**, 63-108 (1974).
199. SHAW, N. *Adv. Microbial Phys.* **12**, 141–167 (1975).
200. SHAW, N. and DINGLINGER, F. *Biochem. J.* **112**, 769–775 (1969).
201. SMITH, P. F. and MAYBERRY, W. R. *Biochemistry* **7**, 2706–2710 (1968).
202. SORKIN, E., ERLENMEYER, H. and BLOCH, H. *Nature* **170**, 124 (1952).
203. SPITZNAGEL, J. K. and DUBOS, R. J. *J. Exp. Med.* **101**, 291–311 (1955).
204. STÄLLBERG-STENHAGEN, S. and STENHAGEN, E. *J. Biol. Chem.* **159**, 255–262 (1945).
205. STODOLA, F. H., DEINEMA, M. H. and SPENCER, J. F. T. *Bacteriol. Revs.* **31**, 194–213 (1967).
206. STODOLA, F. H., LESUK, A. and ANDERSON, R. J. *J. Biol. Chem.* **126**, 505–513 (1938).
207. SUZUKI, T., TANAKA, H. and ITOH, S. *Agr. Biol. Chem.* **38**, 557–563 (1974).
208. SUZUKI, T., TANAKA, K., MATSUBARA, I. and KINOSHITA, S. *Agr. Biol. Chem.* **33**, 1619–1627 (1969).
209. SWEELEY, C. C., BENTLEY, R., MAKITA, M. and WELLS, W. N. *J. Am. Chem. Soc.* **85**, 2497–2507 (1963).
210. TABAUD, H., TISNOVSKA, H. and VILKAS, E. *Biochimie* **53**, 55–61 (1971).
211. TAKAYAMA, K., SCHNOES, H. K., ARMSTRONG, E. L. and BOYLE, R. W. *J. Lipid Res.* **16**, 308–317 (1975).
212. TOCANNE, J. F. *Carbohyd. Res.* **44**, 301–307 (1975).
213. TOCANNE, J. F. and ASSELINEAU, C. *Bull. Soc. Chim. Fr.* 4519–4525 (1968).
214. TOIDA, I. *Am. Rev. Respir. Dis.* **108**, 694–697 (1973).
215. TOIDA, I. *Am. Rev. Respir. Dis.* **110**, 641–646 (1974).
216. TOUBIANA, R., DAS, B. C., DEFAYE, J., MOMPON, B. and TOUBIANA, M. J. *Carbohyd. Res.* **44**, 308–312 (1975).
217. TOUBIANA, R., TOUBIANA, M. J., DAS, B. C. and RICHARDSON, A. C. *Biochimie* **55**, 569–573 (1973).
218. TOUBIANA, R. and TOUBIANA, M. J. *Biochimie* **55**, 575–578 (1973).
219. TREVELYAN, W. E., PROCTER, D. P. and HARRISON, J. S. *Nature* **166**, 444–445 (1950).
220. TULLOCH, A. P. and HILL, A. *Can. J. Chem.* **46**, 2485–2493 (1968).
221. TULLOCH, A. P., HILL, A. and SPENCER, J. F. T. *Can. J. Chem.* **46**, 3337–3351 (1968).
222. TŸORINOJA, K., NURMINEN, T. and SUOMALAINEN, H. *Biochem. J.* **141**, 133–139 (1974).
223. VILKAS, E., ADAM, A. and SENN, M. *Chem. Phys. Lipids* **2**, 11–16 (1968).
224. VILKAS, E. and ROJAS, A. *Bull. Soc. Chim. Biol.* **46**, 689–701 (1964).
225. WALKER, R. J., PROME, J.-C. and LACAVE, C. *Biochim. Biophys. Acta* **326**, 52–62 (1973).
226. WELBY-GIEUSSE, M., LANEELLE, M. A. and ASSELINEAU, J. *Eur. J. Biochem.* **13**, 164–167 (1970).
227. WELSH, K., SHAW, N. and BADDILEY, J. *Biochem. J.* **107**, 313–314 (1968).
228. WIEGANDT, H. *Adv. Lipid Res.* **9**, 249–289 (1971).
229. WILLSTAEDT, H. and BORGGÅRD, M. *Bull. Soc. Chim. Biol.* **28**, 733–735 (1946).
230. WINDER, F. C. and COLLINS, P. B. *J. Gen. Microbiol.* **63**, 41–48 (1970).
231. WINDER, F. C., TIGHE, J. and BRENNAN, P. J. *J. Gen. Microbiol.* **73**, 539–546 (1972).
231a. YAMAGUCHI, M., SATO, A. and YUKUYAMA, A. *Chem. Ind. (London)*, 741–742 (1976).
232. YANAGIHARA, Y., MIFUCHI, I., AZUMA, I. and YAMAMURA, Y. *Jpn. J. Microbiol.* **12**, 103–110 (1968).
233. YANO, I., FURUKAWA, Y. and KUSUNOSE, M. *J. Gen. Appl. Microbiol.* **17**, 329–334 (1971).
234. YANO, I. and SAITO, K. *FEBS-Lett.* **23**, 352–356 (1972).
235. YANO, I., SAITO, K., FURUKAWA, Y. and KUSUNOSE, M. *FEBS-Lett.* **21**, 215–219 (1972).
236. YARKONI, E., BEKIERKUNST, A., ASSELINEAU, J., TOUBIANA, R., TOUBIANA, M.-J. and LEDERER, E. *J. Natl. Cancer Inst.* **51**, 717–720 (1973).
237. YARKONI, E., WANG, L. and BEKIERKUNST, A. *Infect. Immun.* **9**, 977–984 (1974).

References added in proof

Review articles devoted to the cell components of *Mycobacteria*:
RATLEDGE, C. *Adv. Microb. Physiol.* **13**, 115–244 (1976).
BARKSDALE, L. and KIM, K.-S. *Bacteriol. Revs.* **41**, 217–372 (1977).

Review article on acyl-trehaloses:
LEDERER, E. *Chem. Phys. Lipids* **16**, 91–106 (1976).

Structural studies:
Isolation of 6-mycoloyl-6'-acetyltrehalose from the H37 Ra strain of *Mycobacterium tuberculosis*; possible role: TAKAYAMA, K. and ARMSTRONG, E. L. *Biochemistry* **15**, 441–447 (1976).
Structure determination of sulfatide SL-I: GOREN, M. B., BROKL, O., ROLLER, P., FALES, H. M. and DAS, B. C., *Biochemistry* **15**, 2728–2735 (1976).

Biochemical studies:
Accumulation of triglycerides in liver lipids induced by Cord Factor: TOIDA, I. *Amer. Rev. Respir. Dis.* **114**, 767–773 (1976).

Production of sulfatides during cell division in *Mycobacterium avium*: McCARTHY, C. *Infect. Immun.* **14,** 1241–1252 (1976).
Absence of interactions between phleates of trehalose and mitochondria: KATO, M. *Jpn. J. Tuberc. Chest Dis.* **20,** 9–14 (1976).

Biological properties:
Production of granulomas by Cord Factor: REGGIARDO, Z. and SHAMSUDDIN, A. K. M. *Infect. Immun.* **14,** 1369–1374 (1976).
GRANGER, D. L., YAMAMOTO, K.-I. and RIBI, E. *J. Immunol.* **116,** 482–488 (1976).
Stimulation of macrophages by Cord Factor: YARKONI, E., WANG, L. and BEKIERKUNST, A. *Infect. Immun.* **16,** 1–8 (1977).
Nonspecific resistance against infection by *Salmonella typhi* and *S. typhimurium* induced in mice by Cord Factor: YARKONI, E. and BEKIERKUNST, A. *Infect. Immun.* **14,** 1125–1129 (1976).
Nonspecific inhibition of growth of tumor cells by Cord Factor: LECLERC, C., LAMENSANS, A., CHEDID, L., DRAPIER, J.-C., PETIT, J.-F., WIETZERBIN, J. and LEDERER, E. *Cancer Immunol. Immunother.* **1,** 227–232 (1976).
RIBI, E., TAKAYAMA, K., MILNER, K., GRAY, G. R., GOREN, M., PARKER, R., McLAUGHLIN, C. and KELLY, M. *Cancer Immunol. Immunother.* **1,** 265–270 (1976).

Prog. Chem. Fats other Lipids. Vol. 16, pp. 101–124. Pergamon Press, 1978. Printed in Great Britain

MOLECULAR ARRANGEMENT IN CONDENSED MONOLAYER PHASES

Monica Lundquist

*Institute of Medical Biochemistry, University of Gothenburg,
Gothenburg, Sweden*

CONTENTS

I. Introduction	101
II. General Theory	102
III. Methods	104
IV. Materials and Experimental Techniques	104
A. Materials	104
B. The surface balance	104
C. Equipment for the study of polymorphism in the three-dimensional state	105
V. Results and Discussion	105
A. *n*-Alkyl acetates	105
B. Ethyl esters of *n*-aliphatic acids	114
C. *n*-Aliphatic acids	116
D. The formation of racemic compounds in monolayers	118
E. The formation of multifilms at the air–water interface	120
VI. Terminology of Monolayer Phases	120
VII. Summary	120
A. Monolayer phases with a vertical chain arrangement	120
B. Monolayer phases with a tilted chain arrangement	122
References	124

I. INTRODUCTION

The history of that part of surface science that deals with insoluble surface films dates back to ancient time. The calming effect of oil on a rough sea is a classical example of a natural phenomenon that nowadays can be explained by the physical chemistry of surfaces. In 1913 Sir William Hardy[15] pointed out the important fact that monolayers are formed only from molecules consisting of a hydrophobic part and a hydrophilic part, the so-called amphiphilic molecules. With a fundamental work on fatty acid monolayers, Langmuir[22] in 1917 opened a new epoch in monolayer research. He was the first to make an analysis of monolayer properties in terms of molecular structure and molecular arrangement. Some of the basic principles regarding monolayer structure which were put forward by Langmuir are still valid. Thus, it was shown by him that, at the air–water interface, the molecules are oriented with the polar, hydrophilic groups immersed in the water, while the hydrophobic parts are directed towards the air.

With improvement of the technique used in monolayer research and with an increasing knowledge of molecular arrangement in the ordinary three-dimensional state, our understanding of the structure and behavior of molecules in the monolayer state has increased enormously. However, there still remains much to be learned especially about the detailed structure of condensed monolayers and regarding the mechanism of molecular rearrangement in monolayer phase transitions. As there are no methods available for direct structure analysis of a monolayer on a water substrate, knowledge of the monolayer structure has, for the most part, been achieved by a sort of translation and application of the knowledge gained from the study of the ordinary three-dimensional state.

The monolayer behaviour of a great number of substances has been systematically investigated. Several classes of naturally occurring compounds are able to form stable monolayers, e.g. long chain aliphatic compounds with small polar groups, glycerides, phospholipides, sterols and other molecules with condensed ring systems. A large number of studies of polymer and protein films have also been reported. For comprehensive reviews, the reader is referred to the well-known monographs by Adam,[1] Harkins[16] and Gaines[12] and review articles by Dervichian[8] and Stenhagen.[45]

Einar Stenhagen and Stina Ställberg-Stenhagen started monolayer studies about 1940. Originally the aim was to explore the possibilities of the monolayer technique in the determination of organic structures. The application of the monolayer method to such problems is, of course, limited to compounds that form stable monolayers. These authors first applied the monolayer technique in the investigation of film-forming lipids originating from the tubercle bacillus, phtiocerol[47] and phtioic acid.[46] The structure of these compounds was later solved by mass spectrometry.[43] However, the monolayer work carried out by Einar Stenhagen and Stina Ställberg-Stenhagen showed to be important from other aspects, e.g. it led to the discovery of new phases in condensed monolayers.[48] The use of surphase phase diagrams in monolayer research was also introduced by these authors.[45,48]

An analysis of molecular arrangement in insoluble monolayers at the liquid–gas interface has been the subject of recent investigations carried out by the present author. The concept of monolayer or "two-dimensional" polymorphism has been developed and the relation between monolayer polymorphism and polymorphism in the ordinary three-dimensional state has been studied. The investigations have been confined to long-chain aliphatic compounds with small polar groups, principally long-chain esters. The bearing of the results on the understanding of monolayer structure in general will be discussed here. Moreover, the general principles employed and used in the present investigation may be easily adopted to other groups of lipids as well. A detailed account of the results obtained is found in original papers.[28,29,32,33,35,36] The present communication sketches the background of these investigations and gives a brief summary of the results obtained.

II. GENERAL THEORY

Ultimately, the structure and stability of an insoluble monolayer is governed by the delicate balance of the forces of attraction and repulsion between the molecules in the monolayer, as well as between these molecules and the substrate molecules. The most important variables in determining monolayer structure are temperature, pressure, pH and the ionic composition of the substrate.

On the basis of molecular parameters, three major monolayer states are recognized, the gaseous or vapour state, the expanded state and the condensed monolayer state. A close relationship exists between these monolayer states and the corresponding states in the ordinary three-dimensional state. Thus, the gaseous monolayer state, where the molecules are assumed to be floating far apart on the substrate surface, is often compared with the ordinary gaseous state.[1,12,16] The expanded or liquid-expanded monolayer state is generally considered as a very thin liquid phase with the hydrophobic part of the molecules in a randomlike arrangement but with the polar groups anchored to the substrate surface.[1,12,16] In the condensed monolayer state, the molecular parameters are of the same order of magnitude as in the three-dimensional solid state and the molecules must be relatively closely packed. Often several condensed phases are recognized which differ in respect to molecular parameters and physical properties. The concept "monolayer polymorphism" was introduced by Dervichian[7] and has been further developed in the present study.[35,36] It was the aim of the present investigation to show that monolayer polymorphism is equivalent to polymorphism as it occurs in the three-dimensional state. It is assumed here that the molecular arrangement in the condensed monolayer state is analogous to that in the ordinary crystalline state. Differ-

ent condensed monolayer phases then might correspond to different crystalline arrangements of the film-forming molecules.

Long-chain aliphatic compounds with small polar groups have been extensively studied as model substances in this respect, as they are able to form a large number of
different condensed monolayer phases. The character and definition of the phases formed
have been described by a number of authors, cf Harkins[16] and Stenhagen.[45] There
has been much controversy regarding the molecular structure in these phases. One
of the prevailing views has been that of Adam[1] who introduced such concepts as *films
with close-packed heads* and *films with close-packed chains*. Unfortunately these notations,
which imply a somewhat arbitrary interpretation of monolayer structure, are still in
use in monolayer publications and discussions, cf Goddard *et al.*, in their investigations
on fatty acid monolayers.[14] Lyons and Rideal[38] suggested an alternative explanation
for the occurrence of condensed phases with molecular parameters of different magnitude. According to these authors, the hydrocarbon chains are always close-packed in
condensed monolayers, provided the polar groups are small enough. In small area condensed phases, the chains would then have a vertical orientation with respect to the
substrate surface and, in large area condensed phases, the chains would be tilted at
certain preferred angles to the substrate surface. This theory, although put forward
about four decades ago, is in accordance both with present knowledge of the packing
of hydrocarbon chains in the three-dimensional state and with the results obtained
in this investigation.

The relationship between condensed monolayer phases and crystalline modifications
in the three-dimensional state was first clearly demonstrated by Dervichian.[7] He showed
that there exists a close correspondence between certain phase transition temperatures
in triglyceride monolayers and melting points in the three-dimensional state. Dervichian
also drew attention to the apparent correlation between the molecular parameters of
condensed monolayer phases and those of certain crystalline forms in the three-dimensional state. Even though, due to the limited crystallographic data then available, the
actual correlation between the two- and three-dimensional forms was not quite correct,
his general ideas have been confirmed by the present work.

The general theory of the molecular mechanism of phase transitions in monolayers
has been developed by Landau and Lifshitz[26,27] and independently by Dervichian and
Joly.[7,20] The thermodynamic classification introduced by Ehrenfest is also applicable
to monolayer phase transitions. The phase transitions which take place in a monolayer
are ordinary or first order transitions which involve a discontinuity of the molecular
area, A, and subsidiary thermodynamic properties, and second order transitions which
are characterized by a sudden change of the compressibility, K, of the monolayer. In
monolayer pressure-area isotherms, first order transitions are recognized as regions of
constant pressure or plateaus, and second order transitions are apparent as changes
in slope. As follows, from an application of Gibbs adsorption isotherm, a first order
transition involves a discontinuity of the first partial derivative of μ, the chemical potential of the film-forming substance with respect to surface pressure π:

$$\left(\frac{\delta\mu}{\delta\pi}\right)_{p,T} = A.$$

A change of the compressibility without any discontinuity of the area corresponds to
a discontinuity of the second partial derivative of μ:

$$\left(\frac{\delta^2\mu}{\delta\pi^2}\right)_{p,T} = \frac{\delta A}{\delta\pi}; \quad K = \frac{1}{A} \cdot \frac{\delta A}{\delta\pi}.$$

Considering the mechanism of molecular reorientation accompanying the two different
kinds of phase transitions, it has been concluded[20] that ordinary changes of state correspond to the equilibrium between two immiscible phases. In second order transitions
or continuous transitions on the other hand, the rearrangement of the molecules takes
place successively during compression. Thus, as is the case in ordinary bulk systems,

a point of transformation of second order corresponds to the end of the phase transition. According to Joly,[20] the molecules in a monolayer can only exist in a finite number of stable equilibrium states which correlate with a discontinuous series of discrete energy states and correspond to discrete molecular areas. These areas are precisely the transition points of different orders.

A treatise on the thermodynamics of the monolayer state has recently been published by Eriksson.[10]

III. METHODS

An analysis of molecular structure in monolayers, based only on comparative studies of molecular parameters, is apt to be limited to the main features of molecular configuration and arrangement in the film. If it is possible to correlate transition points in the two-dimensional and three-dimensional states, the analysis of monolayer structure is facilitated. However, for a large number of film-forming substances there exists no such relationship, e.g. for highly polar compounds where forces of interaction between the polar end group and substrate molecules play a considerable role. An alternative approach is the purely thermodynamic treatment of monolayer phase relations. In earlier monolayer investigations, the thermodynamic properties of monolayers have, to a large extent, been overlooked. However, in the progress of the present work, it appeared quite necessary to make a serious thermodynamic approach to the properties of monolayer phase relations.

The principal thermodynamic parameters of interest in discussing structural aspects of monolayer phase transitions are the internal energies of transition, ΔU_{tr}, the entropies of transition, ΔS_{tr}, as well as the change of molecular area involved in a transition, ΔA_{tr}. Derived thermodynamic properties are also of interest, e.g. the compressibility and the thermal expansion coefficient. Of course, a thermodynamic analysis provides no direct information of molecular structure. However, a comparison of thermodynamic data obtained for the monolayer state with corresponding data for the three-dimensional state provides a great deal of additional information which is not otherwise obtained.

The present investigation has been directed along the following lines:
(1) Study of the polymorphism in the ordinary three-dimensional state.
(a) X-ray investigation of the crystalline structure of different polymorphic forms. Data could be obtained from the literature for a number of the compounds studied.
(b) Thermal investigation for the establishment of polymorphic patterns, phase transition points and transition energies.
(c) Thermodynamic treatment of the phase relations in the three-dimensional state.
(2) Study of the monolayer polymorphism.
(a) Determination of molecular parameters for the different monolayer phases.
(b) Determination of phase equilibria and construction of phase diagrams.
(c) Thermodynamic treatment of monolayer phase relations.
(3) Correlation of monolayer data with data for the three-dimensional state.

IV. MATERIALS AND EXPERIMENTAL TECHNIQUES

A. *Materials*

Only high purity samples have been used in the monolayer investigations. Very pure normal long-chain acids and alcohols for the preparation of ethyl esters and acetates as well as optically active alcohols and esters have been kindly supplied by Professor E. Stenhagen and his collaborators.

B. *The Surface Balance*

A recording surface balance of the horizontal Langmuir–Adam type has been used. This apparatus has been described elsewhere[5] but a brief description of the instrument used will be given here. The principle of the method consists in measuring the difference in surface tension between a film-covered surface and a clean substrate surface, i.e. by definition the surface pressure, π. The balance used is a special torsion balance equipped with a surface pressure meter based on the Mikrokator principle, a mechanical device originally designed as a precision instrument for length-measurements. The "surface pressure Mikrokator" was designed by Mr. H. Abramsson of Messrs. C. E. Johansson, Eskilstuna, Sweden, in collaboration with Prof. E. Stenhagen[49] and made by the firm mentioned. A hydraulic drive mechanism is used for the continuous compression of the monolayer and a synchronously driven camera for automatical recording of the pressure-area isotherms. The surface trough is placed in a thermostated box and the substrate temperature can be held constant within the limits. of $\pm 0.1°C$. The use of the Mikrokator surface balance allows a rapid and accurate determination of surface pressures. Even small discontinuities in area and compressibility can be detected and the instrument has proved to be specially well suited for the study of phase transitions in monolayers.

C. Equipment for the Study of Polymorphism in the Three-dimensional State

An X-ray diffraction camera for the continuous recording of diffraction-pattern–temperature (DPT) diagrams has been used. The instrument was designed by E. Stenhagen especially for the study of polymorphism in long-chain compounds.[44] The diagrams obtained give information about the chain packing in the different modifications as well as of the existence of single or double layer arrangement and of the angle of tilt of the chains to the end group planes. An estimation of transition temperatures can also be obtained. A Perkin–Elmer differential scanning calorimeter type DSC-1 was used for differential thermal analysis (DTA).[33,35]

V. RESULTS AND DISCUSSION

For the purpose of a careful correlation study between the monolayer state and the ordinary three-dimensional state, it appeared desirable to choose one or more groups of n-aliphatic compounds as model substances. Suitable substances from this viewpoint appeared to be the n-aliphatic esters. These substances are sufficiently polar to spread as monolayers but at the same time the polar forces are weak compared to those of free acids and alcohols. The interaction energy attributed to reorganization of the polar groups in spreading the substance from the three-dimensional crystalline state will, therefore, be comparatively small, which facilitates comparative studies of the present kind. Long-chain ethyl esters and long-chain acetates have been used in the present investigation.

A. n-Alkyl Acetates

1. Polymorphism in the Three-dimensional State

The polymorphism of n-aliphatic compounds has been investigated by a number of authors.[3,39,41,42] The packing principles for these compounds are characteristic of long-chain compounds with a layer structure and the end groups of the molecules arranged in planes.[21] The packing of the hydrocarbon chains, characterized by the "sub-cell" or methylene group packing and the angle of tilt of the chains to the end group planes, is determined primarily by the nature of the end group and the chain length and the thermodynamic condition of the substance. Current nomenclature is based upon the packing of the hydrocarbon chains.[23] A recent survey of different modes of chain packing is given by Larsson.[24] By means of X-ray and thermal analysis, the existence of four different crystalline forms could be demonstrated for the n-alkyl acetates, one tilted

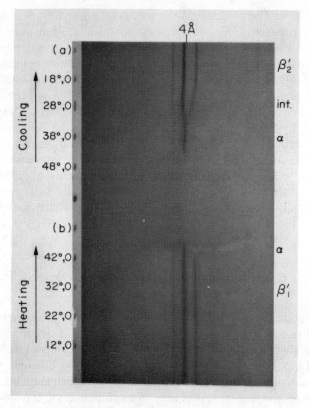

FIG. 1. Diffraction pattern-temperature (DPT) diagrams for *n*-docosyl acetate, (a) on heating, (b) on cooling.

β-form and three vertical forms with different chain packings.[35] The polymorphic behaviour of the even *n*-alkyl acetates investigated can be formulated as follows. The terminology introduced by Lutton[37] and Larsson[23] is used.

$$\beta'_1 \xrightarrow{+\text{heat}} (\alpha \xrightarrow{+\text{heat}}) \text{Liq.}$$

$$\text{Liq.} \underset{+\text{heat}}{\overset{-\text{heat}}{\rightleftarrows}} \alpha \underset{+\text{heat}}{\overset{-\text{heat}}{\rightleftarrows}} \text{interm.} \underset{+\text{heat}}{\overset{-\text{heat}}{\rightleftarrows}} \beta'_2 \xrightarrow{-\text{heat}} \beta'_1.$$

Figure 1 shows the diffraction-pattern–temperature (DPT) diagrams for *n*-docosyl acetate. The diffraction pattern consists of series of reflexions, one series corresponds to the long crystal spacing and another series to the side-spacing pattern. Phase transitions are indicated by sudden displacements of the lines in the diagram. The diffraction pattern disappears on melting. The diffraction pattern of the β'_1-form corresponds to an orthorhombic sub-cell packing and the chains are tilted to the end group plane at an angle of about 47°. Three vertical forms are recognized, the α, the *intermediate* (interm.) and the β'_2 forms. The sub-cell packing in the α-form is hexagonal, it is orthorhombic both in the *intermediate* and the β'_2 forms.

Figure 2 shows calorigrams of *n*-docosyl acetate obtained by differential thermal analysis (DTA). Phase transitions are indicated by pen deflections. The direction of the pen deflection depends on whether the transition is endothermic or exothermic. Calculation of transition energy is based on measurement of the area under a peak. Transition enthalpies (ΔH) and transition entropies (ΔS) obtained from the DTA analysis are plotted against chain length in Figs. 3 and 4, respectively. The functions are linear which has made possible analysis of data in terms of "group contributions".[6]

The bearing of the results on the interpretation of molecular structure in different forms has been discussed.[35] Thus, the $\beta'_1 \rightarrow \alpha$ transition represents the rotational melting[18,41,42] of the hydrocarbon chains. On cooling, a sample in the α-form rotational disorder is probably preserved in the *intermediate* form and is then successively lost

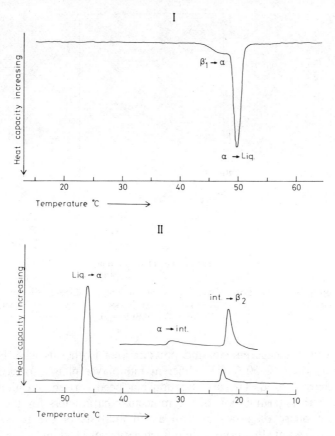

FIG. 2. Calorigrams of *n*-docosyl acetate. Calorigram (I) is obtained on heating a sample in the β'_1-form, (II) on cooling after fusion. The inserted curve is run at a higher sensitivity.

during further cooling. It seems probable that disorder is present also in the β'_2-form to the same degree and is finally lost in the $\beta'_2 \rightarrow \beta'_1$ transition. A similar cooling pattern was suggested for the ethyl esters[33] and might be common for long-chain aliphatic compounds.

2. Monolayer Phase Behaviour

A detailed analysis of the monolayer behaviour of normal alkyl acetates of the C_{18}, C_{20} and C_{22} alcohols has been presented.[35] Most of the information has been gained

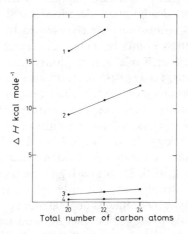

FIG. 3. Enthalpies of transition for even *n*-alkyl acetates in the three-dimensional state plotted against chain length. Curve 1: $\Delta H(\beta'_1 \rightarrow \text{Liq.})$. Curve 2: $\Delta H(\alpha \rightarrow \text{Liq.})$. Curve 3: $\Delta H(\beta'_2 \rightarrow \text{interm.})$. Curve 4: $\Delta H(\text{interm.} \rightarrow \alpha)$.

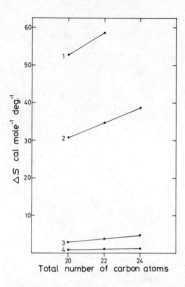

FIG. 4. Entropies of transition of even *n*-alkyl acetates in the three-dimensional state plotted against chain length. Curve 1: $\Delta S(\beta_1' \rightarrow \text{Liq.})$. Curve 2: $\Delta S(\alpha \rightarrow \text{Liq.})$. Curve 3: $\Delta S(\beta_2' \rightarrow \text{interm.})$. Curve 4; $\Delta S(\text{interm.} \rightarrow \alpha)$.

by surface pressure measurements and pressure-area isotherms have been obtained in the temperature range 2–50°C. The different monolayer phases formed have been described and characterized. The investigation has been carried out along the lines presented above with determination of the molecular parameters for the different phases and plotting of phase diagrams as well as an analysis of the phase relationship in terms of thermodynamic functions. The nomenclature of Harkins is used in this paper.[17]

In a series of homologous aliphatic compounds with small polar groups, the different phases exist in different temperature regions according to chain length. An increase of the chain length by one carbon atom has approximately the same effect as a lowering of the temperature about 5°C.[30,45] Figure 5 shows a series of pressure–area isotherms for *n*-docosyl acetate in the temperature range 10–50°C. In this temperature region, condensed films are obtained directly from the vapour phase on compression. Compressible low pressure phases of the type L_2 or *liquid condensed*, according to Harkins,[17] are present in the whole temperature range studied. In docosyl acetate monolayers, three *liquid condensed* phases exist in different temperature regions; the L_2, L_2' and L_2'' phases. Three high pressure, small area phases with low compressibility are present; the S or *solid* phase and the LS or *superliquid* phase described by Harkins and Copeland[17] and the CS or *condensed solid* phase described by Ställberg-Stenhagen and Stenhagen.[45,48] First order phase transitions are represented by plateau regions in the isotherms, second order transition points by kinks. Characteristic molecular areas of the different phases are collected in Table 1 and plotted against temperature in Fig. 6. A surface pressure–temperature (π–T) phase diagram for *n*-docosyl acetate plotted analogously to a three-dimensional phase diagram is shown in Fig. 7. The transition pressures for both first and second order phase transitions are plotted in the diagram as well as monolayer collapse pressures. Second order transitions do not obey the phase rule, however. Of the actual transitions, the $L_2'' \rightarrow CS$, $L_2' \rightarrow S$ and $L_2 \rightarrow LS$ are second order. Three major domains are thus separated from each other by curves representing first order phase transitions; the L_2''–CS, L_2'–S and L_2–LS regions.

Transition energies and entropies for first order phase transitions in the monolayer state could be calculated by a two-dimensional analogue of the Clausius–Clapeyron equation.[9,13,35,40] An analysis of transition energies and entropies in terms of "group contributions" has been made analogously to that for the three-dimensional state. The important knowledge gained by this procedure is the increment per methylene group to the lattice energy and entropy in the different monolayer transitions. It is evident

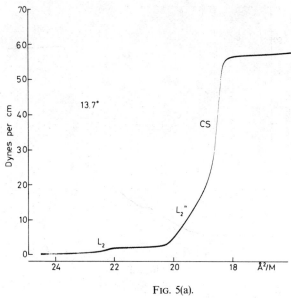

FIG. 5(a).

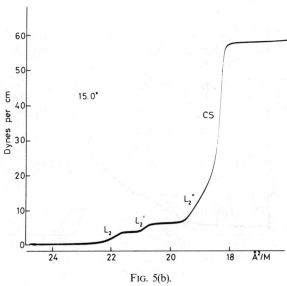

FIG. 5(b).

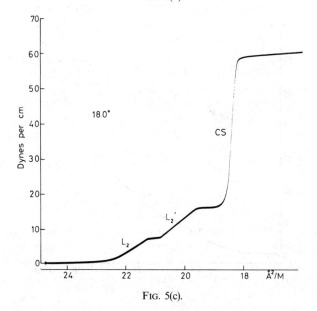

FIG. 5(c).

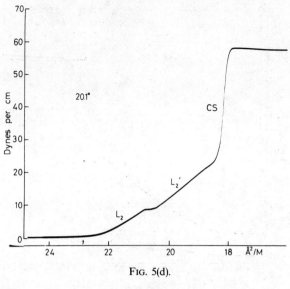

Fig. 5(d).

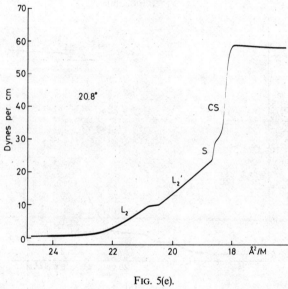

Fig. 5(e).

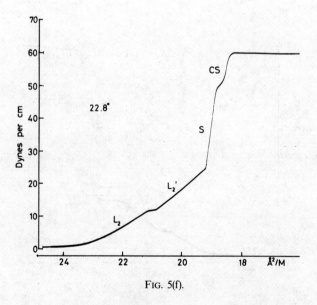

Fig. 5(f).

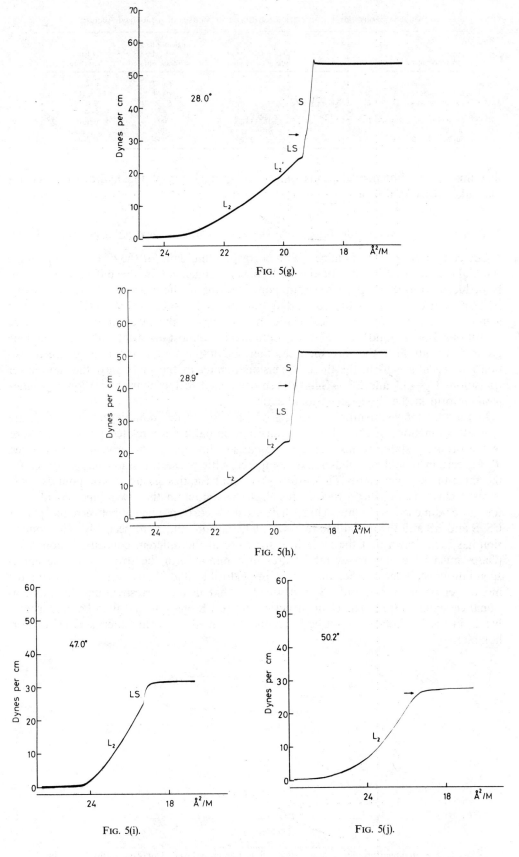

FIG. 5(g).

FIG. 5(h).

FIG. 5(i).

FIG. 5(j).

FIG. 5. π-A isotherms for n-docosyl acetate monolayers on 0.01 N HCl obtained at a compression rate of 4 Å^2/molecule × min in the temperature range 13.7–50.2°C.

TABLE 1. Molecular Parameters for Monolayer Phases of *n*-Docosyl Acetate

Monolayer phase	Area in $Å^2/M$	Area in $Å^2/M$ at the collapse point	Angle of tilt, $v(°)$, of molecules to substrate surface	Chain structure
Liquid condensed, L_2''	19.8–19.9		66–67	$0^\perp(1,0,1)$
Liquid condensed, L_2'	21.6–21.9		60–61	$0_{interm.}(1,1,1)$
Liquid condensed, L_2	22.3–23.5 (40.4°C)		58–59	$H(1,1,1)$
	-24.7 (50.2°C)			
Condensed solid, CS	18.3–18.5	18.1–18.2	90	$0^\perp(0,0,1)$
Solid, S	19.0–19.2	18.9–19.1	90	$0_{interm.}(0,0,1)$
Superliquid, LS	19.4–20.1	19.1–20.0	90	$H(0,0,1)$

that this value ultimately depends on the rearrangement of the hydrocarbon chains that takes place in a transition.

3. *Correlation of Monolayer Data with Data for Three-dimensional Crystalline Forms*

Correlation of the data obtained, with corresponding data for the ordinary crystalline state, allows a good picture of the molecular arrangement in the different phases to be deduced. Thus, the methylene group contributions to the changes in internal energy, ΔU, and entropy, ΔS, in the correlated transitions $L_2' \to L_2''$ and $S \to CS$ are almost identical, indicating that identical chain arrangements take place in these two phase transitions. It was found that the values obtained are almost identical with the methylene group contribution to the corresponding thermodynamic functions for the phase transition *intermediate* $\to \beta_2'$ in the three-dimensional state. In the same way, the monolayer transitions $L_2 \to L_2'$ and $LS \to S$ have been correlated to the phase transition $\alpha \to$ *intermediate* form in the three-dimensional state.

In Fig. 8, changes in internal energy, ΔU, for first order monolayer transitions, are plotted as functions of chain length. Transition enthalpies for related phase transitions in the ordinary state are inserted for comparison. In Fig. 9, the critical temperatures, T_c, for certain monolayer phases, are compared with phase transition temperatures for the three-dimensional state. The results show that for this group of compounds there exists a close relationship between the phase behaviour in the "two-dimensional" and the "three-dimensional" state. Thus, there exists a correspondence between the phases CS, S and LS and the crystalline forms β_2', *intermediate* and α, respectively. The conclusion has been drawn that the molecular structure in the different condensed monolayer phases must have one or several features in common with the structure in the corresponding crystalline modifications. It seems plausible that the subcell structure in the monolayer phases CS, S and LS is identical to that in the corresponding three-dimensional crystalline forms. The chain structure in the CS phase would then be orthorhombic $\perp$, in the LS form it would be hexagonal and in the S form "intermediate" orthorhombic.

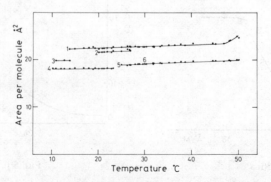

FIG. 6. Area–temperature diagram for *n*-docosyl acetate monolayers. Curves 1, 2 and 3: limiting areas of the L_2, L_2' and L_2'' phases, respectively. Curves 4, 5 and 6: collapse areas of monolayer in CS, S and LS phase, respectively.

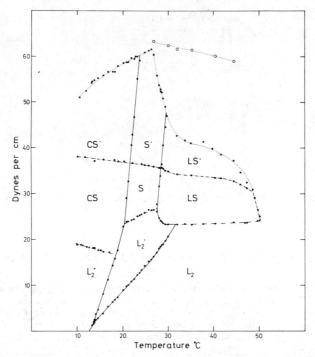

FIG. 7. Monolayer pressure–temperature, π–T, phase diagram for n-docosyl acetate. The curves representing first order phase transition points are full-drawn. The curves representing second order phase transitions are dashed. Equilibrium collapse pressures are indicated by an interpunctated line. Monolayer collapse pressures at a compression rate of 4 Å²/molecule × min are indicated by filled circles and a dotted line and multilayer collapse pressures by open circles and a dotted line. The regions indicated CS′, S′ and LS′ thus represent thermodynamically metastable conditions and are due to a kinetic effect.

The orientation of the molecules in relation to the substrate surface has been determined from values of the characteristic molecular area in the film related to the known cross-sectional area of the hydrocarbon chains in the corresponding packing mode in the three-dimensional state.[31,35] In small-area condensed phases (CS, S and LS), the molecules are oriented vertically to the substrate surface. Sufficient evidence has appeared in this investigation to settle the question of molecular arrangement in condensed monolayer phases whose molecular areas are too large to be consistent with

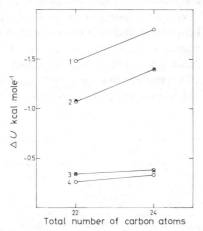

FIG. 8. Changes in internal energy for first order monolayer transitions of n-alkyl acetates as functions of chain length. Enthalpies of transition for related phase transitions in the three-dimensional state are inserted in the diagram. Curve 1: open circles: $\Delta U(L_2' \rightarrow L_2'')$. Curve 2: open circles: $\Delta U(S \rightarrow CS)$; crosses: $\Delta U(\text{interm.} \rightarrow \beta_2')$. Curve 3: open circles: $\Delta U(LS \rightarrow S)$; crosses: $\Delta U(\alpha \rightarrow \text{interm.})$. Curve 4: open circles: $\Delta U(L_2 \rightarrow L_2')$.

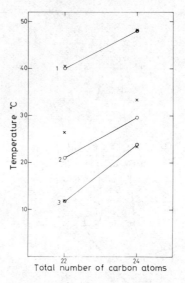

FIG. 9. Critical temperatures, T_c, for certain monolayer phases of n-alkyl acetates as functions of chain length. The transition temperatures for related phases in the three-dimensional state are inserted. Curve 1: circles: T_{cLS},; crosses: melting points of the α-forms. Curve 2: circles: T_{cS}; crosses (above the line): transition temperatures interm. → a. Curve 3: circles: T_{cCS}; crosses: transition temperatures $β'_2$ → interm.

a vertical arrangement of the molecules (L_2, L'_2 L''_2). Evidently there is a regular chain arrangement in these phases. To account for the large molecular areas, the hydrocarbon chains must be tilted to the substrate surface. A tilting of the molecules is accomplished by a displacement along the hydrocarbon chain, corresponding to an integral multiple of a distance equal to one-half zigzag. Accordingly, the angle of tilt can only have a few possible values. The angle of tilt and the mode of arrangement of the hydrocarbon chains being known, it has been possible to give structure symbols for the different condensed monolayer phases, in the same way as for crystalline modifications in the three-dimensional state, see Table 1.

B. *Ethyl Esters of* n-*Aliphatic Acids*

The polymorphic behaviour in the ordinary state has been described recently.[2,3,33] The monolayer behaviour was first described by Alexander and Schulman.[4] The monolayer behavior of ethyl esters in the C_{18}–C_{23} range have been investigated recently by the present author.[36] Three condensed phases with a vertical chain arrangement are formed (the CS, S and LS phases) and two phases with a tilted chain arrangement (L''_2 and L'_2). An expanded phase, *Liquid-expanded*, L_1, or $E^{16,36}$ exists in ethyl ester monolayers. According to Adam, this phase is a so-called vapour-expanded phase.[1] The existence of two kinds of expanded states has been questioned, however.[12,36]

Molecular parameters for the monolayer phases are collected in Table 2. The variance

TABLE 2. Molecular Parameters for Monolayer Phases of the Ethyl Esters of the n-Aliphatic Acids

Monolayer phase	Area in Å²/M	Area in Å²/M at the collapse point	Angle of tilt, $v()$, of molecules to substrate surface	Chain structure
Liquid condensed, L''_2	19.8–20.0	—	66–67	O ⊥ (1,0,1)
Liquid condensed, L'_2	20.4–21.5	—	63–65	$O_{interm.}$(1,0,1)
Condensed solid, CS	18.4–18.6	18.1–18.4	90	O ⊥ (0,0,1)
Solid, S	18.8–19.1	18.5–18.8	90	$O_{interm.}$(0,0,1)
Super-liquid, LS	19.1–20.0	18.9–19.2	90	H (0,0,1)
	(–22.0)	(–17.0–16.0)		
Expanded, E	70–80	28.0–29.0	—	—

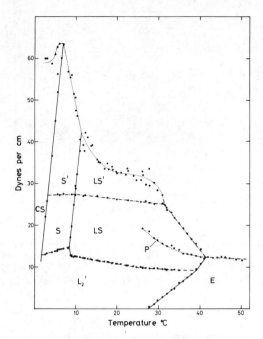

FIG. 10. Monolayer pressure–temperature, π–T, phase diagram for ethyl n-octadecanoate (ethyl stearate). For notations, see text of Fig. 7. A kink point P in the high temperature region of the LS phase, considered as a point of partial collapse, is indicated by a dotted line.

of the values given is caused by thermal expansion. Surface pressure π–T diagrams have been plotted for each homolog studied. Figure 10 and 11 show the phase diagrams for ethyl stearate and ethyl n-nonadecanoate, respectively. The general appearance of the phase diagrams is the same as was found for the n-alkyl acetates. However, there is only one liquid condensed phase related to the phases S and LS, $viz.$ the L_2' phase. The transition region from the expanded to the condensed state has been considered as a distinct monolayer phase by some authors, the I or $intermediate$ phase.[1,17] Accord-

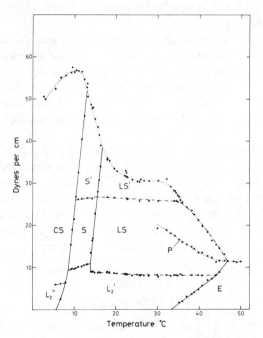

FIG. 11. Monolayer pressure–temperature, π–T, phase diagram for ethyl n-nonadecanoate. For notations see text of Figs. 7 and 10.

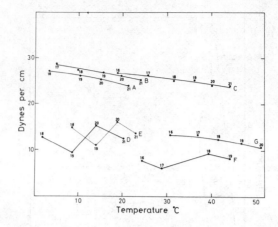

FIG. 12. Critical temperatures and triple points for monolayer phases of ethyl esters with different chain lengths plotted in a π–T diagram. The number of carbon atoms in the acid parts of the esters are indicated. Curve A: T_{ccS}; Curve B: T_{cS}; Curve C: $T_{cLS} = T_{c\ multilayer}$; Curve D: $T_{L_2',CS,S}$; Curve E: $T_{L_2',S,LS}$; Curve F: $T_{L_2',LS,E}$; Curve G: $T_{c\ condensed\ phase}$.

ing to the present investigation, however, the plateau between the expanded state and the condensed state represents ordinary first order transitions, *viz.* the $E \rightarrow L_2'$ and $E \rightarrow LS$ transitions.

Figure 12 shows a π–T diagram where triple points and critical temperatures, T_c, for the different ethyl ester homologues are plotted. It is apparent from the phase diagrams (Figs. 10, 11 and 12) that there is an even–odd alternation in phase transitions related to phases with a proposed tilted chain arrangement. The alternation is most prominent in the surface pressure function. Even–odd alternation in the monolayer state has not been described previously.

The cause for alternation in the ordinary crystalline state is thought to be the different structures in the methyl end group planes for even and odd homologs.[24,39] Also in the monolayer state, alternation was found to be confined to phases with a proposed tilted chain arrangement which might be taken as a further evidence of a regular tilted arrangement in these phases.

Transition energies and entropies have been calculated and the data have been compared with data for the three-dimensional state.[33] As for the *n*-alkyl acetates, it has been possible to show that there exists a close relation between condensed monolayer phases and crystalline forms in the ordinary state.[36] This relationship seems to be confined to the arrangement of the hydrocarbon chains. The sub-cell structure has been determined for the different ethyl ester phases and structure symbols are given in Table 2.

The transitions which take place between the expanded phase E and the condensed monolayer phases L_2' and LS are of special interest. The expanded phase is generally considered as a two-dimensional analogue of the liquid state.[1,16] The transitions $L_2' \rightarrow E$ and $LS \rightarrow E$ would then be related to the melting of crystalline forms in the three-dimensional state. "Melting heats" have been calculated for the monolayer state and the results favour the view that the expanded state might be considered as a two-dimensional liquid with a random arrangement of the hydrocarbon chains.

C. n-*Aliphatic Acids*

The monolayer properties of free fatty acids have been systematically investigated by a number of authors.[1,8,16,30] Monolayer investigations of fatty acids with very long hydrocarbon chains have been carried out by Stenhagen and Ställberg-Stenhagen.[45,48]

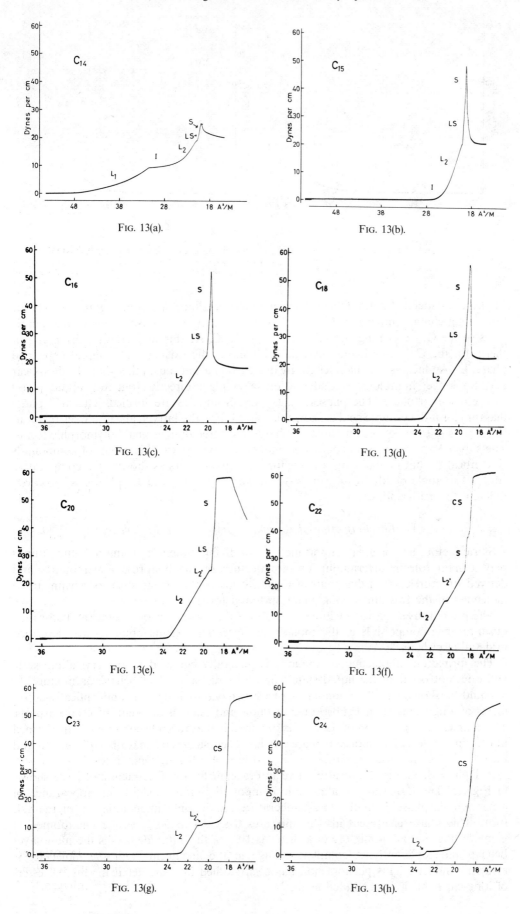

FIG. 13(a).

FIG. 13(b).

FIG. 13(c).

FIG. 13(d).

FIG. 13(e).

FIG. 13(f).

FIG. 13(g).

FIG. 13(h).

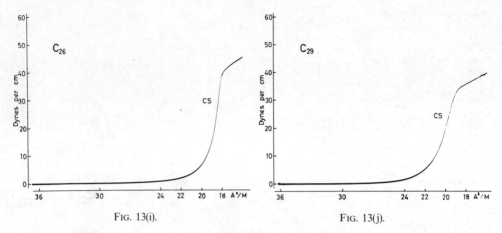

FIG. 13(i).   FIG. 13(j).

FIG. 13. π–A isotherms for straight-chain fatty acids in the undissociated state on 0.01 N HCl at 15°C.

In the undissociated state, the monolayer phase relationship is analogous to that for other long chain compounds. Figure 13 shows a series of π–A isotherms of *n*-aliphatic acids in the C_{14}–C_{29} range on 0,01-N HCl at 15°C. At this temperature, the π–A isotherm for the C_{14} homologue (myristic acid) shows the existence of a liquid expanded phase, L_1. An increase in chain length by one carbon atom causes this phase to disappear and, for longer homologs, a condensation takes place directly from the vapour phase to a condensed phase. The phases L_2, L_2', CS, S and LS are formed. The π–T phase diagram for *n*-docosanoic acid is shown in Fig. 14. For the fatty acids, no direct relationship seems to exist between monolayer phases and three-dimensional polymorphic forms. The sub-cell arrangement in condensed monolayer phases for this group of compounds is identical to that in analogous phases for long chain esters as shown by a comparative study. The angle of tilt to the substrate surface in the L_2 and L_2' phases is, however, different for acids and esters.

D. *The Formation of Racemic Compounds in Monolayers*

Steric factors are of great importance in the living system and among other things play a great role in determining the architecture of the lipoprotein membranes and derived structures. From this point of view, a study of the arrangement of asymmetrical molecules in the two-dimensional state appeared to be of interest.

The present investigation includes a study on the formation of racemic and so-called quasi-racemic compounds in the monolayer state. The results obtained are described in detail elsewhere.[28,29,32]

The formation of racemic compounds is generally confined to the crystalline state and does not occur in the liquid state where racemic substances behave as mixtures.[19] It could be shown that the monolayer phase behaviour of racemic and optically active forms of long-chain methyl substituted alcohols and esters is different in certain aspects. The simplest interpretation of the results obtained is that racemic compounds are formed in certain condensed monolayer phases. It has been suggested that optically active and racemic forms are arranged in monolayer lattices of different symmetries.[29]

π–T phase diagrams of optically active and racemic forms of Tetracosanol-2 are shown in Fig. 15. The formation of a racemic compound is indicated by the appearance of a new phase, phase 3, in the low temperature region only in monolayers of the DL form. This phase is present also in mixtures from 10 to 50% of one enantiomer. It is interesting that an admixture of as little as 10% of the antipode makes the monolayer behave quite differently. A complete change in the arrangement of the molecules is evidently induced. This phenomenon has been studied in some detail for methyl esters of long-chain methyl-substituted acids.[29]

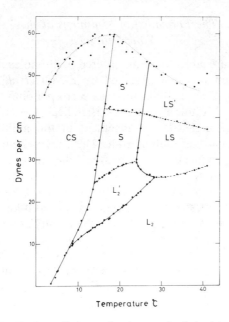

FIG. 14. Monolayer π–T phase diagram of *n*-docosanoic (behenic) acid in the undissociated state. For notations see text to Fig. 7. (In collaboration with E. Stenhagen and S. Ställberg-Stenhagen).

The formation of quasi-racemates in the monolayer state has been described.[31] The term quasi-racemate compound was introduced by Fredga[11] to characterize a molecular compound formally derived from a true racemic compound by some minor change in one of the components. In the present work, quasi-racemic compounds were formed in monolayers of equimolar mixtures of an acetate and a methyl ester of optically active long-chain compounds.

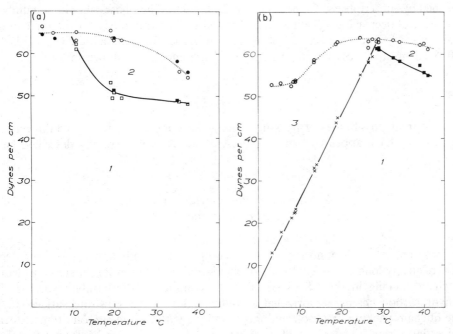

FIG. 15. π–T diagram for (a) D (+)-tetracosanol-2, and (b) DL-tetracosanol-2. First order transitions are represented by full-drawn lines. Monolayer collapse pressures are indicated by open circles and a dotted line.

E. *The Formation of Multifilms at the Air–Water Interface*

Many classes of lipids are able to form ordered multimolecular films on a water subphase. This phenomenon was first observed by Ekvall *et al.*[9] Such multifilms are obtained on compression of a spread film to areas corresponding to two or more layers of close-packed molecules. Multifilm formation is indicated by a second pressure rise in the π–A isotherms. The molecular arrangement in the multifilm seems to follow principles valid for the ordinary crystalline state. Thus, the system "multifilm on water" might be considered as a connecting link between the monomolecular film and the ordinary three-dimensional state.

Multifilm formation has been shown to occur in films of long-chain fatty acids,[34] esters,[35,36] triglycerides,[34] cholesterol[34] and bile acids.[9] Multifilms are easily obtained in films of long-chain aliphatic esters. Normal chain ethyl esters and alkyl acetates both form multifilms from the *super-liquid* or LS phase. The phenomenon is described in some detail elsewhere.[34–36] Figure 16 shows the formation of double and triple layers from the LS phase in ethyl stearate films. A spontaneous respreading of the compressed multifilm to a monolayer takes place if the compressing barrier is withdrawn.

VI. TERMINOLOGY OF MONOLAYER PHASES

At least three different nomenclatures are in use, those introduced by Adam,[1] Harkins[16] and Dervichian.[8] For a comparison of these nomenclatures, the reader is referred to the review article by Dervichian cited above. In view of recent knowledge of the molecular arrangement in monolayer phases, none of these terminologies is free from objection, e.g. the term "liquid condensed" used by Harkins cannot be an adequate notation for a phase with a regular chain arrangement. The introduction of a new terminology has been suggested by the present author.[35] Condensed phases with a crystalline or semicrystalline arrangement of the molecules would simply be denoted *condensed* or C. The terms *condensed vertical*, C_v, and *condensed tilted*, C_t, would denote phases where the molecules are arranged vertically or tilted to the water surface. If there are several *condensed vertical* or *condensed tilted* phases, they would be distinguished by subscripts, e.g. C_{v1}, C_{v2}, etc. or C_{t1}, C_{t2}, etc. The phases should be numbered in the order of decreasing T_c values in analogy with the terminology for the three-dimensional state.[23,37] The classification of monolayer phases for *n*-aliphatic compounds with small polar groups according to the suggested terminology is shown in Table 3.

VII. SUMMARY

The different possibilities of packing of long-chain molecules in a "two-dimensional" array, as they have appeared from the present results, will be briefly described below.

A. *Monolayer Phases with a Vertical Chain Arrangement*

The *condensed solid*, CS or C_{v3} phase: the CS phase is the most condensed phase described in monolayers of normal aliphatic compounds. It only appears for homologues with a relatively long chain length where chain–chain attraction forces are considerable. The chain packing in the CS phase is of the common orthorhombic type, $O\perp$, and the hydrocarbon chains are oriented at right angles to the substrate surface. In this mode of chain packing, the molecules are arranged with all the carbon atoms coplanar to give flat zigzags. The plane of every second chain is perpendicular to the planes of the others so that interlocking of the chains is effective. Figure 17 shows a schematic illustration of the proposed molecular arrangement in the CS phase.

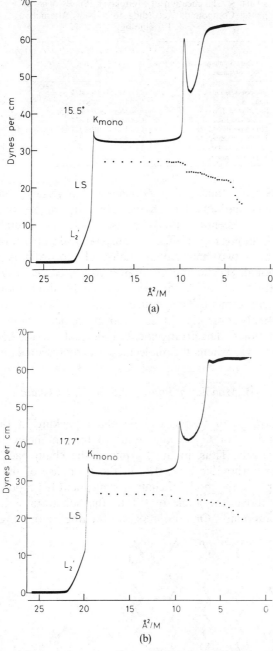

FIG. 16. π–A isotherms for ethyl stearate films on 0.01 N HCl obtained at a compression rate of 4 Å^2/molecule × min. (a) 15.5°, (b) 17.7°. K_{mono} denotes "collapse" of the monolayer or the beginning of the transition to a multilayer. Compression was continued until visible aggregates were formed. Equilibrium pressures are inserted below the continuously recorded curves.

The *solid*, S, or C_{v2} phase: the degree of order in the S phase is also considerable although chain–chain interaction is less pronounced than in the CS phase. The chain arrangement is of an orthorhombic intermediate type. The exact mode of chain arrangement is not known. It seems to represent a transitionary state between the common orthorhombic, $0 \perp$, chain packing and the hexagonal, H, chain packing modes. Thus, the sub-cell structure is of the orthorhombic type and a successive rearrangement of the chains from the common orthorhombic into a more symmetrical mode of chain packing takes place with increasing temperature. Calorimetric measurements have shown that the energy level is relatively high in this mode of chain packing indicating some degree of molecular disorder.

TABLE 3. Classification of Monolayer Phases of *n*-Aliphatic Compounds According to a New Alternative Terminology

Harkins *et al.* Ställberg-Stenhagen and Stenhagen	Lundquist
Condensed solid, CS	Condensed vertical, C_{v3}
Solid, S	Condensed vertical, C_{v2}
Superliquid, LS	Condensed vertical, C_{v1}
Liquid condensed L_2	Condensed tilted, C_{t1}
Liquid condensed, L_2'	Condensed tilted, C_{t2}
Liquid condensed, L_2''	Condensed tilted, C_{t3}
Liquid expanded, L_1, E	Expanded, E
Vapour or gaseous	Gaseous, G

The *super-liquid*, LS or C_{v1} phase: the LS phase is that vertical phase in which chain–chain interaction is less marked. It is characterized by an exceptionally low surface viscosity. The domain of existence of the LS phase is often considerable. Thus, in monolayers of short-chain homologs, it is often the only vertical phase formed. The LS phase is proposed here to be a two-dimensional analog of the ordinary α form. The chain arrangement is hexagonal, H, and the internal energy is large. It is generally assumed that there is some degree of molecular orientational freedom about the long axis in this mode of chain arrangement. The nature of this molecular motion is not known in detail but it is probably some kind of hindered molecular rotation or torsional oscillation or even free rotation.[25] The arrangement of molecules in the LS phase is schematically illustrated in Fig. 18, where the molecules are represented by circular cylinders indicating rotation.

B. *Monolayer Phases with a Tilted Chain Arrangement*

The *liquid condensed*, L_2 or C_t phases: the chain packing in the tilted phases L_2'' or C_{t3}, L_2' or C_{t2} and L_2 or C_{t1} is analogous to that in the related vertical phases CS, S and LS, respectively. Thus, in the L_2'' phase, the chain packing is proposed to be of the common orthorhombic type, $0\perp$, in the L_2' phase of the intermediate type, $0_{interm.}$ and in the L_2 phase the mode of chain arrangement is proposed to be hexagonal. As was pointed out above, the angle of tilt to the substrate surface is different for different groups of compounds. On compression of a tilted monolayer phase, a reorien-

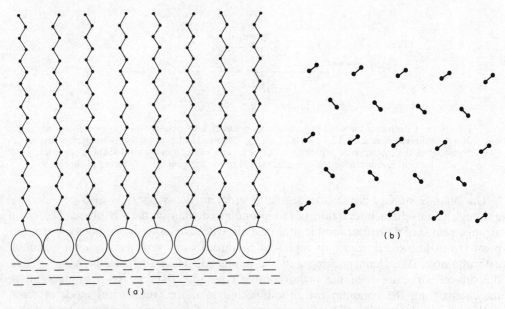

FIG. 17. Proposed molecular arrangement in the CS phase. (a) Projection in a plane perpendicular to the water surface (corresponding to a projection along the crystallographic *a*-axis, cf ref. 3). In (b) the chain arrangement is seen along the direction of the chain axes.

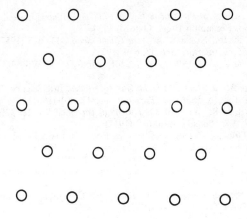

FIG. 18. Proposed molecular arrangement in the LS phase. The chain arrangement is seen along the direction of the chain axes.

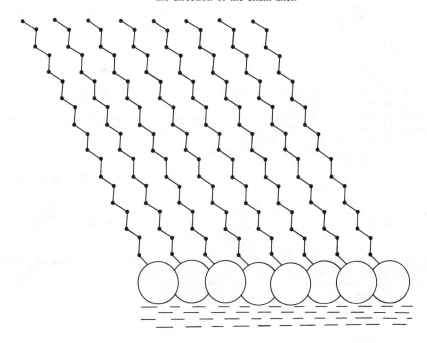

FIG. 19. Proposed molecular arrangement in the L_2'' phase. Projection in a plane perpendicular to the water surface (corresponding to a projection along the crystallographic b-axis). The chain arrangement along the direction of the chain axes corresponds to that in the CS phase, cf Fig. 17 (b).

tation of the molecules into the vertical position takes place. A schematic illustration of the molecular arrangement in the *liquid condensed* phase L_2'' is shown in Fig. 19.

(*Received* 13 *March* 1975)

REFERENCES

1. ADAM, N. K. *The Physics and Chemistry of Surfaces*, 3rd edn, Oxford University Press, London, 1941.
2. ALEBY, S. *Acta Chem. Scand.* **22**, 811 (1968).
3. ALEBY, S. *Solid State Behaviour of Long-Chain Esters*. Abstracts of Gothenburg Dissertations in *Science*, **13** (1969).
4. ALEXANDER, A. E. and SCHULMAN, J. H. *Proc. R. Soc., London* **A161**, 115 (1937).
5. ANDERSSON, K. J. I., STÄLLBERG-STENHAGEN, S. and STENHAGEN, E. *The Svedberg 1884 30/8 1944*, p. 11. Almqvist & Wiksell, Uppsala, 1944.
6. DAVIES, M. and KYBETT, K. *Trans. Faraday Soc.* **61**, 2646 (1965).
7. DERVICHIAN, D. G. *J. Chem. Phys.* **7**, 931 (1939).

8. DERVICHIAN, D. G. *Progress in the Chemistry of Fats and other Lipids*, Eds. HOLMAN, LUNDBERG and MALKIN, Vol. II, p. 193. Pergamon Press, Oxford, 1954.
9. EKWALL, P., EKHOLM, R. and NORMAN, A. *Acta Chem. Scand.* **11**, 693 (1957).
10. ERIKSSON, J. CH. *J. Colloid Interface Sci.* **37**, 659 (1971).
11. FREDGA, A. *The Svedberg 1884 30/8 1944*, p. 261, Almqvist & Wiksell, Uppsala, 1944; *Svensk Kem. Tidskr.*, **67**, 343 (1955).
12. GAINES, G. L. *Insoluble Monolayers at Liquid–Gas Interfaces.* Interscience, New York, 1966.
13. GLAZER, J. and ALEXANDER, A. E. *Trans. Faraday Soc.* **47**, 401 (1951).
14. GODDARD, E. D., SMITH, S. R. and KAO, O. *J. Colloid Interface Sci.* **21**, 320 (1966).
15. HARDY, W. B. *Proc. R. Soc. London*, **A86**, 610 (1912).
16. HARKINS, W. D. *The Physical Chemistry of Surface Films*, 2nd edn Reinhold, New York, 1954.
17. HARKINS, W. D. and COPELAND, L. E. *J. Chem. Phys.* **10**, 272 (1942).
18. HOFFMAN, J. D. and DECKER, B. F. *J. Phys. Chem.* **57**, 520 (1953).
19. HÄGG, G. *The Svedberg 1884 30/8 1944*, p. 140. Almqvist & Wiksell, Uppsala, 1944.
20. JOLY, M. *J. Colloid Sci.* **5**, 49 (1950).
21. KITAIGORODSKII, A. I. *Organic Chemical Crystallography*, Consultant Bureau, New York, 1961.
22. LANGMUIR, I. *J. Am. Chem. Soc.* **39**, 1848 (1917).
23. LARSSON, K. *Acta Chem. Scand.* **22**, 2255 (1966).
24. LARSSON, K. *J. Am. Oil Chem. Soc.* **43**, 559 (1966).
25. LARSSON, K. *Nature* **213**, 383 (1967).
26. LANDAU, L. and LIFSHITZ, E. *Statistical Physics.* Oxford University Press, 1938.
27. LIFSHITZ, E. *Acta Phys. Chim. U.R.S.S.* **19**, 248 (1944).
28. LUNDQUIST, M. *Ark. Kemi* **17**, 183 (1961).
29. LUNDQUIST, M. *Ark. Kemi* **21**, 395 (1963).
30. Lundquist, M. *Fin. Kemistsamf. Med.* **72**, (I) 14 (1963).
31. LUNDQUIST, M. *Surface Chemistry*, Eds EKWALL, GROTH and RUNNSTRÖM-REIO, p. 294. Munksgaard, Copenhagen, 1965.
32. LUNDQUIST, M. *Ark. Kemi* **23**, 299 (1965).
33. LUNDQUIST, M. *Ark. Kemi* **32**, 27 (1970).
34. LUNDQUIST, M. *Proceedings of the Fourth Scand. Symp. on Surface Chemistry*, p. 70. Tylösand (1970).
35. LUNDQUIST, M. *Chem. Scr.* **1**, 5 (1971).
36. LUNDQUIST, M. *Chem. Scr.* **1**, 197 (1971).
37. LUTTON, E. S. *J. Am. Oil Chem. Soc.* **27**, 276 (1950).
38. LYONS, C. G. and RIDEAL, E. K. *Proc. R. Soc. London* **A124**, 333 (1929).
39. MALKIN, T. *Progress in the Chemistry of Fats and other Lipids*, Eds HOLMAN, LUNDBERG and MALKIN, Vol. I, p. 1. Pergamon Press, Oxford, 1952.
40. MOTOMURA, K. *J. Colloid Interface Sci.* **23**, 313 (1967).
41. MÜLLER, A. *Proc. R. Soc. London* **A138**, 514 (1932).
42. RALSTON, A. W. *Fatty Acids and their Derivatives*, p. 363, Wiley, New York, 1948.
43. RYHAGE, R. and STENHAGEN, E. *J. Lipid Res.* **1**, 361 (1960).
44. STENHAGEN, E. *Acta Chem. Scand.* **5**, 805 (1951).
45. STENHAGEN, E. *Determination of Organic Structures by Physical Methods*, Eds BRAUDE and NACHOD, p. 325. Academic Press, New York, 1955.
46. STÄLLBERG, S. and STENHAGEN, E. *J. Biol. Chem.* **139**, 345 (1941).
47. STÄLLBERG, S. and STENHAGEN, E. *J. Biol. Chem.* **143**, 171 (1942).
48. STÄLLBERG-STENHAGEN, S. and STENHAGEN, E. *Nature* **156**, 239, (1945).
49. STÄLLBERG-STENHAGEN, S. and STENHAGEN, E. *Nature* **159**, 814 (1947).

Prog. Chem. Fats other Lipids. Vol. 16, pp. 125–143. Pergamon Press, 1978. Printed in Great Britain

LATERAL PACKING OF HYDROCARBON CHAINS

Sixten Abrahamsson, Birgitta Dahlén, Håkan Löfgren and
Irmin Pascher

Department of Structural Chemistry, Faculty of Medicine, University of Göteborg,
P.O.B. S-400 33 Göteborg 33, Sweden

CONTENTS

I. Introduction	125
II. The Subcell Concept	126
III. Simple Subcells	127
A. Orthorhombic packing O⊥	127
B. Orthorhombic packing O'⊥	129
C. Triclinic packing T‖	130
D. Orthorhombic packing O‖	131
E. Orthorhombic packing O'‖	132
F. Monoclinic packing M‖	133
G. Hexagonal packing H	133
IV. Hybrid Type Subcells	134
A. HS1	134
B. HS2	135
V. Functional Groups in a Hydrocarbon Chain Matrix	137
VI. Cholesterol and Hydrocarbon Chains	139
Acknowledgements	141
References	142

I. INTRODUCTION

Lipid bilayers play an important structural role in biological membranes. They vary in internal order and mobility from liquid-like to almost crystalline states. Phase transitions between these states can easily be induced in intact membranes. Even in liquid bilayers, the lipid molecules are so close that the local packing must resemble that in the solid state though long range order does not exist. Data from X-ray diffraction studies of crystalline lipids are, therefore, highly relevant for discussions on bilayer structure particularly as single crystal analyses provide precise information on the atomic level. Such information will no doubt be needed for an understanding of structural details of membranes and their relation to specific functions.

Over the years a large number of X-ray diffraction studies of lipids with increasing complexity have been performed. Patterns are emerging which can now be used to explain structural features of membrane bilayers.[4]

The interaction between the hydrocarbon chains is of prime importance for the bilayer structure but the molecular arrangement is also highly dependent on polar groups, double bonds, branches etc. When the chain is short and the polar group is large, no regular packing patterns of chains are observed in the solid state.[6] However, when the chains become more dominating, they usually pack with their axes all parallel and with the lateral arrangements of the chains limited to relatively few types of structure. Some of these have been described earlier in this series[9] but since then, new types of chain packings have been discovered and more accurate structural parameters have become available. It seemed, therefore, appropriate at this time to review the subject again.

In a few cases such as soaps,[68] amides,[51,62] and 11-aminoundecanoic acid hydrobromide,[54] the hydrocarbon chains are reported to pack with their axes crossed. As this is unlikely to be the case in membrane bilayers or in technically important bilayer lipid systems, only chain arrangements with parallel axes are discussed here.

II. THE SUBCELL CONCEPT

The lateral chain packing is usually described by a subcell indicating the translation (c_s) between equivalent positions within the periodic carbon chain and in adjacent chains (a_s and b_s). The symbols T, O, M and H are used to indicate triclinic, orthorhombic, monoclinic and hexagonal symmetry respectively and $\parallel$ and $\perp$ mutually parallel and perpendicular zig-zag planes of the chains. A prime (') is used to differentiate subcells with otherwise identical symbols.

Kitaigorodskii[30] and Segerman[52] have derived various chain packing possibilities from close packing considerations. These included previously found subcells but also other types which have not yet been discovered in actual structures. In this paper, only the experimentally observed chain packings will be described. Segerman[52] introduced a new nomenclature based on uniform rows of chains all with parallel zig-zag planes. Whereas this is applicable to simple chain packings, hybrid type subcells found recently cannot be accounted for. The original subcell notation has, therefore, been used here.

Figure 1 shows the structure of 13-oxoisostearic acid[19] with the typical layer arrangement of lipids in the solid state. The chain axes form an angle of 44° (angle of tilt) with the planes containing the end groups (polar and/or methyl groups) of the molecules. The angle of tilt is an indication of the relative displacement of the molecules in the direction of the chain axes. This can also be described, as suggested by Vand,[66] by

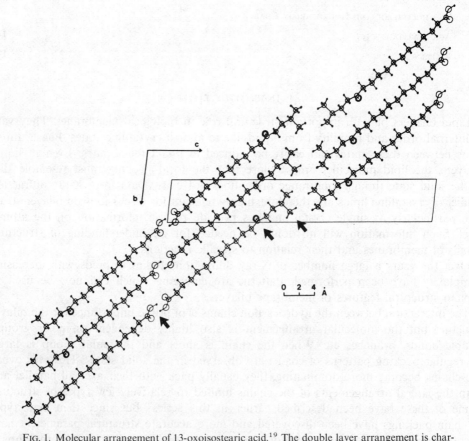

FIG. 1. Molecular arrangement of 13-oxoisostearic acid.[19] The double layer arrangement is characterized by a hydrocarbon matrix with triclinic T$\parallel$ packing into which the carboxylic and carbonyl groups have been accommodated. The methyl branches are given space between the ends of the chains by a translation of the molecules in the chain axes direction. The functional groups are indicated by arrows.

giving the subcell symbol and the subcell indices of the end group plane. The structure in Fig. 1 should then be denoted $T_{\parallel}$ (1, −1,1). As such a classification can only give the general outlines of the structure, it is not commonly used for complex lipid structures.

III. SIMPLE SUBCELLS

A. Orthorhombic Packing $O\perp$

The structure of the orthorhombic subcell was determined early by Bunn[15] in an X-ray diffraction study of polythene. Vainshtein and Pinsker[63,64] obtained very similar results from an electron diffraction study of a mixture of *n*-paraffins. These data agree remarkably well with those obtained from studies on homogeneous *n*-hydrocarbon monomers.[53,59] They indicate virtually identical subcell co-ordinates for $O\perp$ in the monoclinic and orthorhombic crystal forms, respectively, of *n*-hexatriacontane. These are (in fractions of the subcell edges) for the independent CH_2-group:

	x	y	z
C(1)	0.065	0.038	1/4
H(1)	0.049	0.179	1/4
H(2)	0.273	0.010	1/4

The subcell contains four CH_2-groups related by the symmetry *Pbnm*. This leads to an arrangement whereby the zig-zag planes of the chains are mutually perpendicular (Figs. 2 and 3). The relative displacement of two such chains is such that, in a projection down a_s, they appear identical (Fig. 4). Figure 16 summarizes the packing features of all simple subcells. Data for $O\perp$ representing a selection of literature values are given in Table 1. There are variations in the individual values which have various origins.

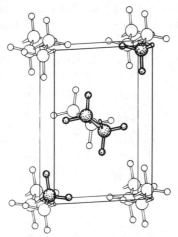

FIG. 2. The orthorhombic packing $O\perp$. The origin of the subcell is at the back, low, left corner. a_s is running horizontally, b_s vertically.

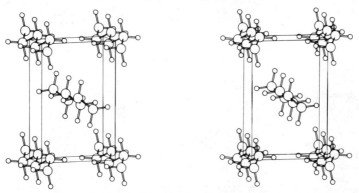

FIG. 3. Stereoscopic pair of $O\perp$.

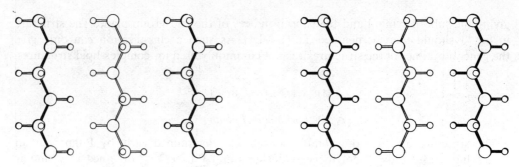

FIG. 4. Projections of the O⊥ (left) and O'⊥ (right) subcells down their 5 Å axes. Whereas the chains of O⊥ are all identical, they are related by mirror planes in O'⊥.

The van der Waals interaction between the chains increases with their length. This is evident from Fig. 5 showing the variation of the unit cell dimension of a homologous series of carboxylic acids.[10] The b-axis which corresponds to a_s of the O⊥ subcell is almost constant whereas a considerable decrease in the a axis and in the monoclinic angle (β) takes place with increasing chain length (b_s is approximately $a \sin\beta$). In the direction of the chain axis there is a linear relationship between the c-axis and the number of carbon atoms.

The subcell responds in an analogous way to temperature increases. The value of a_s is constant (or slightly decreasing) whereas the thermal expansion of the chain packing is accomplished by elongating b_s. For n-octadecanoic acid in the C-form, an expansion of 0.24 Å is observed in b_s for a temperature rise from 20°C to 70°C.[22] A similar effect is also observed for n-tritriacontane (Table 1). Variations in subcell dimensions due to the presence of functional groups are even larger and will be treated later in this paper. If standard dimensions are to be given for O⊥, a hydrocarbon at room temperature with a chain length of about twenty carbon atoms should be used. Smith[55] reported

TABLE 1. Selected Data for the Orthorhombic Packing O⊥

Compound	a_s (Å)	b_s (Å)	c_s (Å)	Area/chain (Å²)	Reference
Polyethylene	4.93	7.40	2.534	18.24	15
n-Paraffins	4.96	7.41	2.54	18.38	63
n-Dodecanoic acid C-form	4.97	7.80	2.52	19.38	69
n-Hexatriacontane	4.95	7.42	2.546	18.36	53
n-Hexatriacontane	4.96	7.42	2.54	18.40	59
n-Tricosane	4.970	7.478	2.549	18.58	55
1,3-Diglyceride of 3-thiado-decanoic acid	4.99	7.46	2.56	18.61	32
L-1-Monoglyceride of 11-bromoundecanoic acid	5.16	7.43	2.47	19.17	34
Triacetylsphingosine	5.00	7.41	2.54	18.53	46
Methyloctadecanoate	5.02	7.35	2.54	18.45	42
n-Octadecanoic acid					
C-form at 20°C	4.958	7.341		18.20	22
70°C	4.923	7.578		18.65	22
n-Tritriacontane < 54.5°C	4.96	7.44	2.545	18.45	49
54.5–65.5°C	4.976	7.567	2.545	18.83	49
65.5–68°C	4.985	7.60	2.545	18.94	49
cis-11,12-Methylene octa-decanoic acid	5.10	7.55	2.53	19.25	18
cis-DL-8,9-Methylenehepta-decanoic acid	5.12	7.70	2.543	19.71	29
$trans$-9,10-Methylene-octadecanoic acid	4.98	8.12	2.60	20.22	13
Methyl-branched polythene with 8 CH₃-groups per 100 carbon atoms	5.00	7.68		19.2	70
12-D-Hydroxyoctadecanoic acid methyl ester	4.86	7.87	2.54	19.12	40

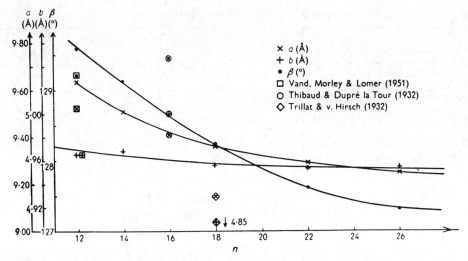

FIG. 5. Variation in some unit cell dimensions with chain length of the crystalline modification C of carboxylic acids.[10] b is identical with the a_s axis of $O\perp$ whereas b_s is approximately $a \sin \beta$ (see refs. 60, 61, 69).

the following dimensions for n-$C_{23}H_{48}$: $a_s = 4.970$, $b_s = 7.478$ and $c_s = 2.549$ Å. The cross section area of one hydrocarbon chain is then 18.58 Å^2.

B. *Orthorhombic Packing $O'\perp$*

The orthorhombic subcell $O'\perp$ was first discovered in 2D-methyloctadecanoic acid.[1] It is similar in shape to $O\perp$ but the relative displacement of the chains along c_s is such that, in a projection down, the 5 Å axis the chains are related by a mirror plane parallel to the chain axes (Fig. 4, 6 and 7). The symmetry elements of $O'\perp$ and $O\perp$ are similar (Pbnm) if in $O'\perp$ the 5 Å axis is chosen as b_s. $O'\perp$ has been found in

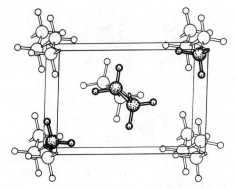

FIG. 6. The orthorhombic $O'\perp$ packing.

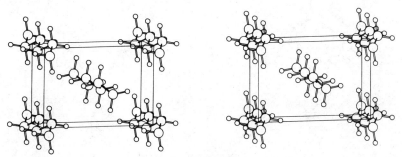

FIG. 7. Stereoscopic pair of $O'\perp$.

TABLE 2. Data for the Orthorhombic Subcell O'⊥

Compound	a_s (Å)	b_s (Å)	c_s (Å)	Area/chain (Å^2)	Reference
2-D-Methyloctadecanoic acid	7.43	5.01	2.50	18.61	1
2-DL-Methyl-7-oxododecanoic acid	7.57	5.06	2.53	19.15	45
n-Dodecylammonium bromide	6.96	5.40	2.54	18.79	39

only a few compounds (Table 2). It is, therefore, difficult to give any standard parameters the most likely values being $a_s = 7.43$, $b_s = 5.01$ and $c_s = 2.54$ Å and

	x	y	z
C(1)	0.029	0.071	1/4
H(1)	−0.025	0.254	1/4
H(2)	0.163	0.086	1/4

The co-ordinates have been calculated assuming a standard C–C distance of 1.515 Å and a C–C–C angle of 114.0°. The chain packing appears to be as effective as O⊥ with a chain area of 18.61 Å^2.

The zig-zag planes of the chains form an angle of less than 45° with the b_s axis which means that they are not perpendicular to each other. The angle between the chain planes is thus 59° in 2D-methyloctadecanoic acid and 67.5° in 2DL-methyl-7-oxododecanoic acid.[45]

C. Triclinic Packing T∥

The triclinic subcell was first analyzed by Vand and Bell.[67] It contains two CH$_2$-groups related by a center of symmetry (spacegroup P$\bar{1}$). All zig-zag planes are parallel. The chain axes of adjacent chains are displaced relative to each other both in directions parallel and perpendicular to the chain planes. (Figs. 8 and 9). The co-ordinates of

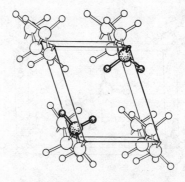

FIG. 8. The triclinic T∥ packing.

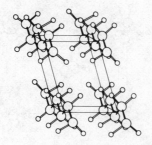 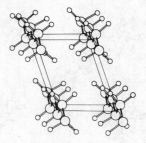

FIG. 9. Stereoscopic pair of T∥.

TABLE 3. Selected Data for the Triclinic Packing $T\parallel$

Compound	a_s (Å)	b_s (Å)	c_s (Å)	α_s (°)	β_s (°)	γ_s (°)	Area/chain (Å^2)	Reference
Trilaurin β-form	4.41	5.40	2.45	74.8	108.5	120.5	19.31	67
11-Bromoundeca-noic acid, D-form	4.40	5.53	2.48	67	105	113	20.45	31
Dodecanoic acid A_1-form	4.25	5.41	2.54	74.9	108.4	120	18.74	37
Potassium palmi-tate (form B)	4.15	5.30	2.557	65.2	108.5	110.4	18.33	24
3-DL-Bromo-octa-decanoic acid	4.27	5.29	2.55	80.4	105.9	119.5	18.89	5
1,3-Diglyceride of 11-bromounde-canoic acid	4.14	5.39	2.54	78	107	120	18.43	28
Isostearic acid	4.28	5.37	2.53	72.3	108.8	117.2	19.01	7
13-Oxoisostearic acid	4.27	5.39	2.55	73.9	108.6	119.6	18.77	19
Tetracosanoyl-phytosphingosine	4.68	5.35	2.54	66	114	122	18.80	21
11-Bromoundecanol	4.50	5.26	2.52	64	114	118	18.17	50
n-Octadecane	4.285	5.414	2.53	80.8	112.2	121.78	18.23	43
Copper (II) decanoate	4.42	5.28	2.56	70.2	109.6	117.5	19.02	38

the independent atoms, referred to c_s, a'_s and b'_s—which latter two are orthogonal to c_s—are:

	x	y	z
C(1)	0.062	0.089	1/4
H(1)	0.353	0.216	1/4
H(2)	−0.056	0.216	1/4

The triclinic packing is together, with the orthorhombic packing $O\perp$ the most common chain arrangement. It is found in a number of different lipids. It is obvious from the data in Table 3 that though the dimensions in the various cases clearly characterize the subcell, there are considerable random deviations in individual parameters. $T\parallel$ thus allows changes required by the actual crystal lattice in the relative orientation of the chains. There is no apparent preferred direction for thermal expansion of the subcell dimensions as in $O\perp$. The subcell in the triclinic modification of n-octadecane[43] may be taken as undistorted dimensions of $T\parallel$.

$$a_s = 4.285 \qquad b_s = 5.414 \qquad c_s = 2.539 \text{ Å}$$
$$\alpha_s = 80.99 \qquad \beta_s = 112.2 \qquad \gamma_s = 121.78°$$

The cross section area is 18.23 Å^2 in this case but, due to the flexibility of $T\parallel$, it can vary considerably and adopt rather large values in some compounds (Table 3).

D. Orthorhombic Packing $O\parallel$

No complete structure determination has been performed of a compound with the orthorhombic $O\parallel$ subcell. von Sydow[58] found this subcell in (−)-2-methyl-2-ethyleicosanoic acid and it also probably exists in α-sodium stearate hemihydrate.[14] The subcell has the dimensions $a_s = 8.150$, $b_s = 9.224$ and $c_s = 2.551$ Å (Figs. 10 and 11). The space-group is $B22_12$[52] and the chain area is 18.77 Å^2.

The following co-ordinates have been calculated:

	x	y	z
C(1)	0.185	0.295	0
H(1)	0.085	0.358	0
H(2)	0.285	0.358	0

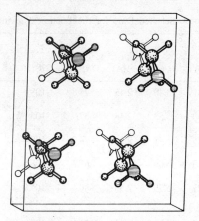

FIG. 10. The orthorhombic O∥ packing.

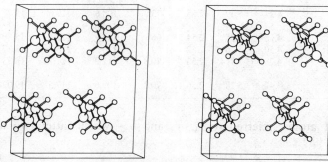

FIG. 11. Stereoscopic pair of O∥.

The axes of neighboring chains are displaced when seen parallel to the zig-zag planes but coincide when seen perpendicular to the chain planes. Chains in contact along a_s have a mutual translation of $1/2\ c_s$.

E. *Orthorhombic Packing O′∥*

The low melting crystalline modification of oleic acid shows another orthorhombic packing with parallel zig-zag planes (O′∥).[8] Here all chains are co-planar in the b_s direction, whereas the axes of neighboring chains are displaced when seen along a_s (Figs. 12 and 13). The dimensions reported are $a_s = 7.93$, $b_s = 4.74$ and $c_s = 2.53$ Å. The spacegroup is P$ma2$[52] and the atomic co-ordinates are:

	x	y	z
C(1)	1/4	−0.015	0.500
H(1)	0.147	−0.138	0.500
H(2)	0.353	−0.138	0.500

H(2) is related to H(1) by a mirror plane. The area per chain is 18.79 Å².

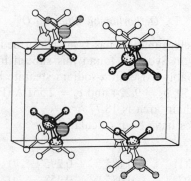

FIG. 12. The orthorhombic O′∥ packing.

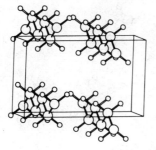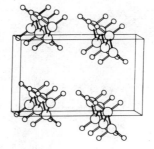

FIG. 13. Stereoscopic pair of O′∥.

F. *Monoclinic Packing M∥*

The monoclinic subcell M∥ was first found in 3-thiadodecanoic acid[11] but has since then been analyzed in two further compounds (Table 4). As in O′∥, there are infinite rows of coplanar chains along b_s (Figs. 14 and 15) but neighboring ones have a $1/2 c_s$ displacement as compared to the case in O′∥. There is a considerable displacement of the chain axes in the direction perpendicular to the zig-zag planes. The most representative dimensions are likely those given by Dahlén *et al.*[20] $a_s = 4.30$, $b_s = 4.74$, $c_s = 2.54$ Å, $\gamma_s = 110.9°$.

The symmetry is $P2_1/m$ and the co-ordinates are:

	x	y	z
C(1)	0.000	0.087	1/4
H(1)	0.203	0.275	1/4
H(2)	−0.203	0.144	1/4

The dimensions give a chain area of 19.04 Å².

G. *Hexagonal Packing H*

Many long-chain compounds adopt a modification just below their melting point for which powder diffraction data imply hexagonal lateral packing (α-phase) of the chain axes. Many authors considered this to be caused by free rotation of the carbon chains, thus behaving as cylinders. The chain axes can either be perpendicular or at an angle to the end group planes. The dimensions of the subcell cannot be obtained with any

TABLE 4. Data for the Monoclinic Subcell M∥

Compound	a_s (Å)	b_s (Å)	c_s (Å)	γ_s (°)	Chain/area (Å²)	Reference
3-Thiadodecanoic acid	4.69	4.56	2.56	122	18.14	11
1-Monoglycerides	4.26	4.83	2.57	114.5	18.72	33
2-DL-Hydroxytetra-decanoic acid	4.30	4.74	2.54	110.9	19.04	20

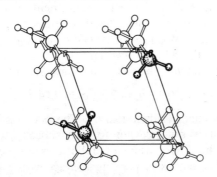

FIG. 14. The monoclinic M∥ packing.

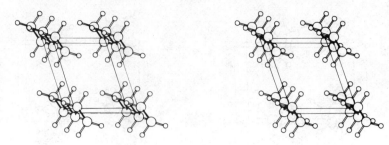

FIG. 15. Stereoscopic pair of M‖.

accuracy. Larsson[35] prepared a multiple film sample of *n*-nonadecane which behaved like a single crystal of the α-phase. He gave the following orthohexagonal axes $a_s = 4.79$ and $b_s = 8.30\,\text{Å}$. Hexagonal symmetry was indeed observed within one lipid layer but adjacent layers showed stacking disorder to give an overall orthorhombic crystal symmetry. This is not unexpected considering the sample used. Piesczek *et al.*[49] reported for the high temperature phase of *n*-triacontane, $a_s = 4.68$ and $b_s = 8.46$ Å. Their analysis indicated clear deviations from hexagonal symmetry.

Even with the chain separation of an ideal hexagonal packing ($b_s = 8.6$ Å), free synchronous rotation of chains with two-fold symmetry is impossible in a defined solid lattice (c.f. ref. 12). Even moderate oscillations about the chain axes will lead to displacement of molecules within the lipid layer. Only in the super liquid state in monolayers[25] in which a vigorous lateral movement in the tightly packed lipid layer is observed, are the chains likely to rotate freely.

It can, therefore, be concluded that the hexagonal symmetry is caused by statistical disorder of the chain planes in the lipid matrix coupled with oscillations which, on the average, have limited amplitudes in the solid state. In the liquid state, these oscillations increase but the interchain distance does not alter much on melting the α-form.[65] This implies that the *trans* zig-zag conformation of the chains is largely maintained in the liquid at least just above the melting point.

IV. HYBRID TYPE SUBCELLS

In the structure determination of a cholesterol ester,[2,3] a twinned O⊥ subcell was observed. This together with the chain packing in a cerebroside[47] seems to be of principal interest for the condensed state of complex lipids (c.f. ref. 4). As these chain arrangements contain features of several simple packings, they will be called hybrid subcells (HS).

A. HS1

This chain arrangement has been found in cholesteryl-17-bromoheptadecanoate[2,3] and in 1,2-dilauroyl-DL-phosphatidylethanolamine.[27] The same chain packing was also discovered in 1,2-dipalmitoyl-DL-phosphatidylethanolamine.[23] The characteristic feature of HS1 (Figs. 17 and 18) is a pleated sheet arrangement of chains all with parallel zig-zag planes. Adjacent sheets show alternate tilt of the chain planes. Hereby carbon chains pack both with parallel and roughly perpendicular chain planes. Dorset[23] called the subcell parallel/perpendicular which, however, is not definite enough a nomenclature.

The subcell is orthorhombic with the following dimensions (c.f. Table 5)

$$a_s = 10.3 \quad b_s = 7.5 \quad c_s = 2.54\,\text{Å}$$

The area per hydrocarbon chain is then 19.3 Å². The packing is less ordered than in simple subcell cases and it is not relevant to give subcell atomic co-ordinates. The chain packing can be thought of as a twinned O⊥ but is best described as a combination of O⊥ and T‖ (Fig. 19). The chains in packing contact with parallel planes have thus a relative displacement according to T‖ (c.f. Fig. 16).

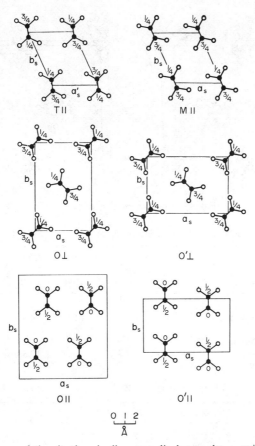

FIG. 16. Cross sections of the simple subcells perpendicular to the c_s axis. a'_s and b'_s of T∥ are thus projections. The relative displacement of the chains are indicated by the co-ordinates 0, 1/4, 1/2 and 3/4 representing fractions of c_s.

B. *HS2*

The chain arrangement in β-D-galactosyl-N-(2-D-hydroxyoctadecanoyl)-D-dihydro-sphingosine[47] is very complex but contains like HS1 pleated sheets of chains with parallel zig-zag planes. The tilt of the chain planes to the sheet planes is different in the two cases. Whereas the chain planes in HS2 (Figs. 20, 21 and 22) are nearly parallel (21–26°) or perpendicular (72–73°) to the sheet planes and intersect at angles between 93–99°, the planes in HS1 are at an angle of about 43° with the sheet planes. The chain planes then form an angle of 86°. The subcell dimensions are $a_s = 9.2$, $b_s = 8.8$, $c_s = 2.55$ Å and $\gamma_s = 98°$. This corresponds to a chain area of 19.1 Å^2. HS2 is even less ordered than HS1 due to the structural dominance of the polar region (Fig. 23).

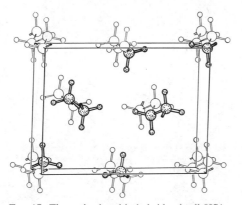

FIG. 17. The orthorhombic hybrid subcell HS1.

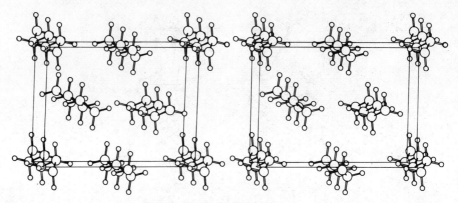

FIG. 18. Stereoscopic pair of HS1.

TABLE 5. Data for the Orthorhombic Hybrid Subcell HS1

Compound	a_s (Å)	b_s (Å)	c_s (Å)	Area/ chain (Å^2)	Reference
Cholesteryl 17-bromoheptadecanoate	10.31	7.50	2.54	19.33	2, 3
1,2 Dilauroyl-DL-phosphatidylethanol-amine	9.95	7.77		19.33	27
1,2-Dipalmitoyl-DL-phosphatidyl-ethanolamine	10.03	7.76		19.46	23

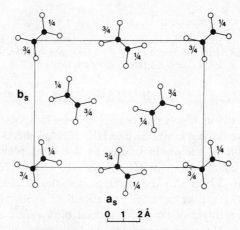

FIG. 19. Cross section of HS1 perpendicular to c_s.

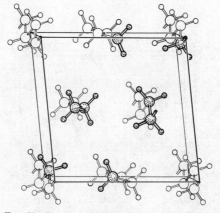

FIG. 20. The monoclinic hybrid subcell HS2.

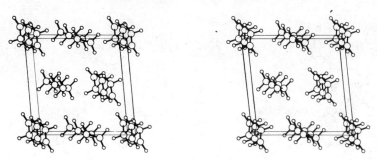

FIG. 21. Stereoscopic pair of HS2.

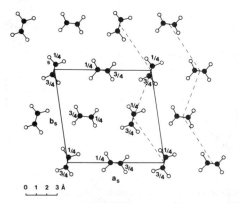

FIG. 22. Cross section of HS2 perpendicular to c_s.

No subcell co-ordinates are therefore given. HS2 shows packing characteristics of several subcells. As indicated in Fig. 24, there exist $O\perp$ and $O'\perp$ building units sharing faces but both with one corner chain in wrong orientation. The lateral arrangement of chains with parallel planes include $T\parallel$, $M\parallel$ and, at a further distance, $O'\parallel$ type packing contacts.

V. FUNCTIONAL GROUPS IN A HYDROCARBON CHAIN MATRIX

As seen in Fig. 1, the displacement of the 13-oxoisostearic acid molecules along the chain axes is such that the carboxylic groups are in packing contact with methylene groups. No major distortions occur—only C(1) is 0.21 Å from the best least-squares plane through C(2) to C(16). The maximum deviation of these atoms from the plane is 0.04 Å. The carbonyl oxygen on carbon-13 is also easily accommodated into the $T\parallel$ packing. Here the sp^2-hybridized C(13) causes distortions. The C(12)–C(13)–C(14) angle can be decreased only to 117.9° and the neighboring chains have to bend somewhat to conform with this angle. Due to the displacement of the molecules, this bend is not local but propagates through the entire chain. The bend takes place *in* the zig-zag plane so that the valence angles at even-numbered carbon atoms are larger than those at odd-numbered atoms.

In DL-2-methyl-7-oxododecanoic acid,[45] the valence angle of the carbonyl carbon atom does not lead to a bend of the chain but to a twist of the chain planes of 7.2° of the parts ($O'\perp$ subcell) on each side of the carbonyl group. The methyl branch, however, is so bulky that it does not fit into the chain matrix. Instead, the molecules of the *iso* compound (Fig. 1) are translated relative to each other so that the methyl end (C(17)) of one molecule just reaches the branch in a neighboring molecule. The branches are thus accommodated in the gap between the methyl ends of the molecules. When forced into a hydrocarbon chain matrix, the methyl group is therefore likely to cause severe local distortions but no detailed structural data is available here. It appears, though, that a limited number of methyl groups can be accommodated into the $O\perp$ packing. Walter and Reding[70] reported that a branched polythene with 8 methyl groups per 100 carbon atoms had a subcell with $a_s = 5.0$ and b_s elongated to 7.68 Å.

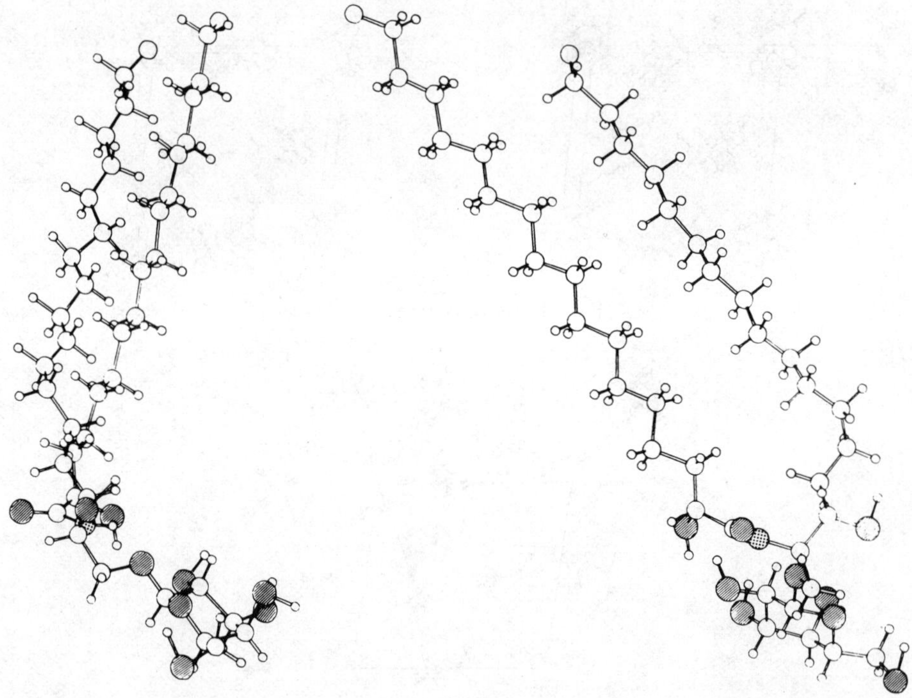

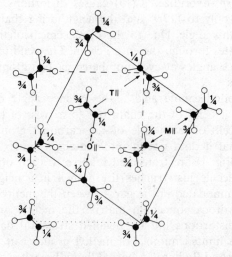

FIG. 23. Molecular conformation of the cerebroside β-D-galactosyl-N-(2-D-hydroxyoctadecanoyl)-
D-dihydrosphingosine as seen along the crystallographic a (right) and b (left) axes.[47]

This is compatible with the effect on the chain packing of hydroxyl groups which also
require considerable space. In 12-D-hydroxyoctadecanoic acid methyl ester,[40] the hydrogen bond system of the hydroxyl group is given space in O⊥ without any local chain
distortions (Fig. 25) by expanding b_s to 7.87 Å whereas a_s is 4.86 Å. In this process,
the angle between the chain planes and the $a_s c_s$ subcell face has increased from 42.3°
in hydrocarbons[59] to 45.8°. The angle between the chain planes are then 84.6° and
91.6°, respectively.

A quite different situation exists in sodium dodecyl sulphate[56] in that the packing
of the bulky polar groups leaves the chains with too much space. The chain packing,
however, adapts in a similar way by keeping one subcell axis nearly constant (5.1 Å)
and expanding the other to 8.2 Å. The arrangement is then very close to a hexagonal
array of chains (b_s = 8.6 Å). Even though the chains have a considerable separation,

FIG. 24. Cross section of HS2 showing features of various simple subcells (c.f. Fig. 16). O⊥
is represented by dashes and O'⊥ by dots. The real subcell is indicated by full lines.

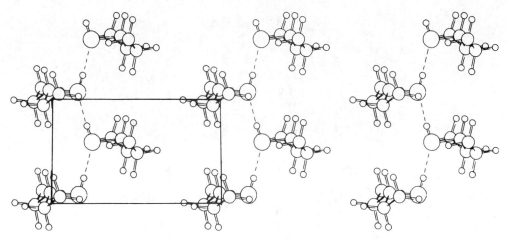

FIG. 25. The chain packing O⊥ with accommodated hydroxyl group hydrogen bonding system in 12 D-hydroxyoctadecanoic acid methyl ester.[40]

a normal *trans* zig-zag conformation is maintained in the structure (Fig. 26) in accordance with what was stated above about chain conformation after melting. There are, of course, disorders in the chains but the arrangement of the chain planes is neither random nor that of O⊥. Instead, the chain planes are oriented mainly parallel to the 5.1 Å axis (Fig. 27).

VI. CHOLESTEROL AND HYDROCARBON CHAINS

The effect of cholesterol on fluidity and packing of phospholipids has been studied extensively but no detailed structural data are available on the co-packing of cholesterol and hydrocarbon chains. Studies of cholesteryl esters have, therefore, been performed in order to obtain such information. Unfortunately, they crystallize with cholesterol skeleta and hydrocarbon chains in separate regions (Fig. 28).[2,3,17]

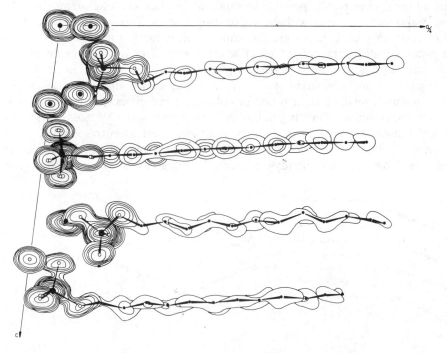

FIG. 26. Electron density map of sodium dodecyl sulphate showing the four molecules in the asymmetric unit.[56]

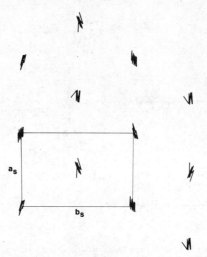

FIG. 27. Projection of the chains in sodium dodecyl sulphate[56] along their axes. The hydrogen atoms are omitted.

The packing of cholesterol skeleta is principally different in cholesteryl esters and cholesteryl sulphate.[57] In the esters, the skeleta, with a cross section area of 36.7 Å^2, are arranged in double layers (Fig. 29) with the projecting methyl groups facing each other. By a relative displacement of the molecules in the direction of maximum extension of the skeleton, the two methyl groups of one molecule can be accommodated in the space between the methyl groups of two molecules in the opposite layer half. The double layers then get smooth contact surfaces. In cholesteryl sulphate, the methyl groups project into the space between two skeleta which then can have no direct contact sideways (Fig. 30). This single layer packing has a similar cross section area (36.8 Å^2) as the double layer arrangement.

If the packing pattern of the hydrocarbon chains in the hybrid cases are compared with the arrangements of cholesterol skeleta, a good fit can be obtained where two chains superpose the area of a cholesterol molecule throughout the lattice (Fig. 31). It thus appears, in principle, possible to randomly replace cholesterol with two hydrocarbon chains and *vice versa* without major disturbances. This would require that both the molecular areas and shapes are the same which is rarely the case as polar groups, chain branches, etc., influence the effective cross section of the long-chain compound.

The only information on cholesterol–hydrocarbon chain co-packing at the present state originates from monolayer studies. It has been shown that monolayers of long-chain compounds retain certain properties of their three-dimensional crystalline states, such as the alternation of melting points in homologous series,[41] cross sectional chain areas and transition temperatures of various phases.[16] Areas corresponding to that of cholesterol $(37–38 \text{ Å}^2)$ have been observed for saturated phosphatidylethanolamines in monolayers[26] but for phosphatidylcholine and sphingolipids, the minimum areas for

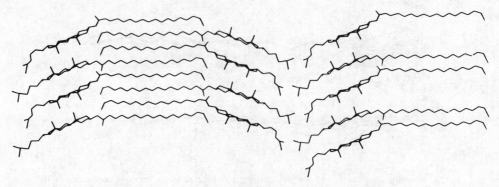

FIG. 28. Molecular arrangement of cholesteryl-17-bromoheptadecanoate.[2,3]

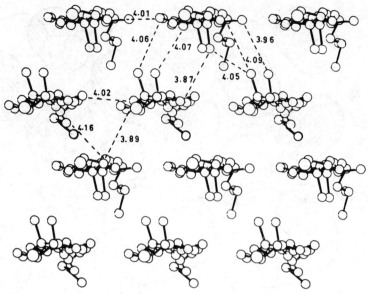

FIG. 29. Packing of cholesterol skeleta in cholesteryl esters as seen along the direction of maximum extension of the skeleton.[2,3]

saturated species are 42 Å^2 [48] and 40 Å^2, respectively (Löfgren, unpublished). The latter areas do not decrease to any extent in systems with cholesterol which is known to have a condensing effect on monolayers of phosphatidylcholines.

However, a monolayer of an equimolecular mixture of phospholipid (distearoylphosphatidylcholine), cerebroside (galactosyl-N-stearoylsphingosine) and cholesterol can be compressed to an area of 37 Å^2 per molecule.[36] This system appears representative for the lipid composition of the outer layer of the myelin membrane[44] if an asymmetric lipid distribution is assumed. This small area can only be explained by a precise stereospecific fit between the three components and shows that cholesterol can be incorporated in a very condensed hydrocarbon chain matrix. It is interesting to note that, despite the close packing, the monolayer is very mobile and shows characteristics of the super liquid state.

Acknowledgements—Grants in support of this Department were obtained from the Swedish Medical Research Council, The Swedish Board for Technical Development, the Wallenberg Foundation and the U.S. Public Health Service (G.M.-11653).

(*Received* 17 *August* 1976)

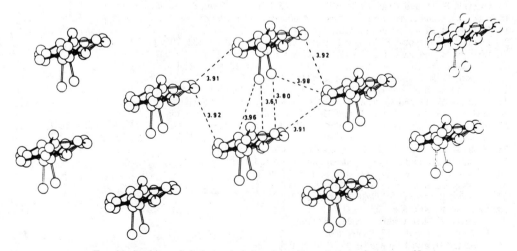

FIG. 30. Packing of cholesterol skeleta in sodium cholesteryl sulphate.[57]

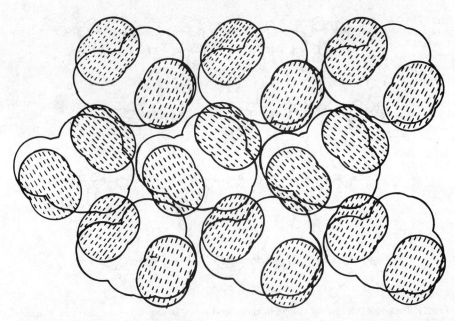

FIG. 31. Superposition of the chain pattern of the HS1 subcell on the steroid skeleta packing in sodium cholesteryl sulphate. The contours are proportional to the van der Waals' radii.

REFERENCES

1. ABRAHAMSSON, S. *Acta Crystallogr.* **12**, 304 (1959).
2. ABRAHAMSSON, S. and DAHLÉN, B. *J. Chem. Soc. Chem. Commun.* **1976**, 117.
3. ABRAHAMSSON, S. and DAHLÉN, B. (1977) *Chem. Phys. Lipids.* In press.
4. ABRAHAMSSON, S., DAHLÉN, B., LÖFGREN, H., PASCHER, I. and SUNDELL, S. In: *Structure of Biological Membranes*, Eds. ABRAHAMSSON and PASCHER, p. 1 (1977). Plenum Press, London.
5. ABRAHAMSSON, S. and HARDING, M. M. *Acta Crystallogr.* **20**, 377 (1966).
6. ABRAHAMSSON, S. and INNES, M. *Acta Crystallogr.* **B30**, 721 (1974).
7. ABRAHAMSSON, S. and LUNDÉN, B.-M. *Acta Crystallogr.* **B28**, 2562 (1972).
8. ABRAHAMSSON, S. and RYDERSTEDT-NAHRINGBAUER, I. *Acta Crystallogr.* **15**, 1261 (1962).
9. ABRAHAMSSON, S., STÄLLBERG-STENHAGEN, S. and STENHAGEN, E. *Progress in The Chemistry of Fats and Other Lipids* Vol. 7. Pergamon Press, London, p. 59 (1963).
10. ABRAHAMSSON, S. and VON SYDOW, E. *Acta Crystallogr.* **7**, 591 (1954).
11. ABRAHAMSSON, S. and WESTERDAHL, A. *Acta Crystallogr.* **16**, 404 (1963).
12. ANDREW, E. R. *J. Chem. Phys.* **18**, 607 (1950).
13. BROTHERTON, T., CRAVEN, B. and JEFFREY, G. A. *Acta Crystallogr.* **11**, 546 (1958).
14. BUERGER, M. J. *Am. Mineral.* **30**, 551 (1945).
15. BUNN, C. W. *Trans. Faraday Soc.* **35**, 482 (1939).
16. BURSH, T., LARSSON, K. and LUNDQVIST, M. *Chem. Phys. Lipids* **2**, 102 (1968).
17. CRAVEN, B. M. and DETITTA, G. T. *J. Chem. Soc. Perkin Trans.* 814 (1976).
18. CRAVEN, B. and JEFFREY, G. A. *Acta Crystallogr.* **12**, 754 (1959).
19. DAHLÉN, B. *Acta Crystallogr.* **B28**, 2555 (1972).
20. DAHLÉN, B., LUNDÉN, B.-M. and PASCHER, I. *Acta Crystallogr.* **B32**, 2059 (1976).
21. DAHLÉN, B. and PASCHER, I. *Acta Crystallogr.* **B28**, 2396 (1972).
22. DEGERMAN, G. and VON SYDOW, E. *Acta Chem. Scand.* **13**, 984 (1959).
23. DORSET, D. L. *Biochim. Biophys. Acta* **424**, 396 (1976).
24. DUMBLETON, J. H. and LOMER, T. R. *Acta Crystallogr.* **19**, 301 (1965).
25. HARKINS, W. D. and COPELAND, L. E. *J. Chem. Phys.* **10**, 272 (1942).
26. HAYASHI, M., MURAMATSU, T. and HARA, I. *Biochim. Biophys. Acta* **255**, 98 (1972).
27. HITCHCOCK, P. B., MASON, R., THOMAS, K. M. and SHIPLEY, G. G. *Proc. Natl. Acad. Sci. U.S.* **71**, 3036 (1974).
28. HYBL, A. and DORSET, D. *Acta Crystallogr.* **B27**, 977 (1971).
29. JEFFREY, G. A. and SAX, M. *Acta Crystallogr.* **16**, 1196 (1963).
30. KITAIGORODSKII, A. I. *Kristallografiya* **2**, 637 (1957).
31. LARSSON, K. *Acta Chem. Scand.* **17**, 199 (1963).
32. LARSSON, K. *Acta Crystallogr.* **16**, 741 (1963).
33. LARSSON, K. *Ark. Kemi* **23**, 29 (1964).
34. LARSSON, K. *Acta Crystallogr.* **21**, 267 (1966).
35. LARSSON, K. *Nature* **213**, 383 (1967).
36. LÖFGREN, H. and PASCHER, I. (1977) to be published.
37. LOMER, T. R. *Acta Crystallogr.* **16**, 984 (1963).

38. LOMER, T. R. and PERERA, K. *Acta Crystallogr.* **B30,** 2912 (1974).
39. LUNDÉN, B.-M. *Acta Crystallogr.* **B30** 1756 (1974).
40. LUNDÉN, B.-M. *Acta Crystallogr.* **B32,** 3149 (1976).
41. LUNDQVIST, M. *Chem. Scripta* **1,** 197 (1971).
42. MACGILLAVRY, C. H. and WOLTHUIS-SPUY, M. *Acta Crystallogr.* **B26,** 645 (1970).
43. NYBURG, S. C. and LÜTH, H. *Acta Crystallogr.* **B28,** 2992 (1972).
44. O'BRIEN, J. S. and SAMPSON, E. L. *J. Lipid Res.* **6,** 537 (1965).
45. O'CONNELL, A. M. *Acta Crystallogr.* **B24,** 1399 (1968).
46. O'CONNELL, A. M. and PASCHER, I. *Acta Crystallogr.* **B25,** 2553 (1969).
47. PASCHER, I. and SUNDELL, S. (1977) *Chem. Phys. Lipids.* In press.
48. PHILLIPS, M. C. and CHAPMAN, D. *Biochim. Biophys. Acta* **163,** 301 (1968).
49. PIESCZEK, W., STROBL, G. R. and MALZAHN, K. *Acta Crystallogr.* **B30,** 1278 (1974).
50. ROSEN, L. and HYBL, A. *Acta Crystallogr.* **B28,** 610 (1972).
51. SAKURAI, T. *J. Phys. Soc. Jpn.* **10,** 1040 (1955).
52. SEGERMAN, E. *Acta Crystallogr.* **19,** 789 (1965).
53. SHEARER, H. M. M. and VAND, V. *Acta Crystallogr.* **9,** 379 (1956).
54. SIM, G. A. *Acta Crystallogr.* **8,** 833 (1955).
55. SMITH, A. E. *J. Chem. Phys.* **21,** 2229 (1953).
56. SUNDELL, S. (1977) To be published.
57. SUNDELL, S. (1977) To be published.
58. VON SYDOW, E. *Acta Chem. Scand.* **12,** 777 (1958).
59. TEARE, P. W. *Acta Crystallogr.* **12,** 294 (1959).
60. THIBAUD, J. and DUPRÉ LA TOUR, F. *J. Chim. phys.* **29,** 164 (1932).
61. TRILLAT, J. J. and HIRSCH, TH. V. *C. R. Acad. Sci. Paris* **195,** 215 (1932).
62. TURNER, J. D. and LINGAFELTER, E. C. *Acta Crystallogr.* **8,** 551 (1955).
63. VAINSHTEIN, B. K. and PINSKER, Z. G. *Dokl. Akad. Nauk SSSR* **72,** 53 (1950).
64. VAINSHTEIN, B. K. and PINSKER, Z. G. *Tr. Inst. Kristallogr. Akad. Nauk. SSSR* **6,** 163 (1951).
65. VAND, V. *Acta Crystallogr.* **6,** 797 (1953).
66. VAND, V. *Acta Crystallogr.* **7,** 697 (1954).
67. VAND, V. and BELL, I. P. *Acta Crystallogr.* **4,** 465 (1951).
68. VAND, V., LOMER, T. R. and LANG, A. *Acta Crystallogr.* **2,** 214 (1949).
69. VAND, V., MORLEY, W. M. and LOMER, T. R. *Acta Crystallogr.* **4,** 324 (1951).
70. WALTER, E. R. and REDING, F. P. *J. Polymer Sci.* **21,** 561 (1956).

Prog. Chem. Fats other Lipids. Vol. 16, pp. 145–162. Pergamon Press, 1978. Printed in Great Britain

LIQUID CRYSTALLINE BEHAVIOR IN LIPID–WATER SYSTEMS

KRISTER FONTELL

Chemical Center, Box 740, S-220 07 Lund 7, Sweden

CONTENTS

I. INTRODUCTION	145
II. METHODS OF INVESTIGATION	147
III. BACKGROUND WORK	147
IV. A MODEL SYSTEM	149
V. OTHER MODEL SYSTEMS	149
VI. LIPID SYSTEMS OF BIOLOGICAL INTEREST	155
VII. CONCLUSIONS	159
REFERENCES	161

I. INTRODUCTION

Lipids have a tendency to form, in combination with water, liquid crystalline phases as well as the thermotropic ones encountered in the nonaqueous state. This capability is due to the molecular structure and nature of lipids. Although they are a heterogeneous group of compounds, including hydrocarbons, dyes, pigments, sterols, phopholipids, glycolipids, detergents etc., a common feature of the lipids is that they contain nonpolar groups of aliphatic and/or aromatic character. In addition many of them have an amphiphilic character in that they possess polar groups some of which form anions or cations. The balance between the lipophilic nonpolar, and the hydrophilic polar properties determines the molecular solubility in water and in fats and thus the behavior in biological systems.

The liquid crystalline lipid phases are characterized by a "liquid" state of the hydrocarbon part of the molecule on the atomic scale while there simultaneously exists a superior order, which is "crystalline" in one, two or three dimensions. This definition is somewhat simplified as there is in Nature place for many molecular conformations intermediate between isotropic amorphous solution and the truly crystalline state. Neither does the definition stress the importance of the water molecules in building up the liquid crystalline aggregates.

In the nonaqueous state, lipids often exhibit liquid crystalline properties at elevated temperatures. When water and/or another solvent is added, one may obtain a lyotropic liquid crystalline system whose properties differ. However, there is never a sharp distinction between the thermotropic and the lyotropic liquid crystalline behavior. For historical reasons, the terminology and way of looking on the phenomena differ and there is little relationship between the two branches of study. This is, for instance, amply exemplified in the recent book *Liquid Crystals and Plastic Crystals.*[29]

The lipids of biological systems are rather complex and it is necessary, for understanding their structures, to study model systems. In biochemistry and biophysics, a common model system is that of lecithin and water. However, lecithins of natural origin may vary in fatty acid composition and, in addition, they may be subject to deterioration with time, especially at elevated temperature. Synthetic lecithins are costly to obtain in large quantities and, with them, there is often a need to work above room temperature to obtain the liquid crystalline state. To understand the fundamentals of the liquid

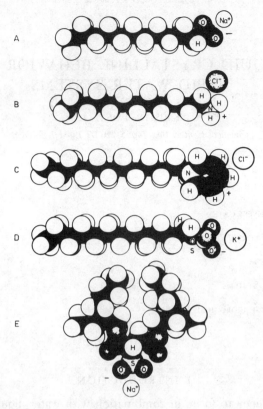

FIG. 1. Molecular models of some surfactant compounds. (After ref. 12). A. Sodium dodecanoate,
B. Dodecyl ammonium chloride, C. Dodecyl pyridinium chloride, D. Potassium dodecyl sulphate,
E. Sodium di-2-ethylhexylsulphosuccinate (Trade name, Aerosol OT).

crystalline behavior of aqueous lipid systems, it therefore was advisable to go one step
further in simplification, to model systems composed of soap and water.

The soaps, in restricted sense, are the alkali-metal salts of higher fatty acids. There
is now in addition a great variety of synthetic surface-active compounds which, together
with the soaps, also are known as surfactants. All these substances have a similar molecular structure (Fig. 1). In addition to the hydrocarbon part, the molecules contain one
or more functional groups that have an affinity for water. There are also surfactants
whose hydrophilic part may consist of nonionizing polar groups.

Several large reviews of the conditions in soap–water systems have been published.
To some extent, they reflect the opinions and methodology of the different research
groups.[13,15,18,22,39,48,55,77,90,91,92] The purpose of the present paper is to attempt to
give an account of the conditions in liquid crystalline phases occurring in soap–water
systems and to connect the conclusions obtained with the conditions existing in aqueous
lipid systems of biological origin.

The lipids may be classified by their physical properties in bulk aqueous system
and at the air–water or oil–water interface.[79] This classification involves first a separation into nonpolar and polar lipids. The nonpolar lipids are insoluble in water and
form either oil droplets or crystals in aqueous systems. The polar lipids may be divided
into *insoluble amphiphilic lipids*, either nonswelling or swelling in water environment,
or in *water-soluble amphiphilic lipids* which may or may not have the capability of
forming liquid crystalline phases. The *soluble amphiphilic lipids* form micelles in relative
dilute aqueous solution and are furthermore capable of solubilizing other classes of
lipids. The soaps belong to the group of water-soluble amphiphiles which have the
capability of forming liquid crystalline phases.

II. METHODS OF INVESTIGATION

The principal methods of investigating the liquid crystalline behavior of soap–water systems are polarizing microscopy and X-ray diffraction. Most of the liquid crystalline phases or mesophases are birefringent and the phases exhibit different characteristic textures in the microscope. However, the familiar terms from thermotropic liquid crystalline studies, *viz.* nematic, cholesteric and smectic, have a very limited applicability and may sometimes lead to confusion and should therefore be avoided. Only the lamellar liquid crystalline lipid–water structures fit in this nomenclature, being smectic, whereas none of the other structures encountered fit into the categories mentioned above. An attempt to classify the liquid crystalline phases occurring in soap–water systems, although now somewhat outdated, has been made by Rosevear.[70]

The inner fine structure is obtained by X-ray diffraction. The techniques employed go back to the work of Luzzati and co-workers.[36,50,52] In addition to a wide diffuse reflection with a position corresponding to a spacing of 4.5 Å, one obtains in the "low angle" region a series of sharp reflections corresponding to crystalline interplanar spacings ranging from 10 to above 100 Å. The nature of the reflection at 4.5 Å suggests a liquid state of the hydrocarbon chains. Electron microscopy has also been used to characterize mesophase structures but the pictures obtained may have been influenced by the technique of preparation. The method is excellent for revealing the morphology of the structures.[28] Information about the liquid state of the hydrocarbon chains may also be obtained by IR, ESR, NMR, DTA (or DSC) and density techniques. Séparation and isolation of different mesophases and studies of the equilibrium conditions may be facilitated by centrifuging technique.[61]

III. BACKGROUND WORK

Much of the basic work on soap–water systems traces back to the school of McBain. A common feature of all temperature-concentration diagrams is that below a certain temperature, the Krafft temperature, described by the T_c curve, the solubility of the soap in water is very low, so that dispersions of soap crystals occur (Fig. 2a). Above the T_c curve, one finds isotropic nonmicellar or micellar solutions and various liquid crystalline phases. In the nonaqueous state as the temperature is increased, a series of different crystalline and liquid crystalline thermotropic phases occur. In an isothermal phase diagram as the soap content is raised, one encounters the sequence *isotropic solution–middle soap–neat soap–nonaqueous or low-water phases*. In a lecithin–water system, one has, in principle, similar conditions. Above the T_c curve, highly diluted aqueous lecithin solutions are in equilibrium with a liquid crystalline lamellar phase (Fig. 2b). This phase is, in turn, on the water-poor side in equilibrium with various nonaqueous or low-water phases (Fig. 3).

Due to the elongated shape of the soap molecules and the layer-like structure of the crystalline compound, it was for a long time believed that all liquid crystalline aggregates encountered in soap–water systems should have lamellar structure. The conclusion was extended to the aggregates of micellar solution systems and this view was maintained for a considerable time although already Hartley had clearly pointed out that the micelles should have a globular shape.[32] For the mesophases, it was realized during the late fifties that other than lamellar liquid crystalline structures were possible. Luzzati and co-workers demonstrated the existence of several nonlamellar liquid crystalline phases in isothermal binary soap–water systems and in nonaqueous systems when the temperature was raised.[48,55] During a transient period, the possibility of a continuous transition from one structure to another inside the same phase was considered but the work of Ekwall *et al.* on phase equilibria in ternary systems of amphiphile and water demonstrated that the various different structures were distinct phases in thermodynamic equilibrium.[14,62,63]

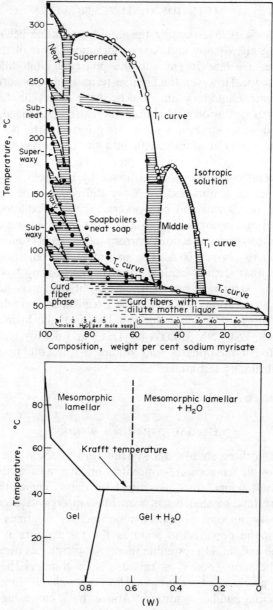

FIG. 2. The binary phase diagrams of (a) sodium tetradecanoate/water system[89] and (b) 1.2 dipalmitoyl-L-phosphatidylcholine/water system.[7,9]

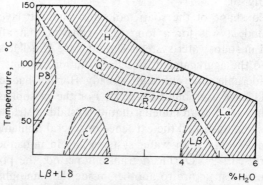

FIG. 3. The "dry" part of the phase diagram of the system egg lecithin/water.[48,56] $L_\alpha L_\beta$, lamellar phase; P_δ, lamellar phase with ribbons; R, mesophase R3m; Q, optical isotropic viscous phase of reversed type, space group Ia3d; H, middle phase of reversed type; C, three-dimensional crystalline phase.

IV. A MODEL SYSTEM

In the ternary system sodium octanoate–decan-1-ol–water at 20°C in addition to two isotropic liquid solution phases (L_1 and L_2 in the terminology of Ekwall *et al.*[13–15,18,22,62,63]) no less than five different liquid crystalline phases occur (B, C, D, E, and F in the same terminology) (Fig. 4). Each of these mesophases has its own macroscopic and microscopic texture and inner fine structure and the phases are separated from another and from the solution phases by two or three phase regions. In the binary soap–water system represented by the bottom axis, there exists, in the concentration region of 46 to 51.5% of soap, a mesophase possessing the texture and inner structure of *middle soap*. In the polarizing microscope, this phase has a fan-like angular texture, of the types 222.5 or 222.6 in the classification of Rosevear.[70] The inner fine structure has by X-ray diffraction been shown to be two-dimensionally hexagonal in that long rod-like aggregate units are parallel and arranged in a hexagonal array. The rod-aggregates are thought to be composed of dualistic amphiphile molecules with the nonpolar hydrocarbon parts forming a nonpolar core and the polar groups facing outwards towards surrounding aqueous polar region. The length of the rod-aggregates exceeds their diameter by several orders of magnitude (Fig. 4, mesophase E). From geometrical considerations, it is clear that the complementary two-dimensional structure of rod-aggregates composed of water molecules and polar groups in a nonpolar hydrocarbon environment is feasible. This structure has been shown to belong to the phase designated F in this model system. Its macroscopic and microscopic texture is also very similar to that of phase E (Fig. 4, mesophase F). Both phases E and F are transparent and have a rather stiff consistency, the apparent viscosities are in the range 20–50 P.[82]

The central part of the phase diagram is occupied by three mesophases designated B, C and D. Their location is between the two middle soap phases, and they are more opaque and are less viscous. The apparent viscosities of the phases C and D are in the range 2–3 P while the apparent viscosity of phase B is about 1.5 P.[82] A common feature is that their crystalline low-angle X-ray reflections are in the sequence 1:2:3. The microscopic appearance of phase D corresponds to the types 122.1, 122.3, 121.1 or 113.1 in Rosevear's classification while the appearance of the two other phases has not been described by him.[70]

The fine structure of phase D is that of "*soap boiler's neat soap*" (neat soap for short) (Fig. 4, mesophase D). It has a lamellar structure in which the lamellae consist of large coherent bilayers of amphiphile alternating with water layers. The structure of *neat soap* is able to incorporate large amounts of water (up to about 82% in this case). This capability depends on a certain molar ratio between soap and alcohol. The uptake of water is "ideal" in that all added water seems to be intercalated between the amphiphilic layers, whose thickness and molecular packing remain unaffected. In other parts of the phase region, the water molecules penetrate between the polar groups and thus change the molecular arrangement of the bilayers.

The proposed structures and even the existence of the two other phases B and C have been questioned. Phase B has been considered to have a loose wavy lamellar structure very similar to that of phase D, perhaps a vesicular type consisting of concentric bilayers.[18] Phase C has been proposed to be composed of long rod-like amphiphile aggregates of more or less quadratic cross-section in a tetragonal array.[18] The existence of the latter phase C as an independent phase has been especially queried,[65,86,90] its whitish appearance suggesting that it could be an "emulsion" of the lamellar phase D dispersed in aqueous micellar solution.[86]

V. OTHER MODEL SYSTEMS

The phase behavior in other amphiphile–water systems resembles that shown and described above, irrespective of whether the soap is an anionic, cationic or a nonionic compound, and irrespective of the polarity of the third lipid component. The conditions

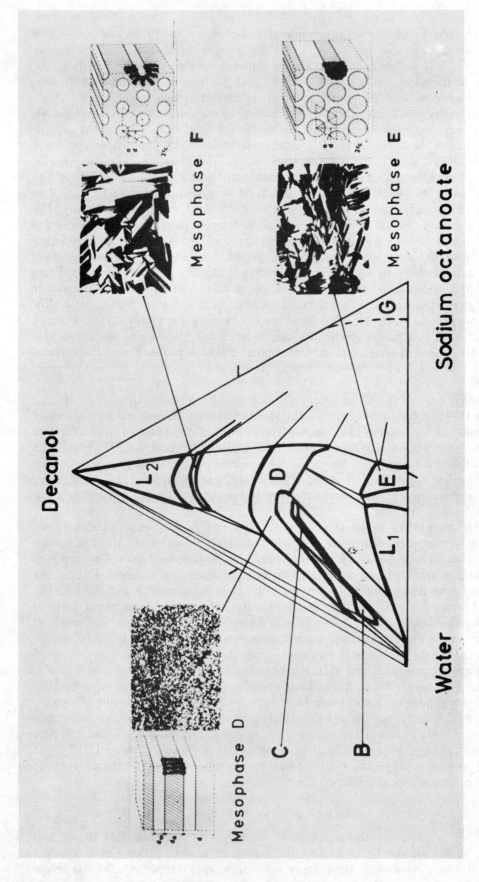

Fig. 4. The isothermal phase diagram for the ternary system sodium octanoate/decan-l-ol/water at 20°C (redrawn after ref. 61). The concentrations are in weight %. The microscopic appearance between crossed polars and the inner fine structure of the three principal liquid crystalline phases are shown. L_1 and L_2, isotropic solution phases; B, C, D, E and F, anisotropic liquid crystalline phases; G, crystalline fibrous soap.

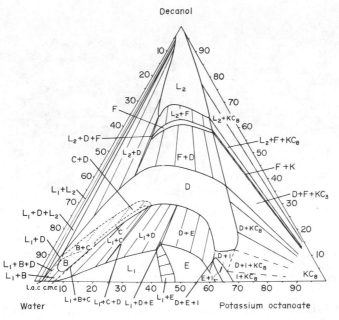

FIG. 5. The isothermal phase diagram for the ternary system potassium octanoate/decan-l-ol/ water at 20°C (19). I, optical isotropic viscous phase; the other notations are the same as in Fig. 4.

for a particular system depend in a predictable manner on the properties of the amphiphile compounds and on the temperature.[13] A prerequisite for the occurrence of the liquid crystalline phases as well as of the solution phase L_2 is often a threshold ratio of soap to water (Figs. 4 and 5).

When the potassium soap is substituted for the sodium soap in the model, the corresponding one obtains a phase map showing the same mesophases and, in addition, in the concentration region above the *middle phase*, an *optical isotropic viscous* phase designated I (Fig. 5).[19] The occurrence of an optical isotropic viscous phase with a location intermediate between the middle phase and the neat phase has been demonstrated also in other potassium soap systems (Fig. 6).[48,51,55] In the same concentration region, some other liquid crystalline phases also occur.[51] For the potassium soap systems, one may thus amplify the sequence of phases to *isotropic solution–middle phase– liquid crystalline phases of miscellaneous types–viscous isotropic phase–neat phase*. However, *optical isotropic viscous phases* have been encountered in other concentration regions of the lipid systems. (For a general survey, see for instance ref. 13).

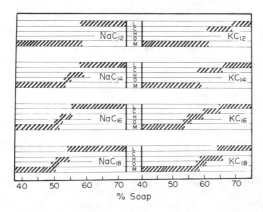

FIG. 6. Range of existence of different mesophases in aqueous alkali soap systems at 100°C.[48,51,55] L, lamellar *neat phase*; C, *optical isotropic viscous phase*; H, *hexagonal complex phase*; M, *middle phase*; O, *deformed middle phase*.

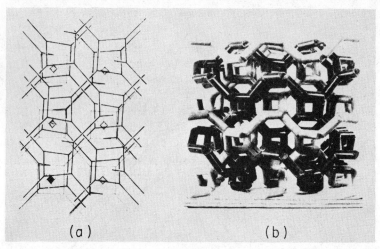

FIG. 7. Schematic presentation of the optical isotropic viscous structure Ia3d. (a) Perspective picture of the axis of the rod-aggregates. The dotted lines are the projections of the axis on the basal plane[54,58] (b) An attempt to visualize the two three-dimensional net-works.[23]

The internal fine structure of these viscous phases has been much debated. In addition to the optical isotropy, a common feature for them is a very stiff consistency. By visual comparison with that of other mesophases, the apparent viscosity must be noticeably higher than 50 P, and even 2500 P has been obtained in the system Aerosol OT water by cone–plate viscometry. When the existence of these phases first were recognized and they were shown to give crystalline three-dimensional low-angle X-ray diffraction patterns, it was thought that the structure consisted of globular aggregates in face-centered close packing, the possibility of the two complementary variants of "oil in water" and "water in oil" being considered.[36,52,53] However, further work showed that, in indexing the diffractograms, the less dense body-centered and primitive cubic structures were often favored.[26,57]

The only *optical isotropic viscous phase* whose structure has been solved with some degree of crystallographic reliability is that of nonaqueous strontium tetradecanoate at 232°C.[54,58] Its structure belongs to the space group Ia3d, No 230 in the International Crystallographic System.[37] According to the proposition of Luzzati *et al.*,[58] the fine structure consists of two interwoven but otherwise independent three-dimensional net-works composed of short rod-like aggregates which are joined three and three at each end (Fig. 7). The core of the rod-aggregates contains the strontium ions and the carboxylate groups and the hydrocarbon chains fill the space between. The complementary

TABLE 1. Dimensions of the Body-centred Cubic Structure Ia3d in Aqueous Amphiphilic Systems[24]

Amphiphile	Type I — Paraffin chains inside the core "oil in water"								Type II — Polar groups inside the rods "water in oil"			
	KC_8	KC_{12}	KC_{14}	KC_{16}	$C_{12}TACl$	$C_{12}TAB$	$C_{16}TAB$	$TAC_{12}I$	Aerosol OT	Lec.	Gal.	SrC_{14}
a (Å)	60.5	78.6	90.7	99.5	79	76.9	98.7	82.9	69	96.2	110.2	62.4
$t°$ (C)	20	100	100	100	22	70	70	20	20	82	64	235
v (cm³ g⁻¹)	(~1)	0.99	1.01	1.03	(~1)	0.99_7	1.02_7	1.08	(~1)	1.025	0.986	1.165
ϕ	0.66	0.64	0.62	0.62	0.85	0.82	0.80	0.70	0.75	0.97	0.88	1
l (Å)	21.3	27.8	32.1	35.2	27.9	27.2	34.9	29.3	24.4	34.0	39.0	22.2
r (Å)	11.2	14.0	15.9	17.5	17.5	15.9	20.2	15.7	8.9	11.1	15.6	—
S (Å²)	46.0	46.8	47.5	48.8	41.1	51.0	49.6	47.5	65	67.0	73.3	—

Notation

KC_n: $CH_3(CH_2)_{n-2}COOK$
SrC_{14}: $(CH_3(CH_2)_{12}COO)_2Sr$
$C_{12}TACl$: $CH_3(CH_2)_{11}N(CH_3Cl$
C_aTAB: $CH_3(CH_2)_{n-1}N(CH_3)_3Br$
$TAC_{12}I$: $CH_3(CH_2)_{10}CONN(CH_3)_3$
Aerosol OT: Sodium di-2-ethylhexyl sulphosuccinate
Lec.: hen egg lecithin
Gal.: galactolipids from maize chloroplasts

a: parameter of the cubic cell
t: temperature
v: specific volume of the sample
ϕ: volume fraction of the amphiphile
l: length of the rods
r: radius of the rods (containing non-polar or polar material, respectively)
S: area per polar group at the interface between polar and non-polar material

structure variant also exists (Table 1). However, the structure Ia3d can account for only two of the at least five different *isotropic viscous phases* observed. For one of the other cubic phases, it has been possible to determine its space group, Pm3n, No 223 in the International Crystallographic System[37] and an internal structure has been proposed, a mixed structure of globular aggregates and a net-work.[84]

The possibility of globular aggregates in cubic close packing cannot be excluded. For a long time, the opinion prevailed that the high resolution NMR spectra obtained for the *viscous isotropic phases* excluded all other structures than those composed of globular aggregates. This obstacle has been circumvented by the recent work of Charvolin and Rigny.[10,11] In his preference for structures of globular aggregates, Winsor draws parallels between the structures of the *plastic crystals* and the *viscous isotropic phases*.[92,93] The *plastic crystals* can have other cubic structures than the close packed one, but these are obviously caused by structural requirements of not strictly symmetric molecules and/or their local polarity.

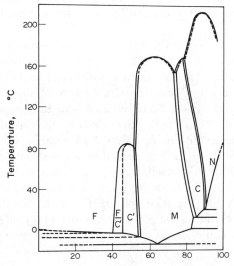

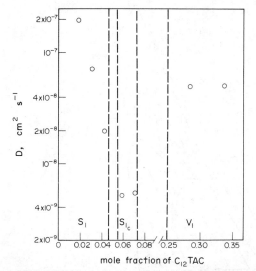

FIG. 8. (a) The binary phase diagram of the system dodecyltrimethyl ammonium chloride/water.[2] F, isotropic solution phase; C', isotropic viscous phase belonging to the space group Pm3n; M, middle phase, C, isotropic viscous phase belonging to the space group Ia3d; N, neat phase. (b) The amphiphile diffusion coefficient, D, as a function of mole fraction of the amphiphile.[4] $S_1 \equiv F$; $S_{1c} \equiv C'$; $V_1 \equiv C$.

That the building principles differ is also demonstrated by measurements of the amphi-philic diffusion by the NMR technique in the binary system dodecyltrimethylammonium bromide–water.[2] In this system, two *isotropic viscous phases* occur, one located on either concentration side of the *middle phase* (Fig. 8a). The translational self diffusion coefficient has been shown to diminish monotonically with decreasing water content through the isotropic micellar solution phase and the neighboring *isotropic viscous phase* whereas, in the second *isotropic viscous phase*, it increases by a factor of 10 and is also consider-ably greater than the value obtained in concentrated micellar solution (Fig. 8b). These findings can be rationalized by the assumption of a continuous network of amphiphiles in the latter *isotropic viscous phase*.[4]

The cubic phase occurring in the system potassium octanoate–water (Fig. 5) seems also to possess a network structure. The translational diffusion coefficients for the cubic phase and the neighboring lamellar neat phase have about the same value after correc-tion for geometrical factors, (unpublished pulsed NMR work, Lindblom, G., Wen-nerström, H., Fontell, K. and Lindman, B.). It may thus be concluded that the diffusional motion in this cubic phase is unrestricted, i.e. the structure must consist of continuous lipid regions, a conclusion which is compatible with the network but not with a globular aggregate structure.

One of the problems in solving the structure of the isotropic viscous phases is that they usually display so few X-ray reflections that it is not possible to index the diffraction pattern unequivocally. Often one obtains just two reflections in the ratio $\sqrt{3}:\sqrt{4}$. This ratio does not constitute proof that the structure is of the face-centered type as is often asserted, for the ratio fits a body-centered structure equally well. The internal fine structures of many of the *isotropic viscous phases* are still unresolved.

A recent survey of the conditions in different amphiphile–water systems is given by Ekwall.[13] In addition to the already presented views, it should be noted that although in some systems the amphiphile itself may be in liquid state, the addition of water may cause formation of a liquid crystalline phase. This behavior is, for instance, observed for the nonionic Triton X-100 (Fig. 99 in ref. 13). This water bonding capacity disappears above the cloud point of the amphiphile. In multicomponent systems, the molecular structure of the "foreign" compound may exert influence on the phase structures formed.

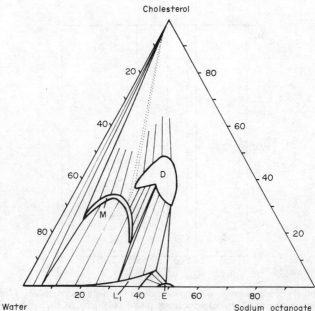

FIG. 9. The isothermal phase diagram for the ternary system sodium octanoate/cholesterol/water at 20°C.[16,60] M, a phase where there exists a "crystalline" short range order; other notations are the same as in Fig. 4.

An example which may be given is the system sodium octanoate–cholesterol–water (Fig. 9). The addition of cholesterol results in the occurrence of a new phase, designated phase M, which in addition to the low angle X-ray reflections displays sharp short spacings, thus demonstrating an ordering influence of the cholesterol molecule upon the "liquid" hydrocarbon chains.[13,16]

One aspect that only occasionally has been studied is the effect of adding neutral salt to a liquid crystalline amphiphile system.[21,25] The addition of alkali halide depresses the critical micillar concentration (c.m.c.) and the other concentration limits of the aqueous micellar soap solutions but, with respect to the liquid crystalline phases, there is often only vague talk about "salting in" and "salting out". The term "coacervation" introduced by Bungenberg de Jong has also bearing upon this group of phenomena.[5] However, the influence of neutral salt must have an immense technical and biological significance.

To sum up, studies of conditions in soap–water systems show an abundance of different liquid crystalline phase structures and, in an equilibrium system, the phases are separated by two- and multiphase regions. The most common phases are the lamellar *neat phase* and the two-dimensional hexagonal *middle phases*, but one has always to recognize the possibility of the existence of other liquid crystalline phases and that some of them may be rather viscous and may be in equilibrium with an aqueous solution. Because the soaps belong to the group of *soluble amphiphiles*, the aqueous solutions in equilibrium with the liquid crystalline phases are micellar. In multi-component systems, and depending upon the polarity of the other lipids, the equilibria may involve dilute aqueous solutions below the critical micelle concentration of the soap (Figs. 3 and 4). It is important to realize that the alkali soaps require a threshold amount of water before the liquid crystalline state is obtained. This amount seems to be what is required for the hydration of the counter ion and for the alkali soaps, the amount required increases from potassium through sodium to lithium.[17] The water molecules are an integrated part of the aggregates forming the mesophase structure. At high water content, there exists in addition "excess" water which, although it is not bound to any particular aggregate unit, participates in the liquid crystalline structure.[13]

VI. LIPID SYSTEMS OF BIOLOGICAL INTEREST

Liquid crystalline phases occur in many biological lipid systems. In the previous sections, the system of lecithin and water has been mentioned, and the basic similarity between soap–water and the lecithin–water phase diagrams has been stressed. In other biological lipid systems, the phase conditions are similar. The same types of phases with the same internal fine structures are often encountered. The lecithins belong to the group of *swelling amphiphiles* in the classification of Small.[79] Virtually insoluble, lecithin swells in water to form a well-defined lamellar liquid crystalline phase as soon as the temperature is above the Krafft point of the lecithin. Below the T_c curve, more or less crystalline gel structures occur in equilibrium with aqueous solution (Figs. 2b and 3).

Much of the basic work on the phase structures of lecithin systems has been done by the group of Luzzati,[48,49,55,69,83] although other groups have also contributed.[7,9,33,43–45,78,80,88.] Early work was to a large extent focused on egg-yolk lecithin, but lately synthetic lecithins have also been studied. The conditions in egg-yolk lecithin systems perhaps more closely resemble those of naturally occurring membrane systems. On the dry side of the lecithin–water system, one encounters several different phases (Fig. 3). The building units of their internal fine structure are often rod-like aggregates, either infinitely long or of finite length and joined in threes or fours together, to form network structures.[48,55] Due to the low water content, these aggregates are considered to be of the "water in oil" type. These "dry" phases have been considered to be of little interest for the understanding of the conditions in biological systems.

At higher water contents, the predominant phase above the T_c curve is the lamellar phase.[48,55] For the egg-yolk lecithin system, the conditions in this phase are considered to be "ideal" above a certain water content, that is, the behavior is the same as in the soap–long chain alcohol–water systems at a critical ratio of soap to alcohol. Thus, the thickness of the lipid lamellae and the molecular packing inside them seems not to be influenced by the water content of the phase, and the lamellar liquid crystalline lecithin phase may take up about 45% water. By addition of a small amount of a charged lipid, anionic or cationic, one can further increase the water content of the lipid system and still retain a homogeneous phase with lamellar structure. Thus, a system with very thick water layers intercalated between the lipid lamellae can be obtained.[31]

In excess of water, this lamellar phase of lecithin (and other swelling biological amphiphiles) are easily dispersed in the aqueous equilibrium solution forming milky systems. The aggregates of such a system, the liposomes, are closed multilayered structures. The liposomal systems may be clarified by ultrasonic treatment whereby the liposomes are subdivided into vesicles. These are aggregates that are considered to consist of a single amphiphile bilayer around a solvent-filled cavity and are, in turn, dispersed in the aqueous solution. Experience has shown that formation of vesicles is facilitated by the presence of a small amount of a *soluble amphiphile*, for instance lysolecithin or a detergent. The vesicular systems are optically clear and may remain stable for a considerable time. Nevertheless they are not in thermodynamic equilibrium and they should not be confused with the micellar solutions occurring in soap–water systems, such as the aqueous solutions containing "normal" micelles or the solutions containing "reversed" micelles. The vesicular systems are more related to the "transparent emulsions" or "microemulsions" of the type studied by Schulman and subsequent workers in the field.[1,3,34,66,71–73] However, one has to keep in mind that some of the systems known as "microemulsions" in reality are thermodynamically stable solutions containing "reversed micelles".[20]

The resemblance between the bilayer structure of vesicular systems and the bilayers of biological membranes has motivated the use of vesicular systems as model systems. The egg-yolk lecithin–water system has been especially favored. In such membrane-like structure, as in liquid crystalline lamellar and nonlamellar phases, the motion of the individual amphiphile molecules and their dependence on the addition of "foreign" molecules and on changes in temperature has been studied by the use of electron spin resonance (ESR) spin-labelling and NMR techniques (for a recent review, see ref. 42). The local environment of a particular amphiphile molecule seems to be, to an astonishingly large degree, independent of the addition of "foreign" molecules or the actual phase structure. Thus, for example, the lipid translational diffusion coefficients measured in different mesophase structures are of about the same magnitude (Lindblom, G. and Wennerström, H., paper presented at the VIIth International Conference on NMR in Biological Systems, September 19–24, 1976, St. Jovite, Canada).

While the conditions above the transition temperature seem to a large extent to be similar for all lecithins, there are noticeable differences in phase behavior below the T_c curve for systems containing synthetic lecithins. The C_{16} and C_{18} saturated lecithins have been shown to differ from the C_{12} and C_{14} ones (83, and unpublished NMR results of ion binding by Lindblom, G.). In addition, the existence of a pretransition temperature for the C_{16} saturated lecithin has been demonstrated by calorimetric methods. Some authors consider the pre-transition peak to reflect a change from tilted stiff hydrocarbon chains to perpendicular ones inside the lipid lamellae.[35,67] As a word of caution, it should be stressed that it is not always certain that a phase remains homogeneous on crossing the T_c curve. A two-phase zone may exist although the preparation judged by the naked eye still has the appearance of an homogeneous phase.

The Luzzati group has extended its work to a large number of other swelling lipids of biological interest. The systems studied include phosphatidylethanolamines, phosphatidylinositols, cardiolipins, sphingomyelins, cerebrosides, galactolipids, etc.[48] Many of these lipids have often an ill-defined chemical composition. They may have a varying hydrocarbon chain distribution and the degree of unsaturation may also vary. One would, therefore, expect the occurrence of a great number of phases. Actually, the phase diagrams often are rather simple. This circumstance depends on the fact that, above the transition temperature, the hydrocarbon chains behave like a miscible liquid. Therefore the systems are insensitive towards chemical heterogenity both with respect to the lipid chain conformation and to the addition of "foreign" molecules. However, the molecular structure of the foreign molecule may have an influence. For instance, the stiff skeleton of the cholesterol molecule may exert an ordering influence on the hydrocarbon chain conformation. On the other hand, cholesterol is sometimes considered to have a fluidizing influence.[8] Of course, the term "fluidity" says very little about the specific molecular motions involved. However, recent studies by NMR technique have shown that the addition of cholesterol to gel phase specimens of the dipalmitoyl lecithin or dimyristoyl lecithin systems resulted in an *increased lateral diffusion* of the lecithin molecules.[87]

Another aspect worth considering is that the hydrocarbon chains of a heterogeneous lipid may have different conformations in different domains of the structure.[6,30,55,59,64,68,85] This heterogenity may involve entire chains or only part of a chain. The properties of the chains in different states may depend on the actual "crystallographic" lattice and/or on the temperature and composition, for T_c is very badly defined in multicomponent lipid systems. The predominant liquid crystalline phase in these systems is the lamellar one but other phase structures may exist. The lipid derivative monogalactosyl diglyceride obtained from pelargonium leaves possesses, according to Shipley, a hexagonal middle phase structure of the "reversed type".[75] Digalactosyl diglyceride, which also occurs in pelargonium leaves, has been shown to have a lamellar structure. Both these structures contain up to about 20% water and they are in equilibrium with dilute aqueous solution. It is the low water content which motivates the assumption of the hexagonal phase of the reversed type. These results are supported by recent studies of the related lipid derivative monoglucosyl diglyceride obtained from *Acholeplasma laidlawii*. A hexagonal phase has been observed at 40°C and this phase also contains up to about 20% water (unpublished results, Wieslander, Ulmius, Fontell and Lindblom). Another group of swelling insoluble lipids of interest is the monoglycerides.[38,47] Many of their aqueous systems display lamellar liquid crystalline phases, but very viscous phases, both the hexagonal middle phase type and the optical isotropic viscous types have been observed (Figs. 10a, b). These viscous phases may contain only 25% water and still be in equilibrium with very dilute aqueous solution phases.

The bile acid alkali salts are capable of forming micellar aqueous systems but they do not form liquid crystalline phases alone. However, in the presence of other soluble amphiphiles of suitable molecular structure, they have this capability. Some examples of such systems are presented in Figs. 11 a–d.[40,41,79] The gross appearance of the liquid crystalline phases encountered is similar to that previously described for the common soap–water systems, but the internal fine structure must be somewhat modified to accommodate the stiff phenantrene skeleton of the bile acid molecule. Small has put forward some suggestions for the mechanisms involved.[79,80] Again, we may have a system in which a very viscous phase, the hexagonal middle phase in the system sodium cholate–lecithin–water, is in equilibrium with a dilute aqueous solution (Fig. 11d). The other viscous phase of the system, the *viscous isotropic phase*, has been shown by recent NMR studies to contain continuous lipid regions, thus supporting the idea of a network structure for a cubic phase.[46]

Biological membrane systems contain, in addition to the lipids, protein molecules. The molecular arrangement of the liquid crystalline structures in biological membranes is not yet clarified and a coherent picture of the conditions prevailing in them has not been developed. A rather recent survey of structure studies in such systems is given by Shipley[74] and the importance of the physical state of the lipids for the development of atherosclerosis has been discussed by Small and Shipley.[81] Figure 12 shows recent visualizations of the membrane structures.[27,76]

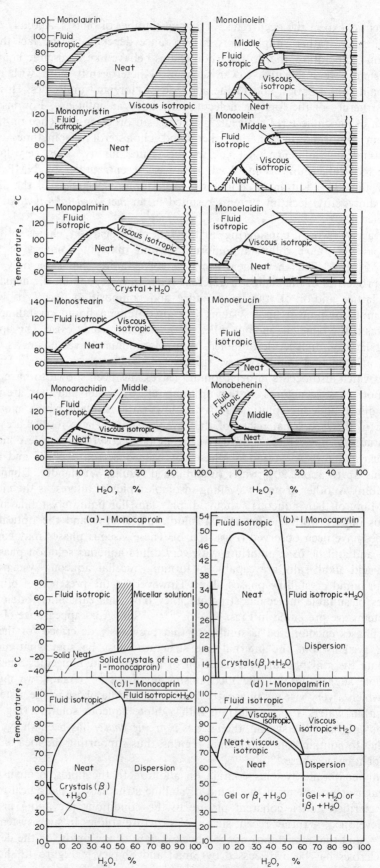

FIG. 10. The binary phase diagrams of (a) ten and (b) four different monoglyceride/water systems studied by Lutton[47] and Larsson,[38] respectively.

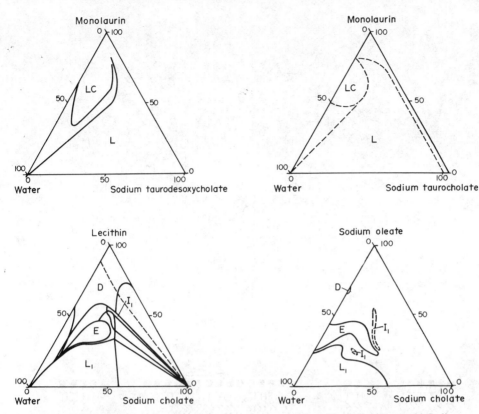

FIG. 11. Isothermal phase diagrams of the ternary systems bile acid salt/lipid/water. (a) Sodium taurodesoxycholate/monolaurin/water at 45°C.[40] (b) Sodium taurocholate/monolaurin/water at 45°C.[41] (c) Sodium cholate/lecithin/water at 22°C.[79] and (d) Sodium cholate/sodium oleate/water at 37°C.[79] LC, liquid crystalline phase of undefined internal structure, probably neat phase; other notations are the same as in Fig. 5.

VII. CONCLUSIONS

There are great similarities in the liquid crystalline behavior between simple soap–water systems and systems containing lipids of biological interest. At sufficiently high temperature and water content, liquid crystalline phases may occur in all lipid systems. In addition, phases having more ordered molecular conformations may be observed at the low water and low temperature part of the phase diagram. One property common to all lipid–water systems is the segregation of the hydrophilic and lipophilic moieties into distinct regions separated from each other by internal interfaces formed by the polar groups of the lipid molecules. Many of the characteristic properties of the lipid–water systems are direct consequences of this segregation into polar and nonpolar regions.

One may distinguish between two main categories or liquid crystalline phases; viz. (1) the lamellar structures consisting of large coherent amphiphile lamellae separated by water layers and (2) rod-like aggregates which are either infinitely long or of finite length and joined together in threes or fours to form networks. Of the latter structures, there are two topologically distinct variants in that the hydrocarbon chains may be inside and the polar medium groups outside or vice versa, that is, "oil in water" or "water in oil". With the exception of the *viscous isotropic phases*, one of these variants will form a continuous matrix into which the opposite structure elements are fitted. In the *viscous isotropic phases*, the two structure elements are continuous and form two intertwined networks through the three-dimensional structure. However, there is at least, in principle, also the possibility that some of the *viscous isotropic phases* are formed by globular aggregates of either type, "oil in water" or "water in oil" in cubic packing.

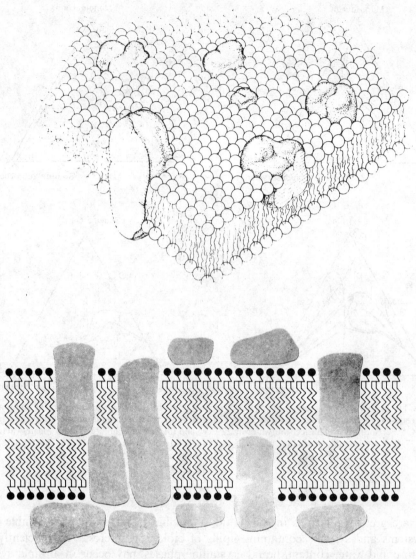

FIG. 12. Two recent visualizations of the biological membrane structure by (a) Singer and Nicholson[76] and (b) Fox,[27] respectively. Phospholipid molecules provide the basic bilayer structure while the protein molecules may be either attached to the bilayer surface, be interdigitated between the phospholipid molecules or be extended through the bilayer.

There are differences in size and shape of the structure elements of a liquid crystalline lipid system. The hydrocarbon chains are separated from the water regions by internal interfaces covered by polar groups. Thus the area per polar group at these interfaces will be an operational parameter indicating the molecular packing at these interfaces and the possible molecular conformations.

The liquid crystalline state is characterized by the "liquid" nature of the hydrocarbon chains and a "crystalline" long range order. The long range organization determines the "crystallographic" lattice and symmetry of the structure while the short range organization is determined by the conformation of the hydrocarbon chains. This conformation is, in turn, dependent on the temperature and/or the presence of other "foreign" molecules. The presence of "water of hydration" is often required in order to render the formation of a liquid crystalline structure possible.

The lamellar liquid crystalline phase is the most commonly occurring and, in this phase, the average hydrocarbon chain orientation is perpendicular to the lipid–water interface. It is unlikely that the conformations of the hydrocarbon chains in the other liquid crystalline phases are profoundly different. From the chemical stand-point, the hydrocarbon chains behave as a miscible liquid. Therefore, the lipid systems are not sensitive to chemical heterogeneity and are able to take up large amounts of "foreign" molecules without losing their basic structure. This behavior goes a long way to explain the role of lipids in biological membranes.

As a working hypothesis, it is often assumed that the lipids of biological membranes are in a liquid crystalline "fluid" state. This assumption is not necessarily true. It is still unsettled whether biological membranes are also capable of functioning when the lipids are in a gel-like conformation. Although the lamellar liquid crystalline structure is the predominant one in the biological context, the possibility of the formation of viscous structures of rather low water content in equilibrium with very dilute aqueous systems cannot be excluded, implying that, when the molecular (or micellar) solubility of the lipid in water is surpassed, these viscous phases may be formed. One may also speculate about the possible occurrence of local viscous domains, cubic or hexagonal ones, in an otherwise lamellar membrane structure.

REFERENCES

1. Ahmad, S. J., Shinoda, K. and Friberg, S. *J. Colloid Interface Sci.* **47**, 32 (1974).
2. Balmbra, R. R., Clunie, J. S. and Goodman, J. F. *Nature* **222**, 1159 (1969).
3. Bowcott, J. E. and Schulman, J. H. *Z. Elektrochem.* **59**, 283 (1955).
4. Bull, T. and Lindman, B. *Mol. Cryst. Liq. Cryst.* **28**, 155 (1974).
5. Bungenberg de Jong, H. C. *La Coacervation, les Coacervates et leur Importance en Biologie* Vol. 397, Actualities Scientifiques et Industrielles, Herman et Cie, Paris (1936).
6. Caron, F., Mateu, L., Rigny, P. and Azard, R. *J. Molec. Biol.* **85**, 279 (1974).
7. Chapman, D. *Introduction to Lipids* p. 13, McGraw-Hill, London (1969).
8. Chapman, D., Byrne, P. and Shipley, G. G. *Proc. R. Soc.* **290A**, 115 (1975).
9. Chapman, D., Williams, R. M. and Ladbrooke, B. D. *Chem. Phys. Lipids* **1**, 445 (1967).
10. Charvolin, J. and Rigny, P. *J. Magn. Reson.* **4**, 40 (1971).
11. Charvolin, J. and Rigny, P. *J. Chem. Phys.* **58**, 3999 (1973).
12. Ekwall, P. in *Festskrift Carl Kempe*, p. 335, Almqvist & Wiksell, Stockholm (1965).
13. Ekwall, P. *Advances in Liquid Crystals*, Vol. 1, p. 1, Ed., Brown, G. H., Academic Press, New York (1975).
14. Ekwall, P., Danielsson, I. and Mandell, L. *Kolloid-Z.* **169**, 113 (1960).
15. Ekwall, P., Danielsson, I. and Stenius, P. *Chemistry and Colloids MTP Int. Rev. Sci.*, Phys. Chem. Ser. 1, Vol 7, p. 9, Ed. Kerker, M., Butterworths, London (1972).
16. Ekwall, P. and Mandell, L. *Acta Chem. Scand.* **15**, 1407 (1961).
17. Ekwall, P. and Mandell, L. *Acta Chem. Scand.* **22**, 699 (1968).
18. Ekwall, P., Mandell, L. and Fontell, K. *Molec. Cryst. Liq. Cryst.* **8**, 157 (1969).
19. Ekwall, P., Mandell, L. and Fontell, K. *J. Colloid Interface Sci.* **31**, 508 (1969).
20. Ekwall, P., Mandell, L. and Fontell, K. *J. Colloid Interface Sci.* **33**, 215 (1970).
21. Ekwall, P., Mandell, L. and Fontell, K. *J. Colloid Interface. Sci.*, in press.
22. Ekwall, P. and Stenius, P. *Chemistry and Colloids, MTP Int. Rev. Sci.* Phys. Chem. Ser. 2, Vol. 7, p. 215, Ed. Kerker, M., Butterworths, London (1975).
23. Fontell, K. *J. Colloid Interface Sci.* **43**, 156 (1973).
24. Fontell, K. in *Liquid Crystals and Plastic Crystal*, Table 4.1, Vol. 2, Chapter 4, Ellis Horwood, Chichester (1974).
25. Fontell, K. *Colloidal Dispersions and Micellar Behaviour*, ACS Symposium Series, No. 9, p. 270 (1975).
26. Fontell, K., Mandell, L. and Ekwall, P. *Acta Chem. Scand.* **22**, 3209 (1968).
27. Fox, C. F. *Biophysical Chemistry*, readings from *Sci. Am.*, p. 91, Eds., Bloomfield, V. A. and Harrington, R. E., Freeman, San Francisco (1975).
28. Goodman, J. F. and Clunie, J. S. in *Liquid Crystals and Plastic Crystals*, Vol. 2, Chapter 1, Ellis Horwood, Chichester (1974).
29. Gray, G. W. and Winsor, P. A. (Eds.) *Liquid Crystals and Plastic Crystals*, Vols. I and II, Ellis Horwood, Chichester (1974).
30. Gulik-Krzywicki, T., Rivas, E. and Luzzati, V. *J. Molec. Biol.* **27**, 303 (1967).
31. Gulik-Krzywicki, T., Tardieu, A. and Luzzati, V. *Molec. Cryst.* **8**, 285 (1969).
32. Hartley, G. S. *Aqueous Solutions of Paraffin Chain Salts*, Paris (1936).
33. Hitchcock, P. B., Mason, R., Thomas, K. M. and Shipley, G. G. *Proc. Natl. Acad. Sci. USA* **71**, 3036 (1974).
34. Hoar, T. P., and Schulman, J. H. *Nature (London)* **152**, 102 (1942).
35. Hui, S. W. *Chem. Phys. Lipids* **16**, 9 (1976).
36. Husson, F., Mustacchi, H. and Luzzati, V. *Acta Cryst.* **13**, 668 (1960).
37. *International Tables for X-ray Crystallography*, Vol. II, Birmingham, 1959.
38. Larsson, K. *Z. Phys. Chem.* (Frankfurt am Main) N. F. **56**, 173 (1967).
39. Lawrence, A. S. C. *Mol. Cryst. Liq. Cryst.* **7**, 1 (1969).
40. Lawrence, A. S. C. *Mol. Cryst. Liq. Cryst.* **7**, Fig. 37, p. 53 (1969).
41. Lawrence, A. S. C. Boffey, B., Bingham, A. and Talbot, K. *Intern. Congress Surface Active Substances*, 4th (1964) Vol. 2, Fig. 16, p. 681, Gordon and Breach (1967).

42. LEE, A. G., BIRDSALL, N. J. H. and METCALFE, J. C. in *Nuclear Magnetic Relaxation and the Biological Membrane, Methods in Membrane Biology* Ed. KORN, E., Plenum Press, New York (1974).
43. LEVINE, Y. K. Ph.D. Thesis, London, 1970.
44. LEVINE, Y. K. in *Progress in Biophysics and Molecular Biology* Eds. BUTLER, J. A. V. and NOBLE, D., Vol. 24, p. 3, Pergamon Press, Oxford (1970).
45. LEVINE, Y. K. and WILKINS, M. H. F. *Nature* (*London*), *New Biol.* **230,** 69 (1971).
46. LINDBLOM, G., WENNERSTRÖM, H., ARVIDSON, G. and LINDMAN, B. *Biophys. J.,* **16,** 1287 (1976).
47. LUTTON, E. S. *J. Am. Oil Chem. Soc.* **42,** 1068 (1965).
48. LUZZATI, V. *Biological Membranes,* p. 1., Ed. CHAPMAN, D., Academic Press, New York (1968).
49. LUZZATI, V., GULIK-KRZYWICKI, T. and TARDIEU, A. *Nature* **218,** 1031 (1968).
50. LUZZATI, V., MUSTACCHI, H. and SKOULIOS, A. *Nature* **180,** 600 (1957).
51. LUZZATI, V., MUSTACCHI, H. and SKOULIOS, A. *Discuss. Faraday Soc.* **25,** 43 (1958).
52. LUZZATI, V., MUSTACCHI, H., SKOULIOS, A. and HUSSON, F. *Acta Cryst.* **13,** 660 (1960).
53. LUZZATI, V. and REISS-HUSSON, F. *Nature* **210,** 1351 (1966).
54. LUZZATI, V. and SPEGT, P. A. *Nature* **215,** 710 (1967).
55. LUZZATI, V. and TARDIEU, A. *Ann. Rev. Phys. Chem.* **25,** 79 (1974).
56. LUZZATI, V. and TARDIEU, A. *Ann. Rev. Phys. Chem.* **25,** Fig. 2, p. 80 (1974).
57. LUZZATI, V., TARDIEU, A. and GULIK-KRZYWICKI, T. *Nature* **217,** 1028 (1968).
58. LUZZATI, V., TARDIEU, A., GULIK-KRZYWICKI, Y., RIVAS, E. and REISS-HUSSON, F. *Nature* **220,** 485 (1968).
59. LUZZATI, V., TARDIEU, A. and TAUPIN, D. *J. Molec. Biol.* **64,** 269 (1969).
60. MANDELL, L. *Finska Kemistsamf. Medd.* **72,** 51 (1963).
61. MANDELL, L. and EKWALL, P. *Acta Polytech. Scand. Chem. Incl. Metall. Ser.* **74,** I, p. 1 (1968).
62. MANDELL, L., FONTELL, K. and EKWALL, P. *Advan. Chem. Ser.* **63,** 89 (1967).
63. MANDELL, L., FONTELL, K., LEHTINEN, H. and EKWALL, P. *Acta Polytech. Scand. Chem. Incl. Metall. Ser.* **74,** I–III (1968).
64. MATEU, L., LUZZATI, V., LONDON, Y., GOULD, R. M., VASSEBERG, F. G. A. and OLIVE, J. *J. Molec. Biol.* **75,** 697 (1973).
65. PERSSON, N. P., FONTELL, K., LINDMAN, B. and TIDDY, G. J. T. *J. Colloid Interface Sci.* **53,** 461 (1975).
66. PRINCE, L. M. *J. Colloid Interface Sci.* **52,** 182 (1975).
67. RAND, R. P., CHAPMAN, D. and LARSSON, K. *Biophys. J.* **15,** 1117 (1975).
68. RANK, J. L., MATEU, L., SADLER, D. M., TARDIEU, A., GULIK-KRZYWICKI, T. and LUZZATI, V. *J. Molec. Biol.* **85,** 249 (1974).
69. REISS-HUSSON, F. *J. Molec. Biol.* **25,** 363 (1967).
70. ROSEVEAR, F. B. *J. Am. Oil Chem. Soc.* **31,** 628 (1954).
71. SCHULMAN, J. H. and FRIEND, J. A. *J. Colloid Sci.* **4,** 497 (1949).
72. SCHULMAN, J. H. and RILEY, D. P. *J. Colloid Sci.* **3,** 383 (1948).
73. SCHULMAN, J. H., STOCKENIUS, W. and PRINCE, L. M. *J. Phys. Chem.* **63,** 167 (1959).
74. SHIPLEY, G. G. in *Biological Membranes,* Vol. 2, p. 1, Eds. CHAPMAN D. and WALLACH, D. F. H., Academic Press, London (1973).
75. SHIPLEY, G. G., GREEN, J. P. and NICHOL, B. W. *Biochim. biophys. Acta* **311,** 531 (1973).
76. SINGER, S. J. and NICHOLSON, S. L. *Sciences* **175,** 720 (1972).
77. SKOULIOS, A. *Adv. Colloid Interface Sci.* **1,** 79 (1967).
78. SMALL, D. M. *J. Lipid Res.* **8,** 551 (1967).
79. SMALL, D. M. *J. Am. Oil Chem. Soc.* **45,** 1 (1968).
80. SMALL, D. M. and BOURGES, M. *Molec. Cryst.* **1,** 541 (1966).
81. SMALL, D. M. and SHIPLEY, G. G. *Science* **185,** 222 (1974).
82. SOLYOM, P. and EKWALL, P. *Rheol. Acta* **8,** 316 (1969).
83. TARDIEU, A. Ph.D. Thesis, Paris-Sud, 1972.
84. TARDIEU, A. and LUZZATI, V. *Biochim. biophys. Acta* **219,** 11 (1970).
85. TARDIEU, A., LUZZATI, V. and REMAN, F. C. *J. Molec. Biol.* **75,** 711 (1973).
86. TIDDY, G. J. T. *J. Chem. Soc., Faraday Trans.* **1,** 68, 369 (1972).
87. ULMIUS, J., WENNERSTRÖM, H., LINDBLOOM, G. and ARVIDSON, G. *Biochim. biophys. Acta* **389,** 192 (1975).
88. VANDENHEUVEL, E. A. *Chem. Phys. Lipids* **2,** 372 (1968).
89. VOLD, R. D., REIVERE, R. and McBAIN, J. W. *J. Am. Chem. Soc.* **63,** 1293 (1941).
90. WINSOR, P. A. *Chem. Rev.* **68,** 1 (1968).
91. WINSOR, P. A. *Molec. Cryst. Liq. Cryst.* **12,** 141 (1971).
92. WINSOR, P. A., *Liquid Crystals and Plastic Crystal* Vol. 1, Chapters 2, 3 and 5. Ellis Horwood, Chichester (1974).
93. WINSOR, P. A. and GRAY, G. W. *Molec. Cryst. Liq. Cryst.* **26,** 305 (1974).

Prog. Chem. Fats other Lipids. Vol. 16, pp. 163–169. Pergamon Press, 1978. Printed in Great Britain

STABILITY OF EMULSIONS FORMED BY POLAR LIPIDS

KÅRE LARSSON

Division of Food Science, Chemical Center, University of Lund, S-220 07 Lund, Sweden

CONTENTS

I. INTRODUCTION   163

II. EXPERIMENTAL   163

III. RESULTS AND DISCUSSION   164
    A. Surface film behavior   164
    B. Emulsion stability and surface film properties of some mixed systems   166
    C. Interfacial structure and emulsion stability   169

ACKNOWLEDGEMENT   169

REFERENCES   169

I. INTRODUCTION

Polar lipids, e.g. phospholipids, monoglycerides and long-chain esters of fruit acids, are important food additives. In these applications, they are usually classified as food emulsifiers although other effects than emulsification often are utilized, such as complex formation with proteins and starch. Emulsions based on polar lipids are also used in the topological administration of drugs. Knowledge of the structure of the lipid film at the oil/water interface is necessary in order to understand the physical properties of these emulsions, such as stability, rheology and diffusion of substances through the interfacial layer. The molecular arrangement of lipids at interfaces will be considered here and a procedure for identification of the structure corresponding to the highest emulsion stability will be demonstrated.

Polar lipids used for formation of emulsions give aqueous phases with water which can be characterized by X-ray diffraction, and they form condensed monolayers at the air/water interface which can be studied by surface balance technique. Boyd and co-workers[2] have studied the rheological properties of surface films, and on this basis, have been able to correlate increasing emulsion stability with increasing surface viscosity and elasticity. The occurrence of multilamellar liquid-crystalline particles above the limit of swelling of monoglycerides and related emulsifiers have been demonstrated,[6] and the importance of such films with regard to emulsion stability has been pointed out by Friberg and Rydhag.[3] In systems corresponding to food emulsions, however, there is often an interfacial film with crystalline hydrocarbon chains.[10] Such a film can be formed by lipid crystals, exposing a hydrophobic surface towards the oil phase and hydrophilic surface groups towards water. This possibility of formation of two alternative surface structures of lipid crystals (with dominating surfaces parallel to the bimolecular unit layers) means that the crystalline state exhibits emulsification properties, a feature which seems to have been neglected previously. The interfacial film can also be formed by lipid bilayers with crystalline hydrocarbon chains alternating with water layers, a structure characterizing the gel state.[6] The hydrocarbon chains in the gel phase possess some degree of disorder, and such a multilamellar interfacial film is therefore more flexible than that formed by true crystals. This type of structure seems to correspond to the highest emulsion stability as will be demonstrated below.

II. EXPERIMENTAL

A continuously recording surface balance of the vertical Wilhelmy type was used for the recordings of surface pressure *vs* molecular area[1] (IIA). The lipids used were dissolved in hexane–ethanol (9:1) and spread by an "Aglo" microsyringe. The nomenclature of surface film phases according to Harkins[4] was used. X-ray diffraction data were obtained as previously described.[6] The food emulsifiers used were kindly supplied by Dr. Krog, AS Grindstedvaerket, Denmark, and the lecithin samples were prepared according to standard methods and checked by thin layer chromatography (TLC). The emulsions were prepared by mixing in a standard shaking equipment (gyrotoric type, New Brunswick) for 20 min. The emulsion stability was expressed as the time required for visible phase separation. The following emulsifiers were studied: L-α-dimyristoyl- and dipalmitoyl lecithin, distilled monoglycerides of fully hydrogenated lard (termed C_{16}/C_{18} monoglycerides in the following text), tetraglycerol monostearate, sodium stearoyl-2-lactylate, 1-monolinolein, sorbitan monolaurate (SPAN 20), and polyoxyethylene ($n = 20$) sorbitan monooleate (TWEEN 80).

III. RESULTS AND DISCUSSION
A. *Surface Film Behavior*

Food emulsifiers form three-dimensional aqueous phases with excess water of either the liquid-crystalline or the *gel* type.[6] The relations between the molecular arrangements in surface films and in three-dimensional phases of such amphiphiles have not been considered earlier, taking our present knowledge on liquid-crystalline structures into account. These relations in the case of two synthetic lecithins, L-α-dimyristoyl lecithin (DML) and L-α-dipalmitoyl lecithin (DPL), will first be discussed. It should be mentioned that numerous surface film studies of lecithins have been published, but the structural relation between three-dimensional phases and monolayer phases have not been analyzed. An isotherm of DML recorded at 14.8°C is shown in Fig. 1. A liquid condensed phase of the L1-type is formed at low pressure with a molecular area of about 58 Å² at highest condensation. At further compression, a condensed phase of the L2-type is formed by a first-order phase transition and it has a collapse point corresponding to an area of about 46 Å² per molecule. The structural relations between surface film phases and three-dimensional phases were discussed in an earlier paper.[8] The lamellar liquid-crystalline phase and the liquid-condensed phase show the same molecular dimensions and similar mobilities, whereas the molecular mobility and the cross-section of the phases L2 and LS indicate that they correspond to the *gel*-phases in three-dimensional systems. In the case of vertical molecules, the surface film is of the LS type existing at rather high pressures, whereas tilted molecules in same chain packing give surface film phases of the L2-type. The high molecular mobility as well as the high compressibility of the tilted L2 phase can be explained by the hexagonal chain packing allowing rotational movement of the molecules. A confirmation of these proposed structural relations between surface film phases and three-dimensional aqueous phases is

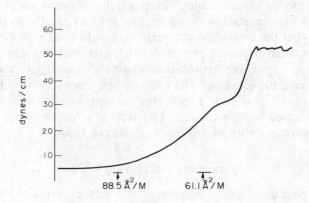

Fig. 1. Pressure–area isotherm of dimyristoyl lecithin at 14.8°C.

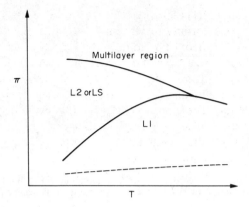

FIG. 2. Schematic illustration of the pressure–temperature phase diagram of a typical food emulsifier. The broken line indicates the lower pressure limit for formation of condensed monolayers.

given by the temperature ranges in which these phases exist. The DML sample used showed a transition *gel* to liquid-crystalline phase (crystalline $\rightleftharpoons$ melted chains) at 22°C. The surface film isotherms recorded at intervals of 3°C showed that the highest temperature at which the L2 phase exist is in the range 20–23°C. The collapse behavior, with the film going over the edges of the surface balance trough, makes an exact determination impossible.

The monolayer phase behavior of DPL has recently been described.[2] Below the chain melting temperature at about 42°C, there are two gel phases in aqueous systems, one with vertical chains (42–34°C), and the other with tilted chains (below 34°C).[12] The same phases seem to occur in the monolayers—an L2 phase which can be observed up to about 34°C and an LS phase up to about 42°C.

A pressure–temperature phase diagram characteristic for most food emulsifiers is shown schematically in Fig. 2. These surface film phase relations were observed in the case of the lecithins described above and in systems of other emulsifiers which will be considered further below. The equilibrium spreading pressures were also determined and were always found to be equal to the pressure at which the L1 phase and L2 or LS phase coexist, and above the critical temperature for existence of the L2 or LS phases the equilibrium spreading pressure is equal to the collapse pressure. One finding of the present work is that emulsions involving the food emulsifiers examined have an oil/water interface of liquid-crystalline type or of *gel* type depending upon whether the temperature is above or below the highest temperature at which the L2 or LS phase can exist.

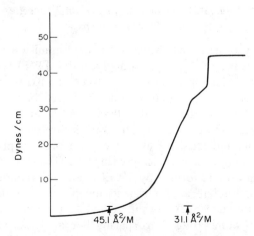

FIG. 3. Pressure–area isotherm of C_{16}/C_{18} monoglycerides at 18°C.

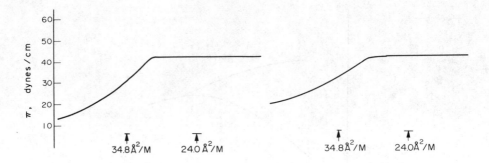

FIG. 4. Pressure–area isotherm recorded at 11.8°C of monolinolein to the left and to the right of the same amount monolinolein mixed with hexadecane in molecular proportions 1:10.

The pressure–area isotherm of a C_{16}/C_{18} monoglyceride sample is shown in Fig. 3. As in the lecithin samples, an L1-phase is first formed, which at about 30 Å^2 per molecule is transformed into a phase which, according to molecular area (about 27 Å^2 per molecule at collapse), must have tilted molecules. The high compressibility shows that this is not an L2-phase but a true solid condensed phase. This can be correlated with the three-dimensional state, where the *gel*-phase formed at cooling is very unstable and is transformed into the β-crystalline form.

Pressure–area isotherms of monolinolein alone and mixed with hexadecane are shown in Fig. 4. It can be seen that the paraffin molecules are solubilized in the L1 phase of monolinolein, and at higher pressure, they are successively squeezed out from the film so that the film consists of monolinolein only at the collapse point. Hexadecane and triolein were examined also in mixtures with the other food emulsifiers and they were always squeezed out from the L2, LS or solid-condensed phases showing that the phases which are more condensed than the L1 phase cannot accommodate the oil molecules in its close-packing. It should also be pointed out that monolinolein forms a cubic liquid-crystalline phase in excess of water[7]—a structure with no counterpart in a monomolecular surface film. Another difference between surface films and three-dimensional phases is observed when the hydrocarbon chains are so short so that no condensed surface films are formed. 1-Monolaurin, for example, forms a lamellar liquid crystal with water above 43°C, but no condensed monomolecular films can be obtained. By the addition of sodium chloride, however, the stability of the monolayer is increased, so that area–pressure isotherms can be recorded.[11] It has also been observed that the transition temperature between the lamellar liquid crystal and crystals dispersed in water is increased by the addition of sodium chloride to the binary system 1-monolaurin–water.[5]

B. Emulsion Stability and Surface Film Properties of some Mixed Systems

The emulsion stability of a triolein–water 1:10 (*w/w*) emulsion with 1% (*w/w*) of an emulsifier mixture consisting of C_{16}/C_{18} monoglycerides–TWEEN 80 is illustrated in Fig. 5. The surface film behavior of this monoglyceride mixture is shown in Fig. 4. TWEEN 80 does not form a condensed monolayer. It can be seen that the emulsion stability is much higher at room temperature than at 60°C, and furthermore, the most stable emulsions are obtained at a weight ratio TWEEN/monoglycerides of about 1:1. Microscopic examinations indicated the existence of an ordered phase at the oil/water interface, detectable by the occurrence of birefringence. The emulsions formed by the 1:1 emulsifier mixture and those formed by TWEEN and the monoglycerides separately were centrifuged (2000 *g* for half an hour), and the phase separating between the oil and water layers was examined by X-ray diffraction. It was found that the 1:1 mixture gives a *gel*-phase with a dominating short-spacing line at 4.2 Å, whereas the emulsion with monoglycerides alone gives a β-crystal form (dominating short-spacing at 4.6 Å)

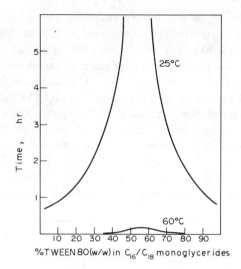

FIG. 5. Emulsion stability according to the time for separation in a system consisting of 10% (w/w) triolein, 89% water and 1% of an emulsifier C_{16}/C_{18} monoglycerides–TWEEN 80 in various proportions. The stability *vs* the weight ratios of the two emulsifiers are shown at 60°C (upper curve) and at 25°C (lower curve).

and that with TWEEN alone gives a liquid-crystalline phase (liquid chains as evident from the 4.5 Å halo). The long-spacing of the *gel*-phase could not be seen on the X-ray photograph, indicating that it was larger than about 300 Å, which was the resolution of the low-angle Luzzati camera used. The first order of such lamellar phases are usually very strong whereas higher orders are quite weak and sometimes missing. This indicates that the presence of the TWEEN molecules, although nonionic, results in swelling of the monoglycerides above the usual limit of about 20 Å in water layer thickness (c.f. Ref. 9).

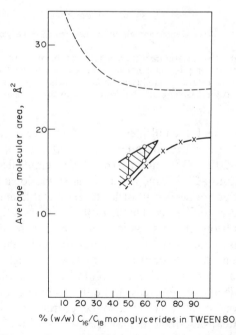

FIG. 6. Surface film behavior as evident from the average molecular areas at monolayer phase transition or collapse *vs* composition for mixtures of C_{16}/C_{18} monoglycerides–TWEEN 80. The broken line shows the molecular areas of the liquid-condensed phase L1 at highest condensation, and the solid line curve shows the collapse of the solid-condensed phase. The shaded region corresponds to the existence of an L2 phase.

The surface film behavior of this binary emulsifier system is shown in Fig. 6. The characteristic molecular areas at phase transitions within the monolayer are given as a function of the composition. The isotherms correspond to the air/water interface, and it was checked that the presence of hexadecane in excess did not change the transition points. In the composition range corresponding to the highest emulsion stability, an L2 phase is formed according to compressibility and the high molecular mobility. It was not possible to obtain good dimensions for this phase at high amounts of TWEEN due to limitations in the compression range of our surface balance. There is obviously complete mutual solubility in the L1-phase, whereas the solid-condensed monoglyceride phase can dissolve very small amounts of TWEEN. The surface film data and the X-ray diffraction suggest independently that the high emulsion stability in the case of the 1:1 mixture is due to the occurrence of a *gel*-phase forming the oil/water interface.

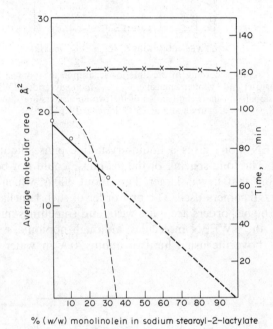

% (w/w) monolinolein in sodium stearoyl-2-lactylate

FIG. 7. Emulsion stability and surface film behavior of binary emulsion mixtures between sodium-stearoyl-2-lactylate and 1-monolinolein examined as described in Figs. 5 and 6. The broken lines illustrate the emulsion stability at 25°C, whereas the upper solid line corresponds to the L1-phase and the lower one to the L2-phase.

The critical amount of the 1:1 emulsifier required to give a stable emulsion was also tested. A mixture of 20% (w/w) triolein in emulsifier/water was kept homogeneous during cooling from 65°C to 20°C. Only about 0.6% (w/w) emulsifier is needed in order to get a stable emulsion, and at 0.3% the emulsion is stable for a few hours. Much larger amounts of the emulsifier are needed in order to get a stable emulsion above the transition temperature liquid-crystal $\geq$ *gel*. This is probably due to the lateral van der Waals interaction in the emulsion bilayers, which must be much higher due to the closer packing in the *gel*-phase compared to the liquid-crystalline phase.

Another binary system of food emulsifiers is shown in Fig. 7. The surface film data indicate that there is almost no molecular solubility of monolinolein in the *gel*-phase of sodium stearoyl-2-lactylate, whereas there is complete mutual solubility in the L1-phase. The emulsion stability could also be correlated with the existence of the *gel*-phase as in the system described above.

An emulsion system consisting of tetraglycerol monostearate as one emulsifier and SPAN 20 as the other was also examined in the same way. SPAN 20 gives a lamellar liquid-crystalline phase in water and liposomes in excess of water just like the monoglycerides. The only surface film phase it forms is the L1-phase. Emulsions with limited stability could only be obtained at small amounts of SPAN (up to about 2:8 in weight ratio). An L2 surface film phase was found also in this system in the case of the highest emulsion stability. Although there are indications of an interfacial film with the structure of the *gel*-phase in these emulsions, the emulsion stability is much smaller than observed in the system TWEEN 80–C_{16}/C_{18} monoglycerides. This is believed to be related to the viscoelastic properties of the interfacial film. The *gel*-phase formed in the tetraglycerol monostearate–SPAN system was not found to give swelling above the water layer thickness of about 20 Å, which is the usual limit in nonionic systems. The existence of a *gel*-phase with very thick water layers between the bilayer units should be expected to give more flexible interfacial films than those with low water content (cf. Ref. 9).

C. *Interfacial Structure and Emulsion Stability*

As suggested in a previous work[10] and confirmed by the results presented here, the state of the hydrocarbon chains is an important factor in emulsions. In the case of ordered chains, the hexagonal chain packing seems to give the highest emulsion stability according to the model systems examined here. One reason could be that it is disordered enough to allow deviations from a planar layer arrangement. If an interfacial film with the *gel* type of structure is compared with a film of the lamellar liquid-crystalline type, it is obvious that the molecular association in the plane of the film is stronger due to the crystalline close-packing in the bilayers, and the higher emulsion stability obtained by such films is certainly related to this. The liquid structure of the chains in the multilamellar liquid crystalline particles (or films) on the other hand can often solubilize a large proportion of nonpolar lipid molecules, like fats and oils, and this structure can therefore exhibit a higher emulsion *capacity* than those with crystalline chains. It is thus useful to know the phase transitions of the interfacial film, particularly the liquid crystal → gel transition. A simple technique to obtain this information is to use the surface balance.

Acknowledgment—This work was made possible by the inspiration and support from the late Professors Stina and Einar Stenhagen.

REFERENCES

1. ANDERSSON, H. J. I., STÄLLBERG-STENHAGEN, S. and STENHAGEN, E. *The Svedberg*, Almqvist-Wiksell, Stockholm (1944).
2. BOYD, J., PARKINSON, C. and SHERMAN, P. *J. Coll. Interface Sci.* **41**, 359 (1972).
3. FRIBERG, S. and RYDHAG, L. *Kolloid Z. Z. Polym.* **244**, 233 (1971).
4. HARKINS, W. D. *The Physical Chemistry of Films*, 2nd edn., Reinhold Publishing Corp., New York (1954).
5. KROG, N. and LARSSON, K. *Chem. Phys. Lipids* **2**, 129 (1968).
6. LARSSON, K. *Z. Phys. Chem.* (*Neue Folge*) **56**, 173 (1967).
7. LARSSON, K. *Chem. Phys. Lipids* **9**, 181 (1972).
8. LARSSON, K. *Surface and Colloid Science*, Vol. 6, p. 261, Ed. E. MATIJEVIC, Wiley, New York (1973).
9. LARSSON, K. and KROG, N. *Chem. Phys. Lipids* **10**, 177 (1973).
10. LARSSON, K. *Chem. Phys. Lipids* **14**, 233 (1975).
11. MERKER, D. R. and DAUBERT, B. F. *J. Am. Chem. Soc.* **80**, 516 (1958).
12. RAND, R. P., CHAPMAN, D. and LARSSON, K. *Biophys. J.* **15**, 1117 (1975).

Prog. Chem. Fats other Lipids. Vol. 16, pp. 171–177. Pergamon Press, 1978. Printed in Great Britain

DERMATOPHYTE LIPIDS

JAN VINCENT

*Department of Structural Chemistry, Faculty of Medicine, University of Göteborg,
Göteborg, Sweden*

CONTENTS

I. INTRODUCTION	171
II. TOTAL LIPIDS IN DERMATOPHYTES	171
III. FATTY ACIDS	173
IV. STEROLS	175
V. PHOSPHOLIPIDS	176
ACKNOWLEDGEMENTS	176
REFERENCES	176

I. INTRODUCTION

This review concerns the distribution of dermatophyte lipids, their possible physiological role and their importance in pathogenesis of dermatophytosis. The text is divided into two parts: the first part covers the increasing amount of information regarding the total lipid composition of dermatophytes with special reference to the various factors influencing the lipid content; the second part covers the three major lipid classes occurring in dermatophytes, i.e. fatty acids, sterols and phospholipids. Although lipids were among the first natural products investigated,[10] they were neglected until the last decades when sophisticated analytical techniques became available. This has, of course, also stimulated more extensive investigations on dermatophyte lipids.

Dermatophytes are a highly specialized group of filamentous fungi, causing lesions of the skin and its appendages in man and animals. Dermatophytes, according to their imperfect forms, include species of three genera: *Epidermophyton*, *Microsporum* and *Trichophyton* (Fungi imperfecti). All known perfect forms belong to genera *Arthroderma* and *Nannizzia* (Ascomycetes).[27] Beside the chemotaxonomical implications, this review may provide useful information for further investigations concerning the role that dermatophyte lipids play in the inflammatory skin reactions in dermatophytosis.

II. TOTAL LIPIDS IN DERMATOPHYTES

Even before dermatophyte lipids were known, their lipolytic activity was described.[8,28] The use of classical staining methods first demonstrated the presence of lipids in dermatophytes.[1,9] Later, mycelial extractions were carried out for evaluation of total lipids in order to elucidate their relationship in the pathogenesis of dermatophytosis.[2,3,17,21,22,36,41,44,51] The total lipid abundances in dermatophytes are summarized in Table 1.

Regarding the relatively large range of abundances, it is apparent from the table that consideration must be given to growth conditions, age and the analytical techniques.

Dermatophytes demonstrate a relatively low growth activity when compared with the majority of saprophytic molds.[40] For example, the mycelia of *E. floccosum* contain total lipids as a function of age, in the range 7.0–19.8% of the dry weight, which is also true for most dermatophytes studied. The total lipids of *E. floccosum* significantly increase with age, except during the early growth period when there is a decrease, mainly at the expense of the triglycerides.[41] Thus, the energy requirements of the developing mycelia would appear to be met, at least in part, by fatty acid oxidation. The

TABLE 1. Total Lipids of Dermatophytes

Genus	Species	Total lipid (% dry wt.)	Reference
Epidermophyton	floccosum	7.0–19.8[a]	41
	floccosum	14.77[b]	3
Microsporum	audouinii	14.63[b]	3
	canis	10.81[b]	2
	canis	26.77[b]	3
	gypseum	13.57[b]	3
Trichophyton	ajelloi*	15.5[c]	22
	ajelloi (mutant)	20.5[c]	22
	gallinae	12.98[b]	3
	mégninii**	8.7[c]	51
	mentagrophytes	15.3[d]	36
	mentagrophytes	18.13[b]	3
	mentagrophytes	8.2[e]	17
	mentagrophytes***	18.8[c]	51
	rubrum	18.49[b]	3
	rubrum	11.2[e]	17
	rubrum	7.7[c]	44
	schönleinii	19.48[b]	3
	schönleinii	8.3[e]	17
	tonsurans	10.85[b]	3
	verrucosum****	43.7[c]	51
	violaceum	10.10[b]	3
	violaceum	16.5[c]	51

Reported as: *Arthroderma uncinatum (stat. ascosp.); **T. rosaceum; ***T. gypseum, asteroides; ****T. faviforme, ochraceum. Growth conditions: [a]Sabouraud dextrose broth, 30°C; [b]asparagine liquid medium, 20°C; [c]Sabouraud dextrose broth, 20°C; [d]dextrose (1%) and peptone (1%) broth, 20°C; [e]Grütz-III-medium, 20°C.

lipid accumulation in older mycelia as a store of energy is confirmed also by cytological studies.[47] The decrease of total lipids in E. floccosum during germination and in the early phase of growth may be a consequence of their utilization as a prime carbon source and substrate for polysaccharide synthesis.[12]

Environmental conditions such as substrate, temperature, agitation, etc. have been recognized to have a particular influence on the total lipid content of dermatophytes. Lipid production of dermatophytes, similar to the majority of filamentous fungi, depends on the nature of the carbon source. In this connection, the richest carbohydrates are, in descending order, glucose, sucrose and fructose. Also amino acids, present in certain concentrations, such as alanine or phenylalanine, can affect the total lipid content.[39] It is generally agreed that a relatively high carbon–nitrogen ratio in the media is required in order to attain maximum lipid production.

In common with most living organisms, dermatophytes show a tendency to enhance lipid content with increasing environmental temperature (unpublished data). Thus, investigations should be carried out on the species grown under optimal temperatures. Moreover, some microorganisms subcultured at low temperatures show a relatively high degree of unsaturation.[13] This alters some physical properties of the lipids, which can be of importance for the survival of the dermatophytes in unfavorable environments.

Aeration of culture has also a pronounced effect on the general metabolism of fungi and their lipid composition. The degree of aeration required depends on the technique applied, e.g. semisolid media, shake or still cultures. Thus, significant differences in the amount of total lipids in still and shake cultures have been reported.[3]

Early stages of germination are characterized by a dependency on the endogenous substrate. It has been suggested[12] that the macroconidia of M. gypseum are capable of immediate response to a favorable environment and therefore constitute an ideal means for the transmittance of infection. The longevity of the macroconidia may depend more on a high concentration of essential materials in the spore rather than any well

Table 2. Dermatophyte Cell Wall Lipids

Genus	Species	Cell wall lipid (% dry wt.)	Reference
Epidermophyton	floccosum	3.3	37
	floccosum	4.6	33
Microsporum	canis	4.0	37
	gypseum	3.6	37
Trichophyton	mentagrophytes	3.1	37
	mentagrophytes	6.6	32

developed dormancy mechanisms. This confirms the changes of dermatophyte *M. quinceanum* during starvation which indicates that the carbohydrates do not form the principal endogenous reserve. In this case, lipids may represent a primary reserve material because they decrease rapidly from 23% in the spores to 13% in a 24 hr old mycelium. The same picture was true during starvation.[40]

Cell walls from filamentous fungi, particularly dermatophytes, have been extensively studied.[32,33,37] However, relatively little is known about their lipid composition. Lipid abundancies of dermatophyte cell walls are given in Table 2. Shah and Knight[37] showed that these lipids may be of importance in conferring rigidity or protection against drying, since morphological distortion was observed after solvent extraction.

The function of dermatophyte lipids is as yet not well understood. Historically, lipids have been considered as reserve material which may be converted to energy and carbon skeletons during growth and reproduction. This is true for those lipids which accumulate in cytoplasma that are composed primarily of triglycerides. However, as we know more about the structure and biosynthesis of less abundant lipids, it is apparent that they have more specific roles to play. For example, lipids are essential components in the membrane structures when they may serve as protective coatings or may be involved in the immunological reactions in dermatophytosis. According to Peck,[35] lipids of pathogenic fungi have a significant role in the mechanism of infectious disease. Moreover, interest in dermatophyte metabolites leads to the detection of specific antigenic components extracted from their mycelia. Lipid fractions, especially fatty acids, have been reported to possess antigenic activity.[4,18]

III. FATTY ACIDS

Fungal fatty acids and their biosynthesis have been extensively reviewed.[20,38] Generally, dermatophyte fatty acids consist of a homologous series of saturated and unsaturated aliphatic acids ranging from C_6 to C_{22} with a predominance of C_{16} and C_{18} acids. These acids can account for over 80% of the total fatty acid content. The ethylenic bonds in the unsaturated acids are in the *cis* configuration and, where more than one ethylenic bond is present, they occur normally in methylene-interrupted sequence.

Fatty acid distribution in some species of dermatophytes from the genus *Trichophyton*[5,23,29–31,42,48–51] and *Microsporum*[43] has been determined mainly by paper chromatography. The fatty acid composition of the remaining monotype genus *Epidermophyton* was established by combined gas chromatography–mass spectrometry in this laboratory.[41]

The distribution of fatty acids encountered in dermatophytes is given in Table 3. For instance, fatty acids of *E. floccosum* decrease significantly with age, which may be due partly to their conversion to glycerides and partly to their complete degradation. However, the composition of the predominating fatty acids was unaffected by age.[41]

The narrow range of dermatophyte fatty acids confirms that they compose a homogenous group of filamentous fungi. Recently, the enzymatic system responsible for fatty acid synthesis in dermatophyte was investigated.[24] It has been postulated that an equilibrium between malonyl CoA (the intermediate in *de novo* synthesis) and acetyl CoA

TABLE 3. Fatty Acids of Dermatophytes[41]

Genus	Species	Analyzed with	6	8	9	10	11	12	13	13:1	14	14:1	15	16	16:1	17	17:1	18	18:1	18:2	19	20	20:1	20:2	21	22
Trichophyton																										
mentagrophytes	MERKEL (29)[1]	PC						×						×				×	×							
	ZAMIECH (51)[2]	PC												×				×	×							
schönleinii	AUDETTE (5)	GC		0.1				0.1			0.3		0.9	23.8	0.2	1.0		11.2	17.0	45.2						
	MERKEL (30)[3]																	×	×							
tonsurans	ZAMIECH. (48)[4]													×				×	×							
	ZAMIECH. (49)[5]	PC												×				×	×							
quincheanum	MERKEL (31)[6]													×				×	×							
mégninii	ZAMIECH. (50)[7]							×						×				×	×							
verrucosum	ZAMIECH. (51)[8]							×										×	×							
violaceum	ZAMIECH. (51)[9]							×						×				×	×							
rubrum	ZAMIECH. (51)																	×	×							
	WIRTH (42)	GC		1.1	0.5	0.7	0.2	tr	tr		0.8	0.8	2.1	23.8				7.4	13.1	52.4		1.1			tr	tr
	KOSTIW (23)				0.7			0.1	tr	1.0	0.8	0.1	0.5	17.6	2.6	0.8	0.7	16.1	19.2	35.6			0.3	0.2		0.8
Microsporum gypseum	WIRTH (43)	GC		tr	tr						2.7			17.2				7.7	10.4	64.7						
Epidermophyton floccosum	VINCENT (41)	GC-MS		0.1	0.3	0.3	0.2	1.7	0.3		1.4	0.2		18.9	2.3	0.7	tr	9.6	16.2	42.8	0.9	1.1	0.5	0.1		

reported as: [1] *T. gypseum*; [2] *T. gypseum*, asteroides; [3] *Achorion*; [4] *T. plicatile*; [5] *T. sulphureum*; [6] *Achorion*; [7] *T. rosaceum*; [8] *T. rosaceum*; [9] *T. faviforme*, ochraceum.

(the intermediate in chain elongation) may provide the site for the mechanism controlling these two processes.

The taxonomical value of fungal fatty acids is limited.[38] However, as *Phycomycetes* are characterized by a predominant occurrence of γ-linolenic acid, its absence in dermatophytes is of taxonomical significance, which thus confirms the relevance of dermatophytes to *Ascomycetes*. Moreover, the fatty acid composition may be of importance in membrane biological activity[6] and may be involved in the allergic inflammatory delayed skin reactions in dermatophytosis.[4] The Landstainer–Draise test of these compounds confirms that the middle chain (C_{10}–C_{12}) fatty acids are able to sensitize the mammalian skin and elicit the delayed inflammatory skin reactions.[18]

IV. STEROLS

The presence of ergosterol in dermatophytes, such as *T. mentagrophytes*,[5] *T. mentagrophytes* (asteroides)[34] and *T. rubrum*,[45] was primarily reported. Brassicasterol and a trace of squalene have been isolated from *T. rubrum*.[44] The occurrence of squalene in trace quantities in the mycelia of *T. rubrum* is consistent with its known position in the biosynthetic pathway between acetate and sterols, and with the fact that squalene does not normally accumulate, to a significant extent, under aerobic conditions.

Subsequently, Blank *et al.*[7] identified two sterols, brassicasterol and ergosterol, from different species of dermatophytes. Ergosterol was found to be more widely distributed than brassicasterol. These sterols occurred together in some species but the second was only present in very small amounts. In a number of species, ergosterol was present to the exclusion of the other. The distribution of these two sterols among the dermatophytes is rather interesting from the taxonomic point of view (Table 4).

Although there are several reports on the subject, the exact function of sterols in fungi is uncertain.[19] The sterol content of *E. floccosum* reaches a maximum during the early growth period,[41] but its relationship with an eventual germination control or cell permeability remains open for investigation.[16] Summarizing, sterols are required for growth and reproduction as well as being implicated in membrane structures and permeability.

Relationships have been established between the sterols of fungal membranes and the action of polyene antifungal antibiotics.[15,46] For instance, the effect of nystatin is probably due to its insertion among the sterols of the lipid bilayer, leading to molecular rearrangements, loss in rigidity and mechanical failure of the membrane.[26]

TABLE 4. Dermatophyte Sterols[7]

Genus	Species	Sterols (% dry wt.) Brassica-sterol	Ergosterol
Epidermophyton	*floccosum*	—	0.083
Microsporum	*audouinii*	—	0.150
	canis	—	0.016
	gypseum	—	0.026
	quinckeanum	—	0.270
Trichophyton	*ferrugineum*	—	0.250
	mégninii	0.039	—
	*mentagrophytes**	—	0.130
	*mentagrophytes***	—	1.50
	*mentagrophytes****	—	0.280
	rubrum	—	0.056
	tonsurans	—	0.126
	*verrucosum*****	0.010	—
	violaceum	0.015	—

Reported as: **T. granulosum*; ***T. persicolor*; ****T. interdigitale*, *****T. discoides*.

TABLE 5. Phospholipids in Dermatophytes

Genus	Species	Phospholipids (% dry wt.)	Reference
Epidermophyton	floccosum	6.96	3
Microsporum	audouinii	5.02	3
	canis	0.048	2
	canis	4.48	3
	gypseum	4.04	3
Trichophyton	ajelloi*	2.6	22
	ajelloi (mutant)	5	22
	gallinae	4.97	3
	mentagrophytes	3.13	3
	mentagrophytes	0.3	36
	rubrum	4.87	3
	schönleinii	3.59	3
	tonsurans	1.33	3
	violaceum	2.58	3

Reported as: *Arthroderma uncinatum (stat. ascosp.)

V. PHOSPHOLIPIDS

Phospholipids are considered to be involved mainly in membrane functions. Their composition in dermatophytes (Table 5) have been, until recently, rather poorly investigated.[2,3,36] Ghoshal[14] found 41% phosphatidylcholine and some sphingomyelin in *T. rubrum*, but the presence of other phospholipids could not be confirmed. However, in *M. cookei*, five major components as phosphatidylethanolamine, phosphatidylserine, phosphatidylcholine, and phosphatidylinositol were identified.[25]

The phospholipid composition of a mutant strain of the dermatophyte *Arthroderma uncinatum* (stat. ascosp. of *T. ajelloi*) was compared with that of the wild type.[22] Both strains contained phosphatidylcholine, phosphatidylethanolamine, phosphatidylserine, diphosphatidylglycerol, phosphatidylinositol, and phosphatidic acid, but important differences in the amounts of phosphatidylcholine, phosphatidylserine, and phosphatidic acid were observed. In both strains, the predominant fatty acid was 18:2 with 54.0% in the wild type and 46.7% in the mutant. Qualitatively, the same fatty acids, with the exception of the C_{20} acids, were found in all phospholipid classes.

T. rubrum contains phosphatidylinositol, polyphosphatidylinocitol, phosphatidylcholine, phosphatidylserine, phosphatidylethanolamine, phosphatidylglycerol, and phosphatidic acid.[11] The relative proportions of these components in total phospholipid fraction remain unchanged with age.

Acknowledgements—I would like to express my gratitude to Professor Sixten Abrahamsson, Associate Professor Ng. Dinh-Nguyen and Assistant Professor Lars Hellgren for their valuable advice, and to Mr Weston Pimlott for reviewing the text. Special thanks go also to Mrs Greta Sonnhagen for her contributions in the preparation of the entire manuscript. The study was supported by the Swedish Medical Research Council.

REFERENCES

1. AKASAKA, T. *Jpn. Dermatol.* **63**, 477 (1953).
2. AL-DOORY, Y. *J. Bacteriol.* **80**, 565 (1960).
3. AL-DOORY, Y. and LARSH, H. W. *Appl. Microbiol.* **10**, 492 (1962).
4. ANDERSSON, B. Å., HELLGREN, L. and VINCENT, J. *Sabouraudia* **14**, 237 (1976).
5. AUDETTE, R. C. S., BAXTER, R. M. and WALKER, G. C. *Can. J. Microbiol.* **7**, 282 (1961).
6. BERTOLI, E., CHAPMAN, D., GRIFFITHS, D. E. and STARCH, S. J. *Biochem. Soc. Trans.* **2**, 964 (1974).
7. BLANK, F., SHORTLAND, F. E. and JUST, G. *J. Invest. Dermatol.* **39**, 91 (1962).
8. BÖHME, H. *Mycopathologia* **46**, 221 (1972).
9. BURDON, K. L. *J. Bacteriol.* **52**, 665 (1946).
10. CHEVREUL, M. E. *Recherches chimiques sur les corps gras d'origine animale*, Levrault, Paris, 1823.
11. DAS, S. K. and BANERJEE, A. B. *Sabouraudia* **12**, 281 (1974).
12. DILL, B. C., LEIGHTON, T. J. and STOCK, J. J. *Appl. Microbiol.* **24**, 977 (1972).
13. GAUGHRAN, E. R. L. *Bacteriol. Rev.* **11**, 189 (1947).
14. GHOSHAL, J. *Sci. Cult.* **35**, 694 (1969).

15. GOTTLIEB, D., CARTER, H. E., SLONEKER, J. H. and AMMANN, A. *Science* **128**, 361 (1958).
16. GOTTLIEB, D. and VAN ETTEN, J. Z. *J. Bacteriol.* **91**, 161 (1966).
17. GOTZ, H. and PASCHER, G. *Dermatologica* **124**, 31 (1962).
18. HELLGREN, L. and VINCENT, J. *Sabouraudia* **14**, 243 (1976).
19. HENDRIX, J. W. *Ann. Rev. Phytopathol.* **8**, 111 (1970).
20. HILDITCH, T. P. and WILLIAMS, P. N. *The Chemical Constitution of Natural Fats*, 4th Ed., Chapman Hall, London (1964).
21. ITO, Y. and FUJII, T. *Inst. Lombardo* (*Rend. Sc.*) *B* **92**, 313 (1958).
22. KISH, Z. and JACK, R. C. *Lipids* **9**, 264 (1974).
23. KOSTIW, L. L., VICHER, E. E. and LYON, I. *Mycopathologia* **29**, 145 (1966).
24. KOSTIW, L. L., VICHER, E. E. and LYON, I. *Mycopathologia* **49**, 67 (1973).
25. KUDRYK, B. Ph.D. Thesis, St. John's University, New York, 1970.
26. LAMPEN, J. O., ARNOW, P. M., BOROWSKA, Z. and LASKIN, A. I. *J. Bacteriol.* **84**, 1152 (1962).
27. LOEFFLER, W. *Zbl. Bakt. Hyg. I. Abt. Orig. A* **220**, 247 (1972).
28. MALLINCKRODT-HAUPT, A. V. *Zbl. Bakt. Hyg. I. Abt. Orig. A* **103**, 73 (1927).
29. MERKEL, M. *Bull. Acad. Polon. Sci. Cl. 2* **5**, 341 (1957).
30. MERKEL, M. *Bull. Acad. Polon. Sci. Cl. 2* **6**, 417 (1958).
31. MERKEL, M. *Bull. Acad. Polon. Sci. Cl. 2* **7**, 501 (1959).
32. NOGUCHI, T., KITAZIMA, Y., NOZAWA, Y. and ITO, Y. *Arch. Biochem. Biophys.* **146**, 506 (1971).
33. NOZAWA, Y., KITAJIMA, Y. and ITO, Y. *Biochim. biophys. Acta* **307**, 92 (1973).
34. OKAZAKI, K. and TAMEMASA, O. *J. Pharm. Soc. Japan* **75**, 1087 (1955).
35. PECK, R. L. In *Biology of Pathogenic Fungi*, Chronica Botanica, Waltham, Mass. (1947).
36. PRINCE, H. N. *J. Bacteriol.* **79**, 154 (1960).
37. SHAH, V. K. and KNIGHT, S. G. *Arch. Biochem. Biophys.* **127**, 229 (1968).
38. SHAW, R. *Adv. Lipid Res.* **4**, 107 (1966).
39. SINGH, B., GUPTA, K. G. and BHATNAGAR, L. *Indian J. Exp. Biol.* **11**, 555 (1973).
40. SWANSON, R. and STOCK, J. J. *Appl. Microbiol.* **14**, 438 (1966).
41. VINCENT, J. *Zbl. Bakt. Hyg. I. Abt. Orig. A.* **233**, 410 (1975).
42. WIRTH, J. C. and ANAND, S. R. *Can. J. Microbiol.* **10**, 23 (1964).
43. WIRTH, J. C., ANAND, S. R. and KISH, Z. L. *Can. J. Microbiol.* **10**, 811 (1964).
44. WIRTH, J. C., BEESLEY, T. and MILLER, W. *J. Invest. Dermatol.* **37**, 153 (1961).
45. WIRTH, J. C., O'BRIEN, P. J., SCHMITT, F. L. and SOHLER, A. *J. Invest. Dermatol.* **29**, 47 (1957).
46. WOODS, R. A. *J. Bacteriol.* **108**, 69 (1971).
47. ZALOKAR, M. In *The Fungi*, Ed. G. C. AINSWORTH and A. F. SUSSMAN, Vol. 1, Academic Press, New York and London (1965).
48. ZAMIECHOWSKA-MIAZGOWA, J. *Bull. Acad. Polon. Sci. Cl. 2* **6**, 423 (1958).
49. ZAMIECHOWSKA-MIAZGOWA, J. *Bull. Acad. Polon. Sci. Cl. 2* **7**, 445 (1959).
50. ZAMIECHOWSKA-MIAZGOWA, J. *Bull. Acad. Polon. Sci. Cl. 2* **10**, 3 (1962).
51. ZAMIECHOWSKA-MIAZGOWA, J. *Bull. Acad. Polon. Sci. Cl. 2* **12**, 67 (1964).

Prog. Chem. Fats other Lipids. Vol. 16, pp. 179–193. Pergamon Press, 1978. Printed in Great Britain

OPTICALLY ACTIVE HIGHER ALIPHATIC HYDROXY COMPOUNDS SYNTHESIZED FROM CHIRAL PRECURSORS BY CHAIN EXTENSION

K. Serck-Hanssen

Institute of Physiological Botany, University of Uppsala, Sweden

CONTENTS

I. Introduction   179
II. Nomenclature for Chiral Compounds   179
III. List of Products and Main Precursors   180
IV. Comments   190
Acknowledgements   191
References   191

I. INTRODUCTION

The most general method available for synthesis of chiral long-chain compounds is based on chain extension of completely resolved precursors. A large number of optically pure compounds may thus be prepared from a small number of chiral key molecules, and resolution steps, uncertain and often laborious, are few, or may be avoided if an optically active natural product is used as starting material.[86]

Pioneering and extensive work in the field was carried out by the Stenhagens from the middle of the 1940s. They started by synthesizing branched-chain fatty acids,[1,73] and began experiments with hydroxy compounds in 1950. At that time, it had been found convenient to use the anodic Kolbe reaction for chain extension of optically active carboxylic acids with branching beyond the α-position,[1,52,75] and they then tried 3-hydroxybutanoic acid in a Kolbe synthesis.[68] When difficulties arose because of skin formation at the electrodes, the acetate was tried instead, and found to react normally.[68] They initiated its resolution by means of quinine, and then asked me to continue the work. At the same time, they proposed that a key molecule allowing chain extension in both directions should be synthesized, and suggested an acetylated monoester of 3-hydroxyglutaric acid.[68]

This review briefly outlines the preparation of twelve key molecules (**1–12**, enantiomers included) and lists some fifty long-chain hydroxy compounds synthesized from them. A few new compounds (**4d, 11, 11a**) are included, and previously unpublished physical data have been added for some known compounds (**6a, 7a, b, d**). Prominent features and compounds of special interest are commented on.

Other reviews do not overlap with the present one to any great extent. It has most in common with a survey of optically active long-chain natural compounds by C. R. Smith,[70] published in 1970, and with a chapter on use of the Kolbe electrosynthesis for assignment of absolute configuration by J. H. Brewster,[10a] published in 1972. Information from readers concerning errors and new or overlooked compounds and publications belonging to the field under review will be appreciated.

II. NOMENCLATURE FOR CHIRAL COMPOUNDS

The two main systems, D/L and *R/S*, used to specify absolute configuration are partly complementary, and in this review, an attempt has been made to combine them. The essence of the former is that a substituent is labeled D or L according to whether it lies to the right or to the left in a vertical Fischer projection formula[4,42] (horizontal bonds above the plane of the paper, and vertical bonds below) with carbon atom no. 1 at the top.[46,47,74] It is essential that the name of the compound leaves no doubt about which atom is numbered 1, and since this has not always been clearly understood, the point is illustrated by the following example.

The terms methyl hydrogen or monomethyl 3-acetoxyglutarate do not indicate which is the top carbon atom, and therefore the enantiomers (6, 7) have been renamed methyl 3-acetoxy-4-carboxybutanoate,[63,69] a little awkward designation, but clearly telling that the methyl ester group is at the top. It is possible, however, to use the first name if it is agreed that methyl hydrogen (Me H) indicates that COOMe is uppermost, and that hydrogen methyl (H Me) is used when COOH takes that position. This convention is followed here.

Such problems are avoided if the R/S system is used, because its essential part, the sequence rule,[16] is independent of name and projection formula.[4,17] But the R/S system is usually less simple to apply than the D/L system, and sometimes obscures relationships[4,16,17,46] because the configurational label may change as the result of a structural change away from the chiral center. This may be exemplified by the reduction of R-heptadecanediol-1,3D (6e) to S-heptadecanol-3D (6f). Sterically related compounds can therefore not generally be grouped as R or S.

III. LIST OF PRODUCTS AND MAIN PRECURSORS

Formulas to the left are Fischer projections of key precursors. To their right are physical data and summaries of methods of preparation. They are followed by names and formulas (Fischer projections turned clockwise 90°) of hydroxy compounds or derivatives prepared from the key molecule. Only compounds with ten or more carbon atoms are included, and they are arranged in order of increasing chain-length. When not specified, chain lengthenings were effected by anodic Kolbe synthesis,[10a,72,78,79,84,87] and optical rotations were measured in chloroform solutions with yellow sodium light (D-lines, $\lambda = 589$ nm). Melting-points recorded in the Stenhagen laboratories are corrected.

Racemic 3-acetoxybutanoic acid has been prepared from 3-hydroxybutanoic acid (now commercially available) by acetylation with acetic anhydride in pyridine.[57,68] The yield was only about 50%, probably mainly due to anhydride formation.[80,81] If acetyl chloride is used, care should be taken to hydrolyze the acid chloride formed.[29,34] Trifluoroacetic anhydride in acetic acid should also be tried,[44] as should partial hydrolysis of an acetylated ester.[80]

1 (−)R-3D-Acetoxybutanoic acid

COOH
|
CH$_2$
|
H—C—OAc
|
CH$_3$
(−)
↓

$\alpha_D^{21-24} - 2.8 \pm 0.1°$ (undiluted, 1 dm), $[\alpha]_D^{21} - 7.7 \pm 0.4°$ (6% in methanol),[62,65,68] obtained remarkably easily and in high yield (73%) from racemic acid via the morphine salt; no recrystallization is needed.[26,62] Absolute configuration is determined by transformation of **1** to (−)R-ZnCa 3D-hydroxybutanoate[68] (rotation in water) and to (−)R-butanediol-1, 3D (rotation in ethanol).[57]

1a (−)R-9D-Hydroxydecanoic acid:
(rotation in methanol)

H
|
(−)-CH$_3$—C—(CH$_2$)$_7$COOH[28]
|
OH

1b $(-)R$-Eicosanol-2L:

$$(-)\text{-CH}_3(\text{CH}_2)_{17}\overset{\overset{\displaystyle OH}{|}}{\underset{\underset{\displaystyle H}{|}}{\text{C}}}\text{CH}_3{}^{68}$$

1c $(+)R$-2L-Acetoxyeicosanone-4:

$$(+)\text{-CH}_3(\text{CH}_2)_{15}\text{COCH}_2\overset{\overset{\displaystyle OAc}{|}}{\underset{\underset{\displaystyle H}{|}}{\text{C}}}\text{CH}_3 \quad \text{Via malonate}^{65}$$

1d $(-)RS$-Eicosanediol-2L, 4L:

$$(-)\text{-CH}_3(\text{CH}_2)_{15}\overset{\overset{\displaystyle OH}{|}}{\underset{\underset{\displaystyle H}{|}}{\text{C}}}\text{CH}_2\overset{\overset{\displaystyle OH}{|}}{\underset{\underset{\displaystyle H}{|}}{\text{C}}}\text{CH}_3 \quad \text{From }\mathbf{1c}^{65}$$

1e $(+)RR$-Eicosanediol-2L, 4D:

$$(+)\text{-CH}_3(\text{CH}_2)_{15}\overset{\overset{\displaystyle H}{|}}{\underset{\underset{\displaystyle OH}{|}}{\text{C}}}\text{CH}_2\overset{\overset{\displaystyle OH}{|}}{\underset{\underset{\displaystyle H}{|}}{\text{C}}}\text{CH}_3 \quad \text{From }\mathbf{1c}^{65}$$

1f $(-)R$-Tetracosanol-2L:

$$(-)\text{-CH}_3(\text{CH}_2)_{21}\overset{\overset{\displaystyle OH}{|}}{\underset{\underset{\displaystyle H}{|}}{\text{C}}}\text{CH}_3{}^{62}$$

2 $(+)S$-3L-Acetoxybutanoic acid[57,62,68]

$$\begin{array}{c} \text{COOH} \\ | \\ \text{CH}_2 \\ | \\ \text{AcO—C—H} \\ | \\ \text{CH}_3 \\ (+) \\ \downarrow \end{array}$$

$\alpha_D^{23} + 2.8°$ (undiluted, 1 dm). Partly resolved acid from mother liquor of the morphine salt has been completely resolved by three recrystallizations of the quinine salt (yield about 30%). Absolute configuration verified by reduction of **2** to $(+)S$-butanediol-1, 3L.[57]

2a $(+)S$-Eicosanol-2D:

$$(+)\text{-CH}_3(\text{CH}_2)_{17}\overset{\overset{\displaystyle H}{|}}{\underset{\underset{\displaystyle OH}{|}}{\text{C}}}\text{CH}_3{}^{68}$$

Absolute configuration verified by partial resolution.[21]

2b $(+)S$-Tetracosanol-2D:

$$(+)\text{-CH}_3(\text{CH}_2)_{21}\overset{\overset{\displaystyle H}{|}}{\underset{\underset{\displaystyle OH}{|}}{\text{C}}}\text{CH}{}^{62}$$

3   (−)S-Ethyl/methyl hydrogen 2L-acetoxysuccinate

COOEt/Me
|
AcO—C—H
|
CH$_2$
|
COOH
(−)

Ethyl half-ester: $[\alpha]_D^{22}$-29.1° (11% in ethanaol), m.p. 50–51°C.[9,41] Prepared from commercial S-malic acid (levorotatory in dilute aqueous solution) by ethanolysis of its acetylated cyclic anhydride obtained by means of acetyl chloride (overall yield about 60%). The isomeric half-ester is apparently not formed.[9]

↓

3a   (+)S-2L-Hydroxydecanoic acid:

$$
(+)\text{-CH}_3(\text{CH}_2)_7\!-\!\overset{\displaystyle \text{OH}}{\underset{\displaystyle \text{H}}{\text{C}}}\!-\!\text{COOH}^{41}
$$

3b   (+)S-2L-Hydroxydecanedioic acid:
                          (rotation in ethanol)

$$
(+)\text{-HOOC(CH}_2)_7\!-\!\overset{\displaystyle \text{OH}}{\underset{\displaystyle \text{H}}{\text{C}}}\!-\!\text{COOH}^{27}
$$

3c   S-Ethyl 2L-acetoxytetradecanoate:

$$
\text{CH}_3(\text{CH}_2)_{11}\!-\!\overset{\displaystyle \text{OAc}}{\underset{\displaystyle \text{H}}{\text{C}}}\!-\!\text{COOEt}^{55}
$$

3d   (+)SS-2L-Hydroxy-12L-methyltetradecanoic acid:
                          (containing 20–25% of the 13-Me acid)

$$
(+)\text{-CH}_3\text{CH}_2\!-\!\overset{\displaystyle \text{CH}_3}{\underset{\displaystyle \text{H}}{\text{C}}}\!-\!(\text{CH}_2)_9\!-\!\overset{\displaystyle \text{OH}}{\underset{\displaystyle \text{H}}{\text{C}}}\!-\!\text{COOH}^{49}
$$

3e   (+)S-2L-Hydroxyhexadecanoic acid:

$$
(+)\text{-CH}_3(\text{CH}_2)_{13}\!-\!\overset{\displaystyle \text{OH}}{\underset{\displaystyle \text{H}}{\text{C}}}\!-\!\text{COOH}^{30,41} \quad (+)\text{-Methyl ester}^{40,59,76,77}
$$

3f   (−)S-2L-Acetoxyoctadecanoic acid:

$$
(-)\text{-CH}_3(\text{CH}_2)_{15}\!-\!\overset{\displaystyle \text{OAc}}{\underset{\displaystyle \text{H}}{\text{C}}}\!-\!\text{COOH}^{31}
$$

3g S-2L-Hydroxyeicosanoic acid:

$$\text{CH}_3(\text{CH}_2)_{17}\!-\!\overset{\displaystyle \text{OH}}{\underset{\displaystyle \text{H}}{\text{C}}}\!-\!\text{COOH}^{30}$$

3h S-2L-Hydroxytetracosanoic acid:

$$\text{CH}_3(\text{CH}_2)_{21}\!-\!\overset{\displaystyle \text{OH}}{\underset{\displaystyle \text{H}}{\text{C}}}\!-\!\text{COOH}^{30}$$

4   (+)R-Methyl/ethyl hydrogen 2D-acetoxysuccinate

$$\begin{array}{l}\text{COOMe/Et}\\ |\\ \text{H}-\text{C}-\text{OAc}\\ |\\ \text{CH}_2\\ |\\ \text{COOH}\\ (+)\\ \downarrow\end{array}$$

Methyl half-ester: physical data apparently not reported. Prepared from commercial R-malic acid (dextrorotatory in dilute aqueous solution) by methanolysis of its acetylated cyclic anhydride obtained by means of acetyl chloride[36,59] (overall yield about 75%).[33] The isomeric half-ester is apparently not formed.[9]

4a   (−)R-Methyl 2D-hydroxyhexadecanoate:

$$(-)\text{-CH}_3(\text{CH}_2)_{13}\!-\!\overset{\displaystyle \text{H}}{\underset{\displaystyle \text{OH}}{\text{C}}}\!-\!\text{COOMe}^{40,59,76,77}$$

4b   (+)R-2D-Acetoxy-4,4,5,5-$^2\text{H}_4$-hexadecanoic acid:

$$(+)\text{-CH}_3(\text{CH}_2)_{10}(\text{CD}_2)_2\text{CH}_2\!-\!\overset{\displaystyle \text{H}}{\underset{\displaystyle \text{OAc}}{\text{C}}}\!-\!\text{COOH}^{33,36}$$

4c   (+)R-2D-Acetoxyoctadecanoic acid:

$$(+)\text{-CH}_3(\text{CH}_2)_{15}\!-\!\overset{\displaystyle \text{H}}{\underset{\displaystyle \text{OAc}}{\text{C}}}\!-\!\text{COOH}^{31}$$

4d R-2D-Hydroxy-11-methyleicosanoic acid:
(mixture of diastereoisomers)

$$\text{CH}_3(\text{CH}_2)_8\text{CHCH}_3(\text{CH}_2)_8\!-\!\overset{\displaystyle \text{H}}{\underset{\displaystyle \text{OH}}{\text{C}}}\!-\!\text{COOH}^{58}$$

4e R-2D-Hydroxy-7,8,10,11,13,14,16,17-^{2}H$_8$-docosanoic acid:

$$CH_3(CH_2)_4(CHDCHDCH_2)_4(CH_2)_3\overset{\displaystyle H}{\underset{\displaystyle OH}{-C-}}COOH^{35}$$

5 $(-)S$-Methyl hydrogen 2L-methoxysuccinate

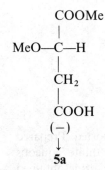

$[\alpha]_D^{20} - 61°$ (7% in acetone), m.p. 46–48°C.[60] Purity not known.[55] Prepared from commercial S-malic acid (yield about 20%) via the methyl ether of the dimethyl ester and methanolysis of the derived cyclic anhydride obtained after alkaline hydrolysis and treatment with acetyl chloride.[55]

5a $(-)S$-Methyl 2L-methoxytetradecanoate:

$$(-)\text{-}CH_3(CH_2)_{11}\overset{\displaystyle OMe}{\underset{\displaystyle H}{-C-}}COOMe^{55}$$

6 $(+)R$-Methyl hydrogen 3D-acetoxyglutarate[50,69]

COOMe
|
CH$_2$
|
H—C—OAc
|
CH$_2$
|
COOH
(+)
↓

$[\alpha]_D^{25} + 6.1 \pm 0.1°$ (20% in chloroform). Obtained from racemic half-ester via the cinchonidine salt recrystallized four times (yield about 30%).[63] Absolute configuration[18] determined by transformation of **6** to $(-)R$-3D-hydroxyhexanoic acid[63,69] and to $(-)R$-3D-hydroxypentanoic acid.[63]

The racemic half-ester[50,69] may be prepared from commercial diethyl 3-hydroxyglutarate (yield about 50%) by methanolysis of the acetylated cyclic anhydride which is obtained by means of acetyl chloride from dipotassium 3-hydroxyglutarate.[63]

6a $(+)R$-Decanolide-4D:
$[\alpha]_D^{22} + 46 \pm 1°$ (2% in ethanol)

$$(+)\text{-}CH_3(CH_2)_5\overset{\displaystyle H}{\underset{\displaystyle O-CO}{-C-}}(CH_2)_2{}^{66}$$

By Arndt/Eistert synthesis[10b] from $(-)R$-3D-AcO-nonanoic acid.[66] Amide from **6a**: m.p. 82.2–82.7°C.[66]

6b $(-)R$-3D-Hydroxydodecanoic acid:

$$(-)\text{-}CH_3(CH_2)_8\overset{\displaystyle H}{\underset{\displaystyle OH}{-C-}}CH_2COOH^3$$

6c   (−)R-3D-Hydroxyhexadecanoic acid:
Dextrorotatory in methanol solution

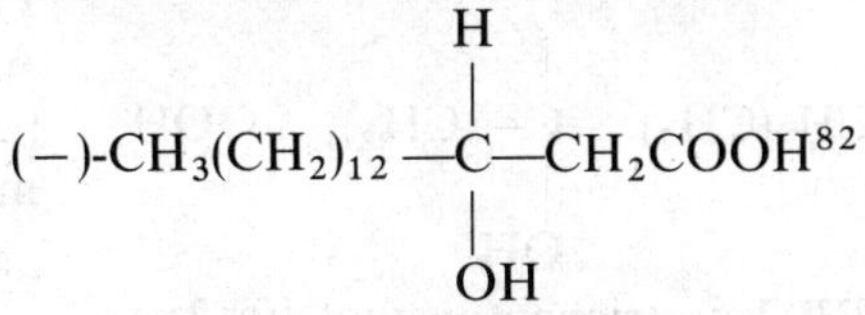

$$(-)\text{-}CH_3(CH_2)_{12}\overset{\displaystyle H}{\underset{\displaystyle OH}{\vert}}\underset{\vert}{C}\text{—}CH_2COOH^{82}$$

Optical purity 72%.
(−)-Methyl ester;[76]
inverted via tosylate.[59]

6d   (+)R-Methyl 3D-acetoxyheptadecanoate:

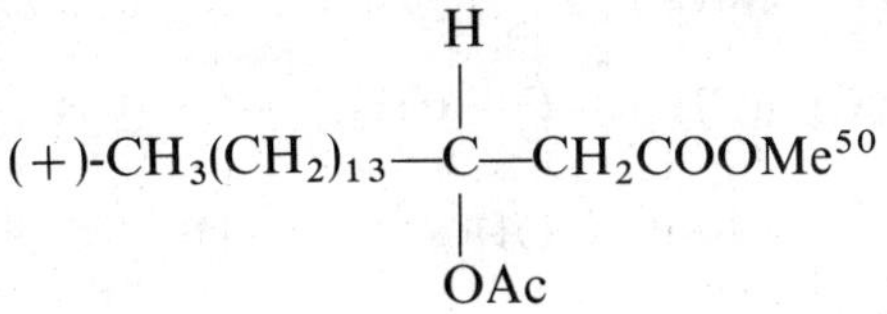

$$(+)\text{-}CH_3(CH_2)_{13}\text{—}\underset{OAc}{\overset{H}{C}}\text{—}CH_2COOMe^{50}$$

6e   (+)R-Heptadecanediol-1,3D:

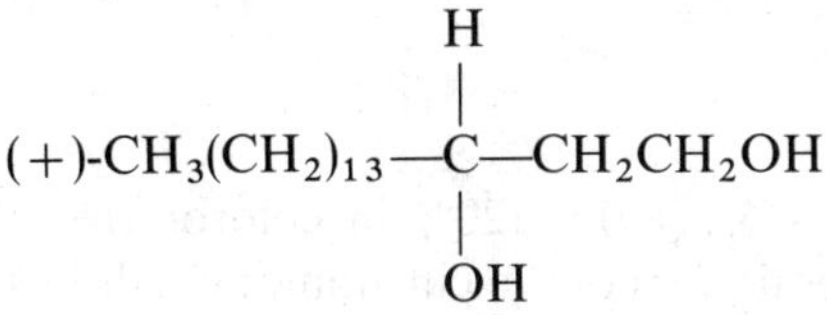

$$(+)\text{-}CH_3(CH_2)_{13}\text{—}\underset{OH}{\overset{H}{C}}\text{—}CH_2CH_2OH \quad \text{From } \mathbf{6d}^{50}$$

6f   (+)S-Heptadecanol-3D:

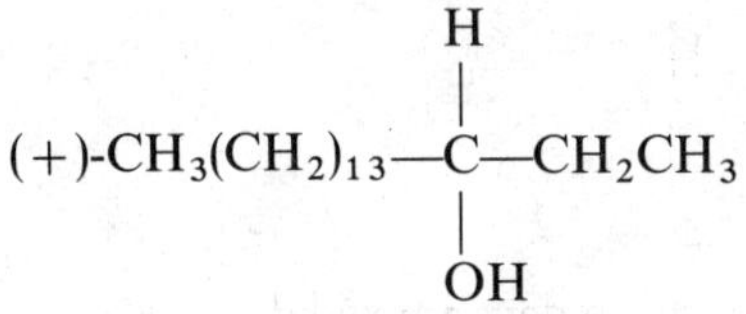

$$(+)\text{-}CH_3(CH_2)_{13}\text{—}\underset{OH}{\overset{H}{C}}\text{—}CH_2CH_3 \quad \text{From } \mathbf{6e}^{50}$$

6g   (−)R-3D-Hydroxyoctadecanoic acid:

$$(-)\text{-}CH_3(CH_2)_{14}\text{—}\underset{OH}{\overset{H}{C}}\text{—}CH_2COOH^{82}$$

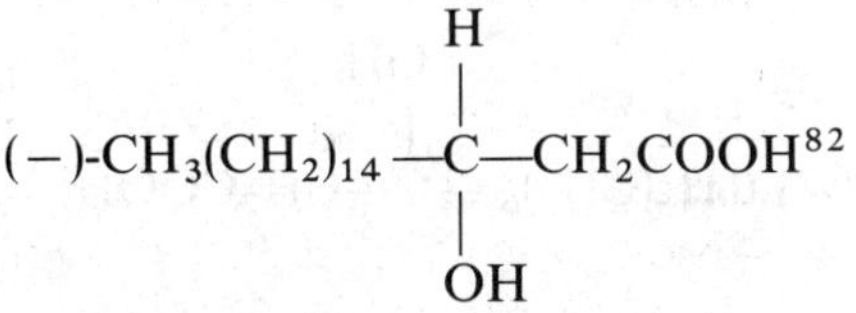

6h R-9D-Hydroxyoctadecanoic acid:

$$CH_3(CH_2)_8\text{—}\underset{OH}{\overset{H}{C}}\text{—}(CH_2)_7COOH$$

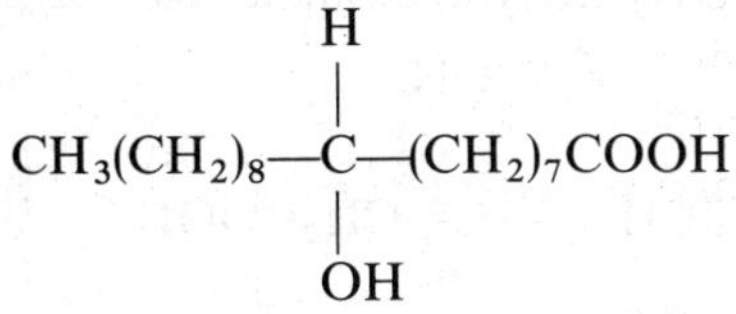

From **6b**[3]
(−)-Methyl ester;[61,80]
inverted via tosylate.[61,80]

6i R-Methyl 11D-hydroxyoctadecanoate:
(from the antipode of **7a**)

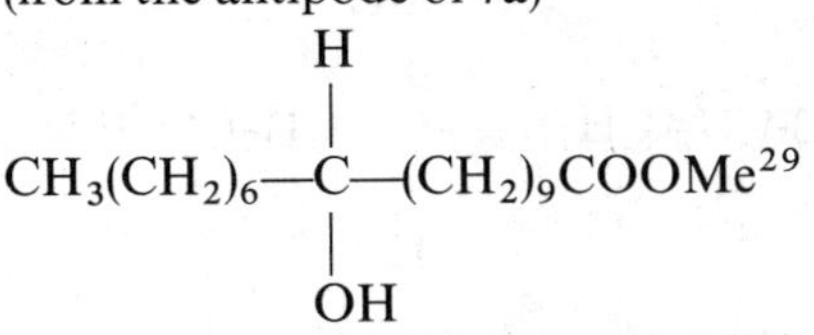

$$CH_3(CH_2)_6\text{—}\underset{OH}{\overset{H}{C}}\text{—}(CH_2)_9COOMe^{29} \quad \text{Inverted via tosylate.}^{29}$$

6j R-Methyl 16D-hydroxyoctadecanoate:

$$CH_3CH_2\text{—}\underset{OH}{\overset{H}{C}}\text{—}(CH_2)_{14}COOMe^{34}$$

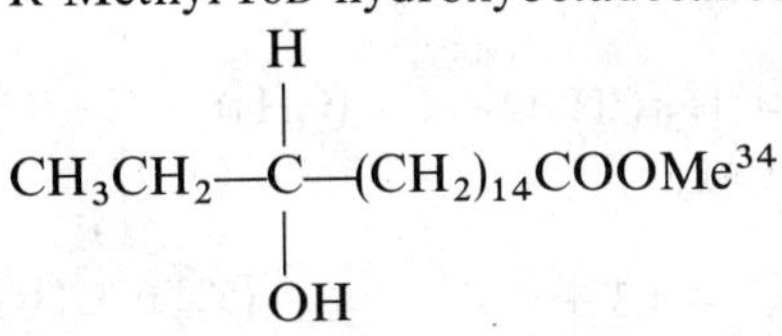

6k   R-13D-Hydroxydocosanoic acid:

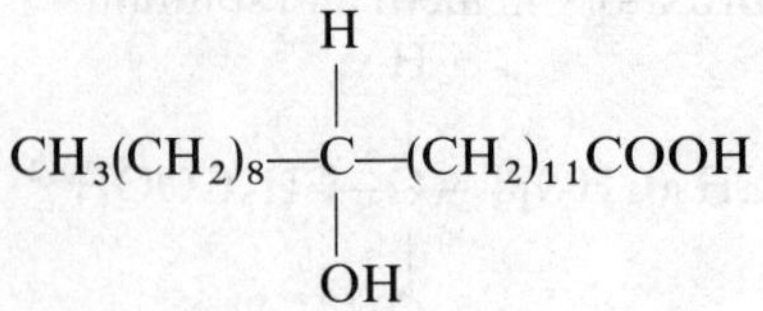

$$CH_3(CH_2)_8-\overset{\overset{\displaystyle H}{|}}{\underset{\underset{\displaystyle OH}{|}}{C}}-(CH_2)_{11}COOH$$

From **6h**[80]
Antipode of
(−)-methyl ester;[80]
made from inverted **6h**.[80]

6l   RR-Tetratriacontanediol-10L,25D:

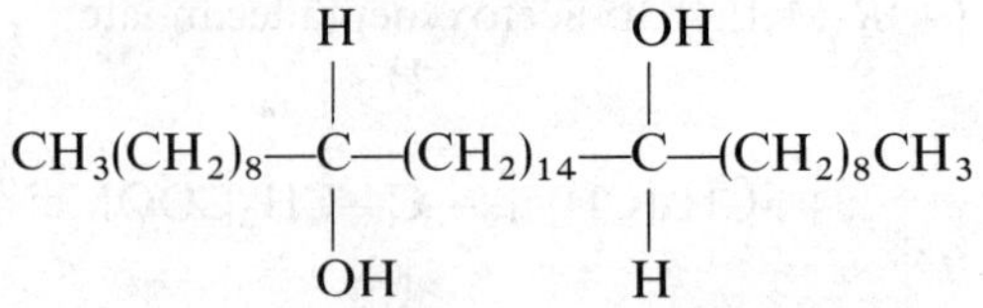

$$CH_3(CH_2)_8-\overset{\overset{\displaystyle H}{|}}{\underset{\underset{\displaystyle OH}{|}}{C}}-(CH_2)_{14}-\overset{\overset{\displaystyle OH}{|}}{\underset{\underset{\displaystyle H}{|}}{C}}-(CH_2)_8CH_3$$ From **6h**[80]

7   (−)S-Methyl hydrogen 3L-acetoxyglutarate[63,69]

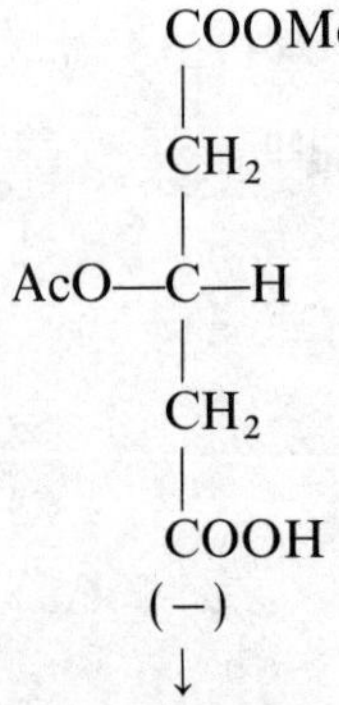

$$\begin{array}{c} COOMe \\ | \\ CH_2 \\ | \\ AcO-C-H \\ | \\ CH_2 \\ | \\ COOH \\ (-) \\ \downarrow \end{array}$$

$[\alpha]_D^{21}$ − 6.2 ± 0.1° (20% in chloroform). Partly resolved acid from mother liquors of the cinchonidine salt has been completely resolved by five recrystallizations of the strychnine salt (yield about 25%).[63]

7a   (+)S-3L-Hydroxydecanoic acid:
              Hydrazide: m.p. 135°, $[\alpha]_D^{21}$ + 15 ± 2°(1% in dioxan)[66]

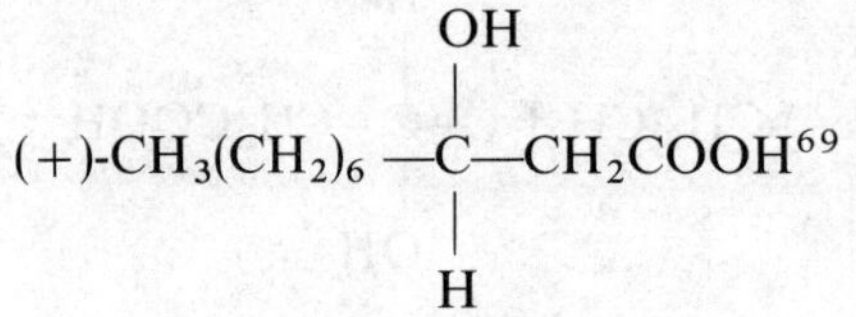

$$(+)\text{-}CH_3(CH_2)_6-\overset{\overset{\displaystyle OH}{|}}{\underset{\underset{\displaystyle H}{|}}{C}}-CH_2COOH^{[69]}$$

Levorotatory in EtOH[69] and in aqueous NaOH; dextrorot. in pyridine.[66]

7b   (+)S-3L-Hydroxydodecanoic acid:

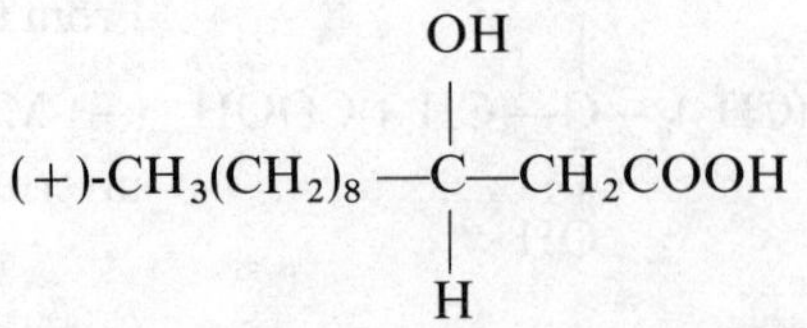

$$(+)\text{-}CH_3(CH_2)_8-\overset{\overset{\displaystyle OH}{|}}{\underset{\underset{\displaystyle H}{|}}{C}}-CH_2COOH$$

m.p. 62.3–62.8°
$[\alpha]_D^{20}$ + 16.0 ± 0.4°
(5% in CHCl$_3$)
$[\alpha]_D^{21}$ − 4.0 ± 0.4°
(5% in HCONMe$_2$)[67]

7c   (+)S-Methyl 3L-hydroxyhexadecanoate:

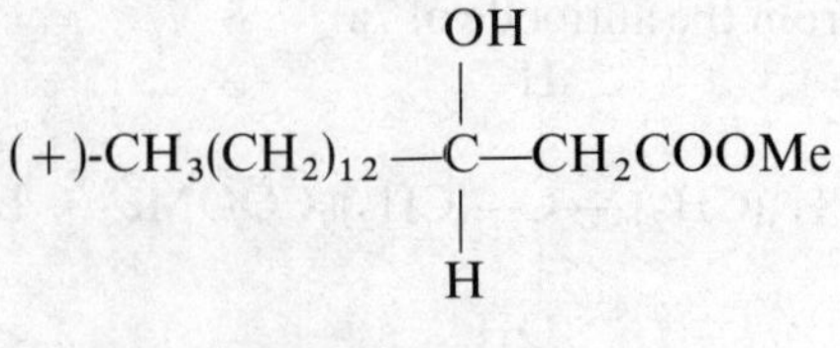

$$(+)\text{-}CH_3(CH_2)_{12}-\overset{\overset{\displaystyle OH}{|}}{\underset{\underset{\displaystyle H}{|}}{C}}-CH_2COOMe$$ Optical purity 74%[76]

7d   (+)SS-Hexadecanediol-7D,10L:

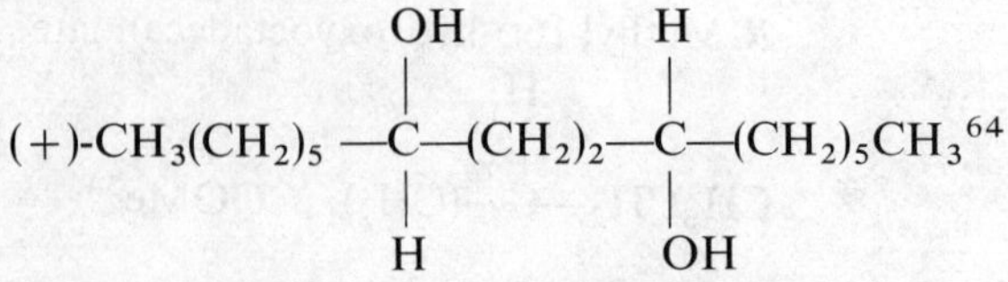

$$(+)\text{-}CH_3(CH_2)_5-\overset{\overset{\displaystyle OH}{|}}{\underset{\underset{\displaystyle H}{|}}{C}}-(CH_2)_2-\overset{\overset{\displaystyle H}{|}}{\underset{\underset{\displaystyle OH}{|}}{C}}-(CH_2)_5CH_3^{[64]}$$

$[\alpha]_D^{25}$ + 4.3 ± 0.4°   (5% in CHCl$_3$)[66]

7e   (+)S-12L-Hydroxyoctadecanoic acid:

$$(+)\text{-}CH_3(CH_2)_5 \overset{\displaystyle OH}{\underset{\displaystyle H}{-\overset{|}{\underset{|}{C}}-}} (CH_2)_{10}COOH$$

Rotation in pyridine and in acetic acid[64]

7f   S-Methyl 15L-hydroxyoctadecanoate:

$$CH_3(CH_2)_2 \overset{\displaystyle OH}{\underset{\displaystyle H}{-\overset{|}{\underset{|}{C}}-}} (CH_2)_{13}COOMe[34]$$

Racemic 2-hydroxy-10-undecenoic acid, m.p. 60.4–60.9°C, has been prepared from commercial 10-undecenoic acid (yield about 13%) via 10,11-dibromoundecanoic acid, methyl 2,10,11-tribromoundecanoate, methyl 2-iodo-10-undecenoate, and methyl 2-acetoxy-10-undecenoate, without purification of the intermediates.[19]

Complete resolution of 2-hydroxy-10-undecenoic acid has been effected by recrystallization of the α-phenethylamine salts five times (yield about 34%). The (−)-amine gave the S-2L-OH isomer, with $[\alpha]_D^{25} + 4.1 \pm 0.2°$ (4% in chloroform) and $[\alpha]_D^{25} - 6.3 \pm 0.3°$ (4% in pyridine), m.p. 65–65.5°C. Numerically equal values were obtained for the R-2D-OH isomer from the salt of the (+)-amine. The absolute configuration was deduced from the rotational data.[19]

8   (−)R-2D-Methoxy-10-undecenol-1[20]

$$\begin{array}{c} CH_2OH \\ | \\ H-C-OMe \\ | \\ (CH_2)_7 \\ | \\ CH \\ \| \\ CH_2 \\ (-) \end{array}$$

$[\alpha]_D^{25} - 22.3 \pm 0.3°$ (4% in chloroform). Prepared by LiAlH₄ reduction of (+)R-methyl 2D-methoxy-10-undecenoate, $[\alpha]_D^{23} + 35.3 \pm 0.3°$ (4% in chloroform) obtained from the hydroxy acid by means of diazomethane in the presence of BF₃ (overall yield about 56%). For malonate chain extension, **8** was transferred to the 1-iodo compound by reaction of the tosylate with NaI (yield 91%).[20]

via ↓ malonates

8a   (+)RR-Methyl 2D,4D-dimethoxy-12-tridecenoate:

$$(+)\text{-}CH_2{=}CH(CH_2)_7 \overset{\displaystyle H}{\underset{\displaystyle OMe}{-\overset{|}{\underset{|}{C}}-}} CH_2 \overset{\displaystyle H}{\underset{\displaystyle OMe}{-\overset{|}{\underset{|}{C}}-}} COOMe[20]$$

8b   (−)SR-Methyl 2L, 4D-dimethoxy-12-tridecenoate:

$$(-)\text{-}CH_2{=}CH(CH_2)_7 \overset{\displaystyle H}{\underset{\displaystyle OMe}{-\overset{|}{\underset{|}{C}}-}} CH_2 \overset{\displaystyle OMe}{\underset{\displaystyle H}{-\overset{|}{\underset{|}{C}}-}} COOMe[20]$$

8c   (+)*RR*-Methyl 2D, 4D-dimethoxyheptacosanoate:

$$
\begin{array}{ccc}
& H & H \\
& | & | \\
(+)\text{-}CH_3(CH_2)_{22}\text{---}C\text{---}CH_2\text{---}C\text{---}COOMe & From \ \mathbf{8a}^{20} \\
& | & | \\
& OMe & OMe
\end{array}
$$

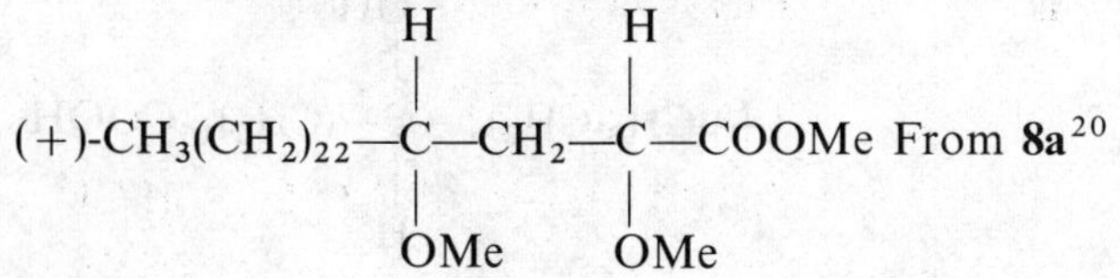

8d   (−)*SR*-Methyl 2L,4D-dim ethoxyheptacosanoate:

$$
\begin{array}{ccc}
& H & OMe \\
& | & | \\
(-)\text{-}CH_3(CH_2)_{22}\text{---}C\text{---}CH_2\text{---}COOMe & From \ \mathbf{8b}^{20} \\
& | & | \\
& OMe & H
\end{array}
$$

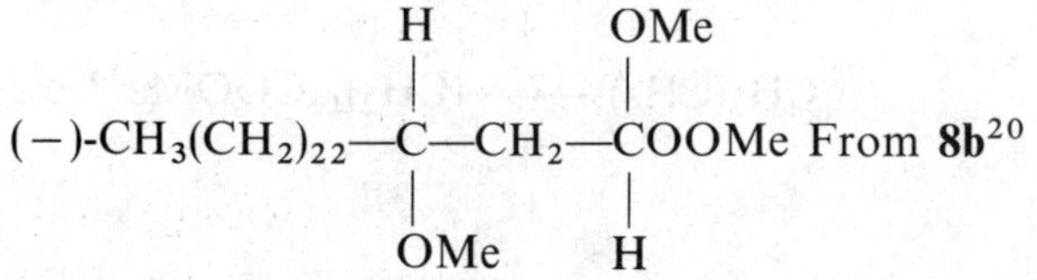

9   (+)*S*-2L-Methoxy-10-undecenol-1[20]

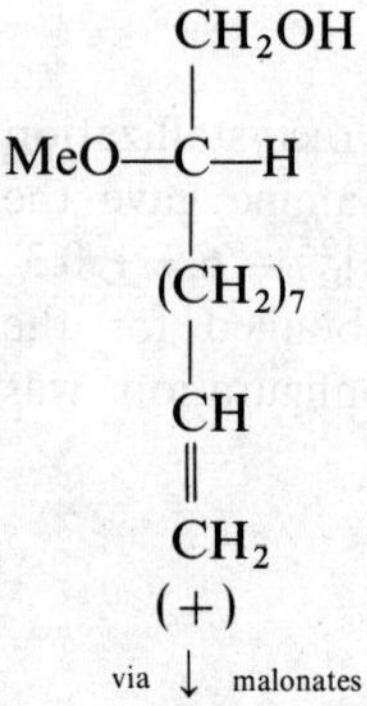

$$
\begin{array}{l}
CH_2OH \\
| \\
MeO\text{---}C\text{---}H \\
| \\
(CH_2)_7 \\
| \\
CH \\
\| \\
CH_2 \\
(+)
\end{array}
$$

$[\alpha]_D^{25} + 22.5 \pm 0.3°$ (4% in chloroform). Prepared by LiAlH$_4$ reduction of (−)*S*-methyl 2L-methoxy-10-undecenoate, $[\alpha]_D^{23} - 35.0 \pm 0.3°$ (4% in chloroform)) obtained from the hydroxy acid by means of diazomethane in the presence of BF$_3$ (overall yield about 54%). For malonate chain extension, **9** was transformed to the 1-iodo compound by reaction of the tosylate with NaI (yield 83%).[20]

via ↓ malonates

9a   (−)*SS*-Methyl 2L,4L-dimethoxy-12-tridecenoate:

$$
\begin{array}{ccc}
& OMe & OMe \\
& | & | \\
(-)\text{-}CH_2{=}CH(CH_2)_7\text{---}C\text{---}CH_2\text{---}C\text{---}COOMe^{20} \\
& | & | \\
& H & H
\end{array}
$$

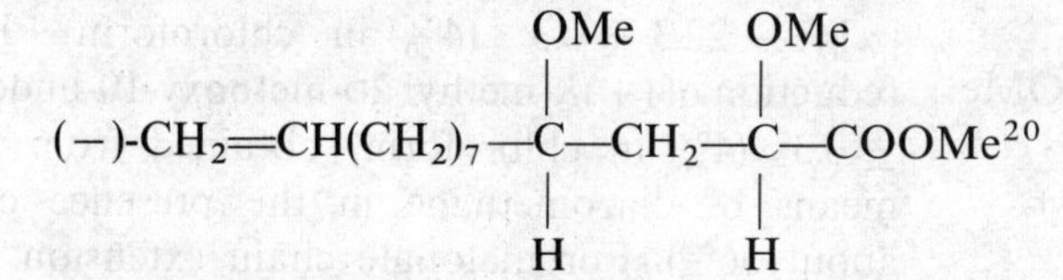

9b   (+)*RS*-Methyl 2D,4L-dimethoxy-12-tridecenoate:

$$
\begin{array}{ccc}
& OMe & H \\
& | & | \\
(+)\text{-}CH_2{=}CH(CH_2)_7\text{---}C\text{---}CH_2\text{---}C\text{---}COOMe^{20} \\
& | & | \\
& H & OMe
\end{array}
$$

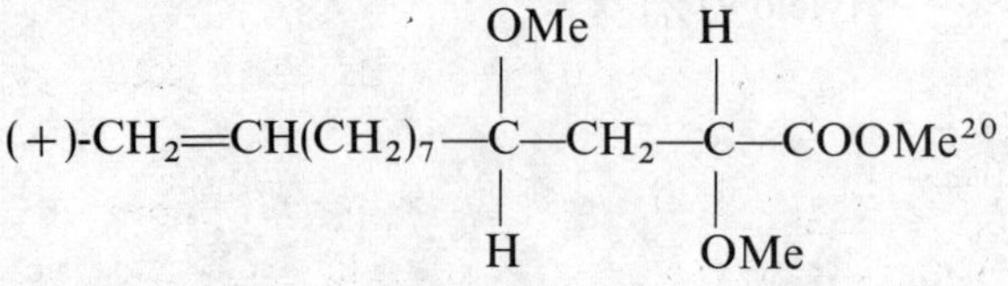

10   Ricinoleic acid: (+)*R*-12D-hydroxy-9*c*-octadecenoic acid

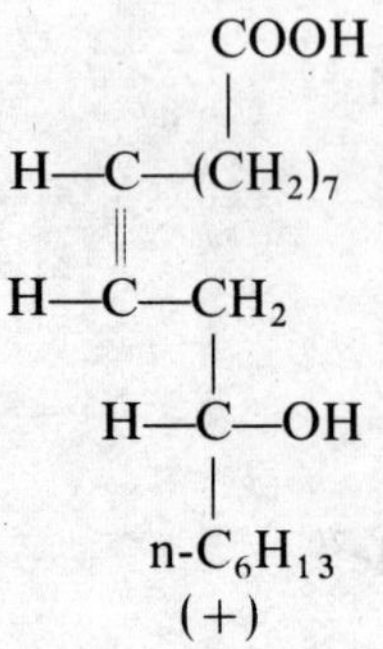

$$
\begin{array}{l}
COOH \\
| \\
H\text{---}C\text{---}(CH_2)_7 \\
\| \\
H\text{---}C\text{---}CH_2 \\
| \\
H\text{---}C\text{---}OH \\
| \\
n\text{-}C_6H_{13} \\
(+) \\
\downarrow
\end{array}
$$

$[\alpha]_D^{25} + 7.79°$ (undiluted, d_4^{25} 0.9416), m.p. 5.0°, 7.7° and 16.0°C.[38] This (+)*R*-12D-hydroxyoleic acid constitutes about 90% of the fatty acids obtained by hydrolysis of caster oil,[5] the commercially available triglyceride seed oil of *Ricinus communis* L. The free acid and several esters are commercial products.

10 **a**
b

R-Homoricinoleic acid
R-Homoricinolaidinic acid:

$$CH_3(CH_2)_5-\overset{\overset{\displaystyle H}{|}}{\underset{\underset{\displaystyle OH}{|}}{C}}-CH_2CH \overset{c}{=} CH(CH_2)_8COOH$$

c: Via nitrile[54]
t: From **10a**[54]

10c

(+)R-Methyl lesquerolate or
(+)R-Methyl 14D-hydroxy-11c-eicosenoate:
(rotation in methanol)

$$(+)\text{-}CH_3(CH_2)_5-\overset{\overset{\displaystyle H}{|}}{\underset{\underset{\displaystyle OH}{|}}{C}}-CH_2CH \overset{c}{=} CH(CH_2)_9COOMe^2$$

By Wittig chain extension

10d

(−)R-Methyl 14D-hydroxyeicosanoate:
(rotation in methanol)

$$(-)\text{-}CH_3(CH_2)_5-\overset{\overset{\displaystyle H}{|}}{\underset{\underset{\displaystyle OH}{|}}{C}}-(CH_2)_{12}COOMe^2$$

From **10c** by hydrogenation

10e

RR-9c,25c-Tetratriacontadienediol-7L,28D:

$$(CH_3(CH_2)_5-\overset{\overset{\displaystyle H}{|}}{\underset{\underset{\displaystyle OH}{|}}{C}}-CH_2CH \overset{c}{=} CH(CH_2)_7)_2^{6,37,45}$$

(structure needs confirmation)

11

COOBe
|
H—C—OAc
|
AcO—C—H
|
COOH
(+)

via ↓ malonate[8]

(+)RR-Benzyl hydrogen 2D, 3L-diacetoxysuccinate[66]

$[\alpha]_D^{21} + 33.1 \pm 0.4°$ (5% in chloroform), m.p. 94°C. Prepared from commercial (+)RR-tartaric acid (yield about 64%) via the diacetate of its cyclic anhydride, which was obtained by means of acetic anhydride + sulfuric acid.[88] The cyclic anhydride reacts smoothly with benzyl alcohol at 100°C to give the half-ester.[66]

11a

(+)RR-Benzyl 2D, 3L-diacetoxy-4-oxoeicosanoate:

$$(+)\text{-}CH_3(CH_2)_{15}CO-\overset{\overset{\displaystyle AcO}{|}}{\underset{\underset{\displaystyle H}{|}}{C}}-\overset{\overset{\displaystyle H}{|}}{\underset{\underset{\displaystyle OAc}{|}}{C}}-COOCH_2Ph^{66}$$

$[\alpha]_D^{21} + 27.4 \pm 0.4°$ (5% in chloroform), m.p. 13° and 40.3–40.7°C.

12   *SRR*-1,2D,3L,4L-Tetrabenzyloxypentanal-5[25]

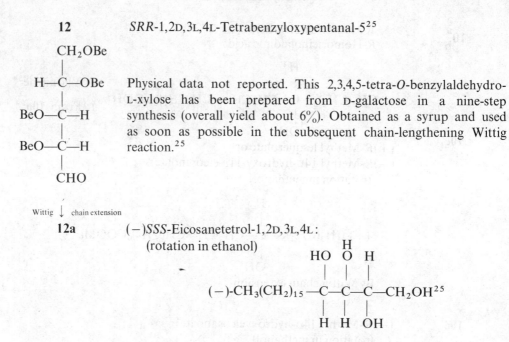

Physical data not reported. This 2,3,4,5-tetra-*O*-benzylaldehydro-L-xylose has been prepared from D-galactose in a nine-step synthesis (overall yield about 6%). Obtained as a syrup and used as soon as possible in the subsequent chain-lengthening Wittig reaction.[25]

Wittig ↓ chain extension

12a   (−)*SSS*-Eicosanetetrol-1,2D,3L,4L :
             (rotation in ethanol)

IV. COMMENTS

More than half of the product molecules listed in this review were synthesized with the aim of determining or verifying the structure and absolute configuration of natural products. This field is also covered by the reviews[10a,70] referred to at the end of the introduction. About half of the other products listed were prepared to serve as substrates (**4b, e, 10a, b**) or as precursors for substrates (**1a, 3a, 4a, b, 6c, i, 7c**) in the elucidation of biochemical reactions. With few exceptions (**10a, b**), the substrate molecules were labeled with heavy hydrogen isotopes, often introduced at the originally hydroxyl-bearing carbon atoms by stereospecific hydrogenolysis of sulfonates. Most of the remaining product molecules were synthesized as model compounds for physical experiments (**1b, d–f, 2a, b, 4d**).

Ten key molecules (**1–10**) and about forty derived compounds have one chiral center only and, except for **12** and **12a**, the others have two. The products with two centers separated by two or more carbon atoms (**3d, 4d, 6l, 6d, 10e**) were obtained from two key molecules by Kolbe synthesis. Those with one carbon atom between the chiral centers (**1d, e, 8a–d, 9a, b**) were prepared by way of malonic esters and chromatographic resolution of resulting diastereoisomeric mixtures. Products with vicinal chiral centers (**11a, 12a**) were synthesized from key molecules with vicinal centers (**11, 12**).

The key molecules which have been used most often are the half-esters of 2-acetoxysuccinic acid and 3-acetoxyglutaric acid. The succinates (**3, 4**), easily prepared from the optically active malic acids of commerce, have so far served as precursors for long-chain α-hydroxy acids or derivatives only, whereas the glutarates (**6, 7**) have been used to synthesize a more varied group of products, of which β-hydroxy acids or their derivatives constitute less than half the members.

In succinate **5**, the hydroxyl group is protected as methyl ether, less reactive than the acetoxyl group of succinates **3** and **4**, and present in some long-chain natural compounds. If demethylation of a product molecule is desirable, this can be done under mild conditions (by I_2/NaBH$_4$) and in high yield without loss of optical purity.[56]

If a small optical impurity can be tolerated, only one enantiomer of a key molecule may be needed, because a derived hydroxy compound can usually be transformed to its antipode with about 10% or less[80] racemization (and roughly 30% yield) by hydrolysis or acetolysis of a sulfonate.[29,61,76,77] The optical purity of some long-chain hydroxy compounds has been determined on the microscale by chromatographic analysis of diastereoisomeric derivatives.[10c,32,34,44] When absolute configuration is determined by comparison of chromatographic retention times of diastereoisomers,[32,34,44,60a,71] reference compounds may be prepared from small amounts of only partly resolved key molecules.

Product molecule **6a** is probably the first simple saturated γ-lactone whose absolute configuration has been firmly established since that of the first member, γ-valerolactone, was determined as $(-)S$-pentanolide-4L in the 1920s.[51] Lactone **6a** is derived from ricinoleic acid (**10**) via the intermediate $(-)R$-3D-hydroxynonanoic acid (levarotatory in chloroform but dextrorotatory in ethanol and in dimethylformamide solutions),[64] the antipode of which has been synthesized by Kolbe chain extension of key molecule **7**.[64] Both lactones support the formulation of the lactone rule [22,39,47,48] that dextrorotatory lactones of this type have the ring to the right in the usual Fischer projection formula. Lactone **6a** and the dextrorotatory homologues with 9, 11 and 12 carbon atoms have been prepared from the corresponding oxo acids by reduction with bakers' yeast.[83] They have molecular rotations (in methanol) close to that of **6a** (in ethanol), indicating high optical purity. Optically impure $(+)$-γ-dodecanolactone has been isolated from palm-kernel oil,[85] and optically impure $(+)$-γ-undecanolactone has been prepared from the fungal metabolite nemotinic acid, $(+)$-$H(C{\equiv}C)_2CH{=}C{=}CHCHOH(CH_2)_2COOH$,[12,13,15] provisionally assigned 4D-O or S configuration on other grounds.[12] In agreement with the lactone rule, lactonization (to nemotin) increases the dextrorotation.[13] Lactonization of the C_{12} homologue odyssic acid (to odyssin) also increases the dextrorotation,[14] showing that the two homologues have the same absolute configuration. This is supported by a close biogenetic relationship.[11] Preparation of an optically active γ-dodecanolactone from odyssic acid or odyssin does not seem to have been reported.

Ricinoleic acid (**10**) has been submitted to Kolbe chain extension with the hydroxyl group unprotected,[6,37,45] apparently giving the normal product **10e** in 67% yield.[37] This suggests that the products (**10a–d**) which have been prepared from ricinoleic acid by multistep syntheses can be obtained in a more simple way by use of the Kolbe reaction.

The hydrogenation product of ricinoleic acid, the cheap $(-)R$-12D-hydroxystearic acid, should also be an attractive starting material, and is expected to undergo the Kolbe reaction unprotected.[7] An isomer, natural $(+)S$-17L-hydroxystearic acid which was, probably unnecessarily, acetylated before anodic condensation with acetic acid to give $(+)S$-octadecanol-2D,[80] was not included in the present list because no chain extension is involved. $(+)S$-Heptadecanol-2D has been prepared by degradation of the same acid.[43]

The key molecules **11** and **12** contain benzyl protecting groups ($PhCH_2$ or Be) removable by hydrogenolysis, an advantage in certain cases.[53] Product **12a** has served as intermediate in a synthesis of 2L-aminoeicosanetriol-1, 3L,4L, the antipode of natural C_{20}-phytosphingosine.[25] Its synthesis could be much simplified if compound **11a** can be reduced to **12a**. Phytosphingosines have also been synthesized from D-glucosamine[24] and from D-galactose[23] by Wittig chain extension of intermediate aldehydes which already contain the nitrogen atom and are therefore not included in this review.

Acknowledgements—To have been allowed to work with the Stenhagens for many years has been a privilege. My new colleagues have by their patience made the writing of this review a pleasure.

REFERENCES

1 ABRAHAMSSON, S., STÄLLBERG-STENHAGEN, S. and STENHAGEN, E. *Prog. Chem. Fats other Lipids* **7**, 1–164 (1963).
2. APPLEWHITE, T. H. *Tetrahedron Lett.* 3391–5, 4160 (1965)
3. BAKER, C. D. and GUNSTONE, F. D. *J. Chem. Soc.* 759–60 (1963).
4. BENTLEY, R. *Molecular Asymmetry in Biology* Vol. 1, Academic Press, New York (1969).
5. BINDER, R. G., APPLEWHITE, T. H., KOHLER, G. O. and GOLDBLATT, L. A. *J. Am. Oil Chem. Soc.* **39**, 513–7 (1962).

6. BÖHME FETTCHEMIE-GESELLSCHAFT (inventor E. Lübbe) *German Pat.* 624 331 (2 pp) (1936); *Chem. Abstr.* **30**, 2504 (1936), *Chem.Zentralbl.* **I**, 3012–3 (1936).

7. BOUNDS, D. G., LINSTEAD, R. P. and WEEDON, B. C. L. *J. Chem. Soc.* 2393–2400 (1953).

8. BOWMAN, R. E. and FORDHAM, W. D. *J. Chem. Soc.* 3945–9 (1952).

9. BRETTLE, R. and LATHAM, D. W. *J. Chem. Soc.* **C**, 906–10 (1968).

10. BREWSTER, J. H. in *Elucidation of Organic Structures by Physical and Chemical Methods* Eds. BENTLEY, K. W. and KIRBY, G. W. 2nd Edn. Part III, Wiley-Interscience, New York (1972). Is Vol. IV of *Techniques of Chemistry*, Ed. WEISSBERGER, A. (a) pp. 119–25, (b) pp. 153–7.

10c. BROOKS, C. J. W., GILBERT, M. T. and GILBERT J. D. *Anal. Chem.* **45**, 896–902 (1973).

11. BU'LOCK, J. D. and GREGORY, H. *Biochem. J.* **72**, 322–5 (1959).

12. BU'LOCK, J. D. and GREGORY, H. *J. Chem. Soc.* 2280–5 (1960).

13. BU'LOCK, J. D., JONES, E. R. H. and LEEMING, P. R. *J. Chem. Soc.* 4270–6 (1955).

14. BU'LOCK, J. D., JONES, E. R. H. and LEEMING, P. R. *J. Chem. Soc.* 1097–1101 (1957)

15. BU'LOCK, J. D., JONES, E. R. H., LEEMING P. R. and THOMPSON, J. M. *J. Chem. Soc.* 3767–71 (1956).

16. CAHN, R. S. *J. Chem. Educ.* **41**, 116–25, 508 (1964).

17. CAHN, R. S., INGOLD, C. K. and PRELOG, V. *Angew. Chem.* **78**, 413–7 (1966); Int. Edn. **5**, 385–415 (1966).

18. See also COHEN, S. G. and MILOVANOVIĆ, A. *J. Am. Chem. Soc.* **90**, 3495–502 (1968).

19. ELIASSON, B., ODHAM, G. and PETTERSSON, B. *Acta Chem. Scand.* **25**, 2217–24 (1971).

20. ELIASSON, B., ODHAM, G. and PETTERSSON, B. *Acta Chem. Scand.* **25**, 3405–14 (1971).

21. ETEMADI, A–H. and GASCHE, J. *Bull. Soc. Chim. Biol.* **47**, 2095-2104 (footnote) (1965).

22. FREUDENBERG, K. *Stereochemie* p. 707–9, F. Deuticke, Leipzig/Wien (1933).

23. GIGG, J. and GIGG, R. *J. Chem. Soc.* **C**, 1876–9 (1966).

24. GIGG, J., GIGG, R. and WARREN, C. D. *J. Chem. Soc.* **C**, 1872–6 (1966).

25. GIGG, R. and WARREN, C. D. *J. Chem. Soc.* **C**, 1879–82 (1966).

26. HAMBERG, M. *Eur. J. Biochem.* **6**, 147–50 (1968).

27. HAMBERG, M. *Analyt. Biochem.* **43**, 515–26 (1971).

28. HAMBERG, M. and BJÖRKHEM, I. *J. Biol. Chem.* **246**, 7411–6 (1971).

29. HAMBERG, M. and SAMUELSSON, B. *J. Biol. Chem.* **242**, 5336–43 (1967).

30. HAMMARSTRÖM, S. *FEBS Lett.* **5**, 192–5 (1969).

31. HAMMARSTRÖM, S. *J. Lipid Res.* **12**, 760–5 (1971).

32. HAMMARSTRÖM, S. *Methods Enzymol.* **35**, 326–34 (1975).

33. HAMMARSTRÖM, S. *Personal communication.*

34. HAMMARSTRÖM, S. and HAMBERG, M. *Anal. Biochem.* **52**, 169–79 (1973).

35. HAMMARSTRÖM, S. and SAMUELSSON, B. *Biochem. Biophys. Res. Comm.* **41**, 1027–35 (1970).

36. HAMMARSTRÖM, S. and SAMUELSSON, B. *J. Biol. Chem.* **247**, 1001–11 (1972).

37. HAUFE, J. and BECK, F. *Chem.-Ing.-Tech.* **42**, 170–5 (1970).

38. HAWKE, F. *J. S. Afr. Chem. Inst.* **2**, 1–5, 125–130 (1949).

39. HIRST. E. L. *J. Chem. Soc.* 4042–58, 4049–50 (1954).

40. HITCHCOCK, C., MORRIS, L. J. and JAMES, A. T. *Eur. J. Biochem.* **3**, 419–21, 473–5 (1968).

41. HORN, D. H. S. and PRETORIUS, Y. Y. *J. Chem. Soc.* 1460–4 (1954).

42. HUDSON, C. S. *Adv. Carbohydr. Chem.* **3**, 1–22 (1948).

43. JONES, D. F. and HOWE, R. *J. Chem. Soc.* **C**, 2801–8 (1968).

44. KARLSSON, K–A. and PASCHER, I. *Chem. Phys. Lipids* **12**, 65–74 (1974).

45. KITAURA, S. *Bull. Inst. Phys. Chem. Res.* (Tokyo) **16**, 765–72 (1937); *Chem. Abstr.* **32**, 4523 (1938); *Chem. Zentralbl.* **I** 390–1 (1939).

46. KLYNE, W. *Chem. Ind.* (London) 1022–5 (1951).

47. KLYNE, W. in *Determination of Organic Structures by Physical Methods* Eds. BRAUDE, E. A. and NACHOD, F. C. Academic Press, New York (1955).

48. KLYNE, W., SCOPES, P. M. and WILLIAMS, A. *J. Chem. Soc.* 7237–42 (1965).

49. LANÉELLE, M. A. *Experientia* **24**, 541–2 (1968).

50. LARSEN, P. K., NIELSEN, B. E. and LEMMICH, J. *Acta Chem. Scand.* **23**, 2552–4 (1969).

51. LEVENE, P. A. and HALLER, H. L. *J. Biol. Chem.* **69**, 165–73, 569–74 (1926).

52. LINSTEAD, R. P., LUNT, J. C. and WEEDON, B. C. L. *J. Chem. Soc.* 3333–5 (1950).

53. See for example LINSTEAD, R. P., WEEDON, B. C. L. and WLADISLAW, B. *J. Chem. Soc.* 1097–1100 (1955).

54. MIZUGAKI, M., UCHIYAMA, M. and OKUI, S. *J. Biochem.* **58**, 273–8 (1965).

55. ODHAM, G., PETTERSSON, B. and STENHAGEN, E. *Acta Chem. Scand.* **B28**, 36–8 (1974).

56. ODHAM, G. and SAMUELSEN, B. *Acta Chem. Scand.* **24**, 468–72 (1970).

57. PAQUETTE, L. A. and FREEMAN, J. P. *J. Am. Chem. Soc.* **91**, 7548–50 (1969).

58. PASCHER, I. *Personal communication.*

59. POLITO, A. J. and SWEELEY, C. *J. Biol. Chem.* **246**, 4178–87 (1971).

60. PURDIE, T. and YOUNG, C. R. *J. Chem. Soc.* **97**, 1524–36 (1910).

60a. RIETSCHEL, E. T. *Eur. J. Biochem.* **64**, 423–8 (1976).

61. SCHROEPFER, G. J. and BLOCH, K. *J. Am. Chem. Soc.* **85**, 3310–1 (1963); *J. Biol. Chem.* **240**, 54–63 (1965).

62. SERCK-HANSSEN, K. *Ark. Kemi* **8**, 401–10 (1955)

63. SERCK-HANSSEN, K. *Ark. Kemi* **10**, 135–49 (1956).

64. SERCK-HANSSEN, K. *Chem. Ind.* (London) 1554 (1958).

65. SERCK-HANSSEN, K. *Ark. Kemi* **19**, 83–96 (1962).

66. SERCK-HANSSEN, K. Unpublished data.

67. SERCK-HANSSEN, K. Unpublished data; but see VINING, L. C. and TABER, W. A. *Can. J. Chem.* **40**, 1579–84 (1962).

68. SERCK-HANSSEN, K., STÄLLBERG-STENHAGEN, S. and STENHAGEN, E. *Ark. Kemi* **5**, 203–21 (1953).

69. SERCK-HANSSEN, K. and STENHAGEN, E. *Acta Chem. Scand.* **9,** 866 (1955).
70. SMITH, C. R. in *Topics in Lipid Chemistry* Ed. GUNSTONE, F. D. Vol. 1, pp. 277–368, Logos Press, London (1970).
71. SMITH, C. R. *J. Chromatogr. Sci.* **14,** 36–9 (1976) (review article).
72. *Specialist Periodical Reports on Electrochemistry, Chem. Soc. (London)* **1,** (1970) and later volumes.
73. STÄLLBERG-STENHAGEN, S. *Ark. Kem. Min. Geol.* **23A,** 1–14 (1946).
74. STÄLLBERG-STENHAGEN, S. *Ark. Kemi* **1,** 187–96 (1949).
75. STÄLLBERG-STENHAGEN, S. *Ark. Kemi* **2,** 95–111, 431–42 (1950).
76. STOFFEL, W., ASSMANN, G. and BISTER, K. *Hoppe-Seyler's. Z. Physiol. Chem.* **352** 1531–43 (1971).
77. STOFFEL, W. and BINCZEK, E. *Hoppe-Seyler's. Z. Physiol. Chem.* **352** 1065–72 (1971).
78. SVADKOVSKAYA, G. E. and VOITKEVICH, S. A. *Russ. Chem. Rev.* **29,** 161–80 (1960).
79. TOMILOV, A. P., MAIRANOVSKII, S. G., FIOSHIN, M. Y. and SMIRNOV, V. A. *Electrochemistry of Organic Compounds,* pp. 366–416, Halsted Press, New York (1972).
80. TULLOCH, A. P. *Can. J. Chem.* **46,** 3727–30 (1968).
81. TULLOCH, A. P., HILL, A. and SPENCER, J. F. T. *Can. J. Chem.* **46,** 3337–51 (1968).
82. TULLOCH, A. P. and SPENCER, J. F. T. *Can. J. Chem.* **42,** 830–5 (1964).
83. TUYNENBURG MUYS, G., VAN DER VEN, B. and DE JONGE, A. P. *Can. Pat.* 648,917 (12pp) (1962); *Nature* **194,** 995–6 (1962); *Appl. Microbiol.* **11,** 389–93 (1963).
84. UTLEY, J. H. in *Technique of Electroorganic Synthesis,* Ed. WEINBERG, N. L., Wiley, New York (1974), Is Part I of Vol. V. of *Techniques of Chemistry,* Ed. WEISSBERGER, A.
85. VAN DER VEN, B. and DE JONG, K. *J. Am. Oil Chem. Soc.* **47,** 299–302 (1970).
86. For an early example see VELICK, S. F. and ENGLISH, J. *J. Biol. Chem.* **160,** 473–80 (1945).
87. WEEDON, B. C. L. *Quart. Rev.* **6,** 380–98 (1952); *Adv. Org. Chem.* **1,** 1–34 (1960).
88. WOHL, A. and OESTERLIN, C. *Ber. Dtsch. Chem. Ges.* **34,** 1139–48 (1901).

NOTE ADDED IN PROOF

S-2L- and *R*-2D-hydroxyhexadecyl acetate have been synthesized from the precursors **3** and **4,** by W. Stoffel and K. Bister,[77] but apparently the work has not been published.

The following publications, dealing with compounds belonging to the subject under review, came to the author's attention after submission of the manuscript.

CHAN, K-K., COHEN, N., DE NOBLE, J. P., SPECIAN, A. C. JR. and SAUCY, G. Synthetic studies on (2*R*,4′*R*,8′*R*)-α-tocopherol. Facile syntheses of optically active, saturated, acyclic isoprenoids via stereospecific [3,3] sigmatropic rearrangements. *J. Org. Chem.* **41,** 3497–3505 (1976).

COKE, J. L. and RICE, W. Y. JR. The absolute configuration of carpaine. *J. Org. Chem.* **30,** 3420–22 (1965).

COKE, J. L. and RICHON, A. B. Synthesis of optically active δ-*n*-hexadecalactone, the proposed pheromone from *Vespa orientalis. J. Org. Chem.* **41,** 3516–7 (1976).

GERLACH, H., OERTLE, K. and THALMANN, A. Synthese des (*R*)-*trans*-11-Hydroxy-8-dodecensäure-lactons (Recifeiolid). *Helv. Chim. Acta* **59,** 755–60 (1976).

IWAKI, S., MARUMO, S., SAITO, T., YAMADA, M. and KATAGIRI, K. Synthesis and activity of optically active disparlure. *J. Am. Chem. Soc.* **96,** 7842–4 (1974).

MASAMUNE, S., YAMAMOTO, H., KAMATA, S. and FUKUZAWA, A. Synthesis of macrolide antibiotics. II. Methymycin. *J. Am. Chem. Soc.* **97,** 3513–5 (1975).

MORI, K. Synthesis of *exo*-brevicomin, the pheromone of western pine beetle, to obtain optically active forms of known absolute configuration. *Tetrahedron* **30,** 4223–7 (1974).

MORI, K. Absolute configuration of (−)-massoilactone as confirmed by synthesis of its (*S*)-(+)-isomer. *Agr. Biol. Chem.* **40,** 1617–9 (1976).

MORI, K. Synthesis of optically active forms of ipsenol, the pheromone of *Ips* bark beetles. *Tetrahedron* **32,** 1101–6 (1976).

MORI, K., ODA, M. and MATSUI, M. Synthesis of (+)-(6*R*:1′*R*)-pestalotin and (+)-(6*R*:1′*S*)-epipestalotin. *Tetrahedron Lett.* 3173–4 (1976).

MORI, K., TAKIGAWA, T. and MATSUI, M. Stereoselective synthesis of optically active disparlure, the pheromone of the gypsy moth (*Porthetria dispar* L.). *Tetrahedron Lett.* 3953–6 (1976).

RAVID, U. and SILVERSTEIN, R. M. General synthesis of optically active 4-alkyl (or alkenyl)-γ-lactones from glutamic acid enantiomers. *Tetrahedron Lett.* 423–6 (1977).

STORK, G. and RAUCHER, S. Chiral synthesis of prostaglandins from carbohydrates. Synthesis of (+)-15(*S*)-prostaglandin A$_2$. *J. Am. Chem. Soc.* **98,** 1583–4 (1976).

STORK, G. and TAKAHASHI, T. Chiral synthesis of prostaglandins (PGE$_1$) from D-glyceraldehyde. *J. Am. Chem. Soc.* **99,** 1275–6 (1977).

These references indicate that more new material contributing to the field of this review can be expected to appear in publications dealing with syntheses of optically active pheromones, macrolides, prostaglandins, and related compounds.

Prog. Chem. Fats other Lipids. Vol. 16, pp. 195–206. Pergamon Press, 1978. Printed in Great Britain

PERDEUTERIATED NORMAL-CHAIN SATURATED MONO- AND DICARBOXYLIC ACIDS AND METHYL ESTERS*

Nguyên Dinh-Nguyên and Aino Raal

Institute of Medical Biochemistry, Faculty of Medicine, University of Göteborg,
S-400 33 Göteborg, Sweden

CONTENTS

I. Introduction   195
II. Syntheses   196
   A. Conversion of a protiated carboxylic acid into perdeuteriated analog (pathway 1)   196
   B. Conversion of a long-chain saturated monocarboxylic acid into lower homologs and dicarboxylic acids (pathway 2)   196
   C. Conversion of a long-chain saturated monocarboxylic acid into lower homologs and alkanes (pathway 3)   196
III. Experimental   196
   A. Oxidative degradation of the perdeuterio-*n*-tetracosanoic acid   197
   B. Separation of the perdeuteriated carboxylic acids   197
      1. Reversed-phase partition chromatography on Lipidex gel   197
      2. Gas–liquid chromatography   198
   C. Mass spectrometric analysis of perdeuteriated methyl monocarboxylates and dimethyl dicarboxylates   198
Acknowledgements   206
References   206

I. INTRODUCTION

The preparation of organic compounds possessing high deuterium content may be performed via syntheses or biosyntheses. Both of these preparation modes are utilized and continuously developed in this laboratory in order to suit our need of deuteriocarbons, perdeuteriated carboxylic acids and amino-acids, and others, for studies of mass spectrometry, infrared spectroscopy, gas–liquid chromatography, and metabolism in several diseases.

In this paper, we describe two synthetic methods which were used to prepare a whole series of normal-chain deuteriocarbons and perdeuteriated mono- and dicarboxylic acids. The first step, which is common in both methods, is to prepare a perdeuteriated long-chain monocarboxylic acid or corresponding sodium salt. This acid, or its methyl ester, is then converted into lower homologs and dicarboxylic acids via permanganic oxidative degradation. The sodium carboxylate is submitted to multiple and successive reactions to convert it into the lower homologs and deuterio-alkanes. As an example of the mentioned methods, we outline here the syntheses of the perdeuteriated monocarboxylic acids C_4–C_{23}, dicarboxylic acids C_3–C_{24} and alkanes C_{13}–C_{23}. All of these were made from *n*-tetradocosanoic acid.

$$CH_3\text{--}(CH_2)_{22}\text{--}COOH \overset{[1]}{\rightarrow} CD_3\text{--}(CD_2)_{22}\text{--}COO^-Na^+$$

$$\xrightarrow{[2]} \begin{cases} CD_3\text{--}(CD_2)_x\text{--}COOH \\ HOOC\text{--}(CD_2)_y\text{--}COOH \\ x = 2\text{--}21,\ y = 1\text{--}22 \end{cases}$$

$$\xrightarrow{[3]} \begin{cases} CD_3\text{--}(CD_2)_{x'}\text{--}COOH \\ CD_3\text{--}(CD_2)_z\text{--}CD_3 \\ x' = 10\text{--}22,\ z = 11\text{--}21 \end{cases}$$

* This paper constitutes Part III of the report series entitled *Perdeuteriated Organic Compounds*. References (10) and (11) indicate Parts I and II, respectively.

II. SYNTHESES

A. Conversion of a Protiated Carboxylic Acid
into its Perdeuteriated Analog (pathway [1])

The preparation of perdeuterio-n-tetracosanoic acid was performed according to the process described in earlier reports.[10–14] This consists of protium–deuterium exchange reaction occurring between sodium n-tetracosanoate and deuterium oxide in the presence of deuterium-reduced Adams catalyst, sodium deuterioxide and deuterium peroxide. Two successive exchanges were conducted at 240 and 270°C, in a magnetically stirred nickel reactor. Details of this synthesis and on important factors influencing the exchange reaction performance have been reported in a previous paper.[11] The perdeuterio-n-tetracosanoic acid and perdeuterio-n-tricosane (a by-product, see ref. 11) had ca 99% deuterium content.

B. Conversion of a Long-chain Saturated
Monocarboxylic Acid into Lower Homologs and
Dicarboxylic Acids (pathway [2])

The oxidative degradation of organic compounds—such as carboxylic acids, esters, and hydrocarbons—which has been developed in this laboratory[6–9] was the convenient process to convert the perdeuteria-n-tetracosanoic acid into perdeuteriated n-monocarboxylic acids C_4–C_{23} and n-dicarboxylic acids C_3–C_{24}. The reaction was performed utilizing potassium permanganate and acetic acid at 95–100°C. The advantage of using acetic acid is that it is a suitable solvent for several organic substances and which prevents the reaction medium from becoming basic. A protiated basic medium at that relatively high temperature (95–100°C) could effect isotopic dilution of the formed deuteriated carboxylates. Full details on this kind of permanganic oxidative degradation will be communicated in a later report. The mono- and dicarboxylic acids resulting from this synthesis had the same deuterium content (ca 99%) as that of the starting perdeuteriated substance.

C. Conversion of a Long-chain Saturated
Monocarboxylic Acid into Lower Homologs and
Alkanes (pathway [3])

Recently it has been found in this laboratory that a normal long-chain sodium monocarboxylate submitted to catalytic degradation involving decarboxylation, disproportionation, hydration and oxidation, etc. gave a whole series of lower homologs and alkanes.[5, 11] This became a convenient method to synthesize normal-chain saturated carboxylic acids and hydrocarbons. Thus, four groups of similar substances composed of six even chain perdeuterio-carboxylic acids C_{12}–C_{22}, six odd-chain homologs C_{13}–C_{23}, six odd chain perdeuterio-alkanes C_{13}–C_{23} and five even homologs C_{14}–C_{22} have been prepared from the sodium perdeuterio-n-tetracosanoate.[11] This catalytic degradation was carried out at 340°C in the same reactor and medium which had been previously utilized to perform the H_2–D_2 exchange (mentioned in Section II. A). The detailed degradation has already been reported in a previous paper.[11] The perdeuterated carboxylic acids and alkanes resulting from the latter synthesis had essentially the same isotopic content as that of the perdeuterio-n-tetracosanoic acid.

III. EXPERIMENTAL

As the experimental methods concerning the protium–deuterium exchange and the catalytic degradation relating to the preparation and isolation of compounds mentioned in the preceding Sections A and C are already published,[11] we describe in the present paper only that relating to the permanganic oxidative degradation of the perdeuterio-*n*-tetracosanoic acid.

A. *Oxidative Degradation of the Perdeuterio-n-tetracosanoic Acid*

For the purpose of gaining as many as possible of the whole series of mono- and dicarboxylic acid, the reation was carried out for 18 hr to induce degradation to several shorter chain carboxylic acids.

In a 250-ml two-necked round-bottomed flask equipped with a reflux condenser and provided with a magnetic stirrer were placed 1 g (2.4 mmoles) of perdeuterio-*n*-tetracosanoic acid (*ca* 99% deuterium content, m.p. 79.8°C) and 30 ml of acetic acid (*p.a.*, E. Merck). The stirring was started, and the temperature of the mixture was raised to 95°C by an oil bath, then 3.5 g (22 mmoles) of powdered potassium permanganate (*p.a.*, E. Merck) were added gradually (over 40 min) to the acidic solution. The stirring was continued, and the temperature was maintained constant at 95°C for 8 hr, then at 114–116°C for 10 hr. Two samples (0.5 ml) were removed after 3 and 8 hr of the reaction to check the extent of the degradation.

Separation of the resulting products was carried out as follows. After cooling to room temperature the resulting dark brown mixture was filtered through paper to separate the insoluble manganese dioxide from the soluble potassium acetate and other carboxylic acids or corresponding potassium salts. An equal quantity of diethyl ether was added to the filtrate, then the mixture was submitted to gentle evaporation in an electrothermal rotary evaporator, and the main part of acetic acid was separated off. The residue was acidified with dilute HCl, and the carboxylic acids were extracted with diethyl ether in the usual manner.

B. *Separation of the Perdeuteriated Carboxylic Acids*

Carboxylic acids constituting the residue (0.45 g) of the above ether extract were converted into methyl esters, by reaction with methanol saturated with HCl gas at room temperature. After removing HCl and the excess methanol, the mixture of methyl esters was dissolved in light petroleum and then chromatographed on a silica gel column to separate methyl monocarboxylates and dimethyl dicarboxylates.

Homologs of each ester group were then isolated via liquid–liquid partition chromatography on columns packed with Lipidex gel. In this paper, we relate only experimental separation techniques applied to the methyl monocarboxylates. The separation of the dimethyl dicarboxylates will be described in a separate report.

1. *Reversed-phase Partition Chromatography on Lipidex Gel*

Column. 120 g of Lipidex TM-5000 (Becker) were slurried in 500 ml of 1,2-dichloroethane/methanol/water (15:80:15, *v/v*). This mixture was kept at room temperature for one day, in order to allow the gel to equilibrate with the solvent, then poured into a glass tube (2 × 130 cm) equipped with Teflon stopcock, sintered glass filter and solvent reservoir. The column was washed with a similar solvent mixture and left in the solvent until just before application of sample. The gel-column height was about 75 cm.

Elution. A solution of 230 mg methyl monocarboxylates in minimum of eluting solvent was added to the gel column, and elution effectuated with 1,2-dichloro-ethane/methanol/water at 13–14 ml/hr in the following ratios and orders: 15:80:15, *v/v*, for C_{14} and lower homologs, and 15:80:10, *v/v*, for C_{15} and higher homologs. Fractions of 20 ml each were collected with an automatic fraction collector, and the fractions were checked by thin layer chromatography (TLC). Methyl C_{14}-carboxylate and lower homologs were recovered each in 60–100 ml of effluent separated by 40–60 ml blank effluent. Methyl

TABLE 1. Quantitative Gas–liquid Chromatographic Separations of Methyl Perdeuterio-
monocarboxylates and of Dimethyl Perdeuterio-
n-dicarboxylates

	Monoesters	Diesters
C_4		⎫
C_5		⎬ 0.10%
C_6		⎭
C_7	1.7 mg	0.33
C_8	7.5	2.02
C_9	13.2	6.41
C_{10}	15.4	9.79
C_{11}	17.0	11.14
C_{12}	18.5	12.83
C_{13}	16.0	10.47
C_{14}	15.0	9.12
C_{15}	13.6	13.51
C_{16}	12.3	5.06
C_{17}	11.2	5.40
C_{18}	11.0	4.72
C_{19}	10.2	5.40
C_{20}	9.0	1.68
C_{21}	7.0	1.35
C_{22}	10.0	⎫
C_{23}	6.1	⎬ 0.67
C_{24}	31.8	⎭

C_{15}–C_{17}-carboxylates were each found each in 100 ml of eluate separated by intervals
of 100 ml. The higher homologs were eluted each in 300 ml fractions separated by
300 ml of effluent. The methyl perdeuterio-*n*-monocarboxylates were isolated from the
above eluates by extraction with light petroleum followed by gentle evaporation. The
pure compounds obtained are listed in Table 1. Because the separation technique involved multiple operations including evaporation, low-mass homologs were lost.

2. Gas–Liquid Chromatography

Separation of small amounts of the methyl esters was conveniently performed by
gas–liquid chromatography, for analysis of the samples removed during the course of
the oxidative degradation, and for preparative separation from the final reaction mixture.
The chromatogram reproduced in Fig. 1 represents two homolog series of methyl monocarboxylates (68%) and dicarboxylates (32%). The amounts of the diesters which were
recovered from the mixture are listed in Table 1. The chromatogram shown in Fig.
2 exhibits a whole series of the methyl perdeuterio-monocarboxylates which were separated from the methyl ester mixture via liquid–liquid partition chromatography on silica
gel (see above). In order to study comparative retention time of these perdeuteriated
monoesters and the nondeuteriated analogs, we chromatographed a mixture composed
of both heavy and light isotopic homologs. The result exhibited by the chromatogram
in Fig. 3 shows that the perdeuteriated esters have lower retention times than the protiated analogs, in concordance with earlier reports;[10,17] if this chromatography was
started at a temperature lower than 95°C, the separation of the perdeuteriated *n*-pentane,
n-hexane and *n*-heptane and their nondeuteriated analogs could occur.

C. Mass Spectrometric Analysis of Perdeuteriated
Methyl Monocarboxylates and Dimethyl Dicarboxylates

The methyl perdeuterio-*n*-monocarboxylates C_4–C_{23} and dimethyl perdeuterio-*n*-
dicarboxylates C_3–C_{24} were submitted to mass spectrometric analysis to determine their
deuterium content and to study their fragmentation patterns. The analyses were accomplished using a combined gas chromatograph–mass spectrometer (LKB 9000) connected
on-line to a computer (Digital PDP 15), so data of ionic fragments were recorded

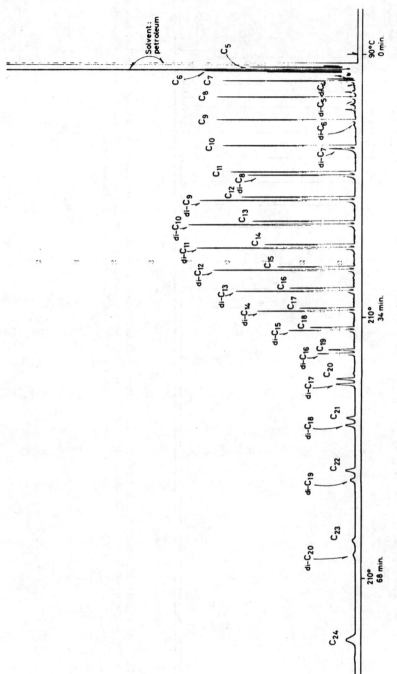

Fig. 1. Gas–liquid chromatogram of a mixture composed of two homolog series of normal-chain methyl perdeuterio-monocarboxylates and dimethyl perdeuterio-dicarboxylates. Conditions: glass capillary column of 24 m in length and 0.28 mm in diameter coated with OV-101 methyl silicone, nitrogen pressure 10 p.s.i., flame ionization detector, temperature programmed from 90 to 210°C at a rate of 3.5°C/min. Gas chromatograph: Perkin–Elmer 900.

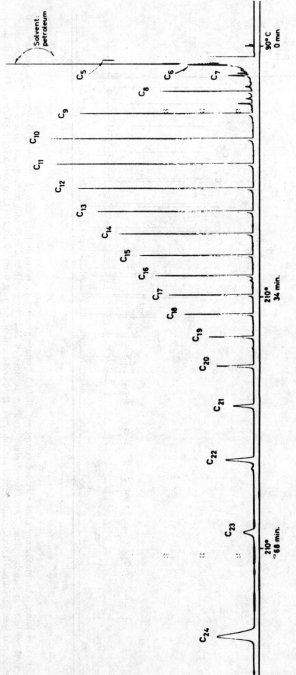

FIG. 2. Gas–liquid chromatogram of a homolog series of normal-chain methyl perdeuterio-monocarboxylates. Conditions: glass capillary column of 24 m in length and 0.28 mm in diameter coated with OV-101 methyl silicone, nitrogen pressure 10 p.s.i., flame ionization detector, temperature programmed from 90 to 210°C at a rate of 3.5°C/min. Gas chromatograph: Perkin–Elmer 900.

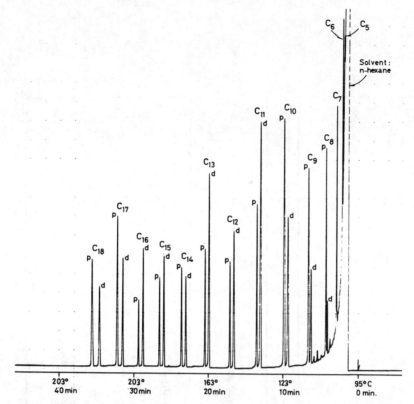

FIG. 3. Gas–liquid chromatogram of a mixture composed of two homolog series of normal-chain methyl perdeuterio-monocarboxylates and nondeuteriated (protiated) analogs. Conditions: glass capillary column of 24 m in length and 0.28 mm in diameter coated with OV-101 methyl silicone, argon pressure 10 p.s.i., flame ionization detector, temperature programmed from 95 to 123°C at a rate of 2.8°C/min and from 123 to 203°C at a rate of 4°C/min. Gas chromatograph: Perkin–Elmer 900. *d* and *p* mark the deuteriated and protiated components, respectively.

by computer and then reconstituted to display mass spectra. Conditions of the gas–liquid chromatographic separation of these esters was as described above, and ionizing potential used was 70 eV.

The deuterium content (%) of each substance was deduced from the heights of molecule-ion peaks, with correction due to the natural abundance (1.1%) of carbon-13. All the esters possessed deuterium contents of over 98.5%. All mass spectra of the present esters are to be published elsewhere. However we show in this paper spectra of some odd monoesters (Fig. 4–6) and diesters (Figs. 7–9). Spectra of the even mono-esters have been published previously.[10,11]

Mass spectra of the normal-chain methyl perdeuterio-monocarboxylates have analogous patterns of those of the nondeuteriated analogs which were investigated by Ryhage and Stenhagen,[18,20] and reviewed and discussed by Budzikiewicz et al.[4] The features of these spectra are the appearance of peaks corresponding to the following ions:

[a] $CH_3OOC-(CD_2)_n-CD_3]^+$ or M^+

[b] $\overset{+}{O}\equiv C-(CD_2)_n-CD_3$ or $M^+ - 31$, where 31: $CH_3O^{\cdot}$

[c]
$$\begin{array}{c} CH_3O \\ \diagdown \\ C \\ \diagup\ \ \ \diagdown \\ CD-(CD_2)_{n-5} \\ \diagup \\ CD_3 \end{array} \overset{\overset{+}{O}}{\underset{}{\diagup\ CD_2}} \quad \text{or } M^+ - 50, \text{ where } 50: \overset{4}{C}D_3-\overset{3}{C}D_2-\overset{2}{C}D_2^{+}$$

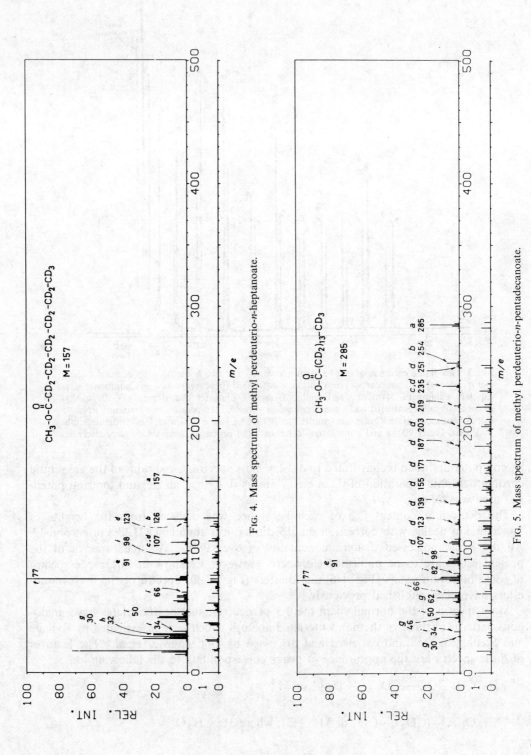

Fig. 4. Mass spectrum of methyl perdeuterio-*n*-heptanoate.

Fig. 5. Mass spectrum of methyl perdeuterio-*n*-pentadecanoate.

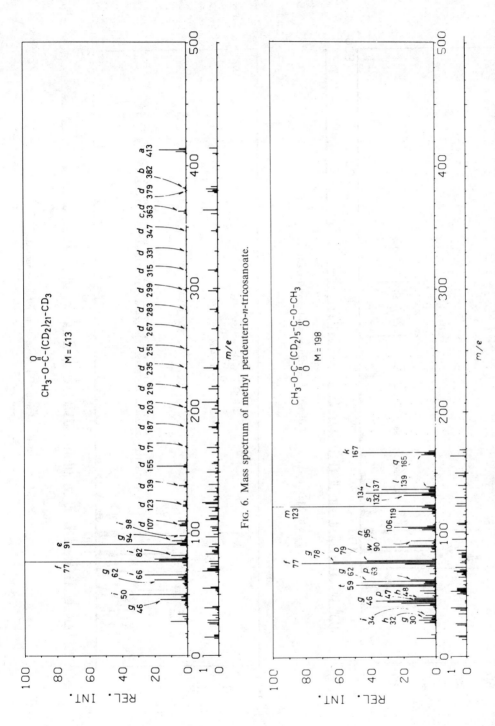

FIG. 6. Mass spectrum of methyl perdeuterio-*n*-tricosanoate.

FIG. 7. Mass spectrum of dimethyl perdeuterio-*n*-heptanedioate. Particular feature of the present spectrum—as of that of C_8 homolog—is the appearance of following peaks: m/e 106 corresponding to ion of type [υ'] (see text) but with a loss of one deuterium atom, m/e 119 corresponding to $M^+ - 79$, and m/e 134 corresponding to $M^+ - 64$.

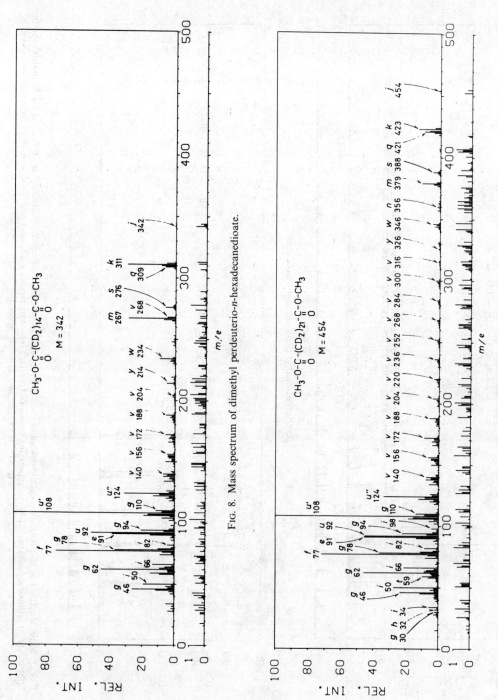

CH₃-O-C-(CD₂)₁₄-C-O-CH₃
      ‖   ‖
      O   O
M = 342

FIG. 8. Mass spectrum of dimethyl perdeuterio-n-hexadecanedioate.

CH₃-O-C-(CD₂)₂₁-C-O-CH₃
      ‖   ‖
      O   O
M = 454

FIG. 9. Mass spectrum of dimethyl perdeuterio-n-tricosanedioate.

[d] $CH_3OOC—(CD_2)_{n,n-1,n-2,...}]^+$

[e] $D\overset{+}{O}{=}C—CD{=}CD_2$, m/e 91
      |
      CH_3O

[f] $DO—C{=}CD_2]^+$, m/e 77
      |
      CH_3O

[g] $C_nD_{2n-1}^+$, where $n = 2, 3, 4, 5, 6, ...$ and m/e 30, 46, 62, 78, 94, ...

[h] $C_nD_{2n}^+$, where $n = 2\text{–}6...$ and m/e 32, 48, 64, 80, 96 ...

[i] $C_nD_{2n+1}^+$, where $n = 2\text{–}6...$ and m/e 34, 50, 66, 82, 98 ...

The normal-chain dimethyl perdeuterio-dicarboxylates also yield mass spectra similar to those of the nondeuteriated analogs. The latter have been previously studied by Ryhage and Stenhagen[19] and by Howe and Williams.[15,16] An outline on these studies has been made by Spiteller and Spiteller.[21] The following presented ions correspond to the characteristic peaks observed on spectra of the deuteriated diesters:

[j] $CH_3OOC—(CD_2)_n—COOCH_3]^+$ or M^+

[k] $CH_3OOC—(CD_2)_n—C{\equiv}\overset{+}{O}$ or $M^+ - 31$, where 31: $CH_3O^·$

[l] $CH_3OOC—(CD_2)_n]^+$ or $M^+ - 59$, or $(M^+ - 31) - 28$, where 28: CO

[m] $CH_3OOC—(CD_2)_{n-1}]^+$ or $M^+ - 75$, or $(M^+ - 31) - 44$, where 44: CD_2CO

[n] $CH_3O—(CD_2)_{n-1}]^+$ or $M^+ - 103$, or $(M^+ - 59) - 44$

[o] $CH_3O—(CD_2)_{n-2}]^+$ or $M^+ - 119$, or $(M^+ - 75) - 44$

[p] $CH_3O—(CD_2)_{n-3, n-4,...}]^+$

[q] $O—COCH_3$
      ⟍ |
        $CD—(CD_2)_{n-1}—C{\equiv}O^{+\cdot}$ or $M^+ - 33$, where 33: CH_3OD

[r] $O—COCH_3$
      ⟍ |
        $CD—(CD_2)_{n-1}]^{+\cdot}$ or $M^+ - 61$, or $(M^+ - 33) - 28$

[s] $OC—(CD_2)_n—CO]^+$ or $M^+ - 66$, or $(M^+ - 33) - 33$

[t] CH_3OOC^+, m/e 59

[u]
```
              OD]+
              |
              C
            /   ‖
        D2C      CD, m/e 92
         |       |
        D2C———CD2
```

[u']
```
              OD]+
              |
              C
            /   ‖
        D2C      CD, m/e 108
         |       |
        D2C      CD2
          \      /
           CD2
```

$$\text{OD}]^+$$
$$\text{C}$$

[u''] D_2C CD, m/e 124

D_2C CD_2

CD_2—CD_2

[v]   Ions of $m/e = 92 + x16$, where $x > 2$

[w] $M^+ - 108$

[y] $M^+ - 128$

[e] $\overset{+}{\text{DO}}{=}\text{C—CD}{=}\text{CD}_2$, m/e 91
$$\text{CH}_3\text{O}$$

[f] $\text{DO—C}{=}\text{CD}_2\,]^+$, m/e 77
$$\text{CH}_3\text{O}$$

[g] $C_n D_{2n-1}^+$, where $n = 2, 3, 4, 5, 6, \ldots$ and m/e 30, 46, 62, 78, 94, $\ldots$

[h] $C_n D_{2n}^+$, where $n = 2$–$6 \ldots$ and m/e 32, 48, 64, 80, 96 $\ldots$

[i] $C_n D_{2n+1}^+$, where $n = 2$–$6 \ldots$ and m/e 34, 50, 66, 82, 98 $\ldots$

Recently it has been shown that deuteriated mono- and dicarboxylic acid pyrrolidides yield more simple mass spectra than those of corresponding methyl esters.[1,2] This prompted us to investigate low resolution mass spectra of the above described deuteriated compounds in the form of pyrrolidide derivatives.[3]

Acknowledgements—This work was supported by grants from *Statens Naturvetenskapliga Forskningsråd* (The Swedish Natural Science Research Council) and from *Styrelsen för Teknisk Utveckling* (The Swedish Board for Technical Development), which are gratefully acknowledged.

REFERENCES

1. ANDERSSON, B-Å., DINGER, F. and DINH-NGUYÊN, NG. *Chem. Scr.* **8**, 200 (1975).
2. ANDERSSON, B-Å., DINGER, F. and DINH-NGUÊN, NG. *Chem. Scr.* **9**, 155 (1976).
3. ANDERSSON, B-Å., DINGER, F., DINH-NGUYÊN, NG. and RAAL, A., *Chem. Scr.* **10**, 114 (1976).
4. BUDZIKIEWICZ, H., DJERASSI, C. and WILLIAMS, D. H. *Interpretation of Mass Spectra of Organic Compounds*, Chapter 1, Holden-Day, San Francisco (1965).
5. DINH-NGUYÊN, NG. *Swed. Pat. Appl.* 76-10801-8.
6. DINH-NGUYÊN, NG. and RAAL, A. *Acta Chem. Scand.* **23**, 1442 (1969).
7. DINH-NGUYÊN, NG. and RAAL, A. *J. Res. Inst. Catal., Hokkaido Univ.* **17**, 171 (1969).
8. DINH-NGUYÊN, NG. and RAAL, A. *Swed. Pat.* 349, 291; *Br. Pat.* 1,276,378; *U.S. Pat.* 3,717,664.
9. DINH-NGUYÊN, NG. and RAAL, A. *Acta Chem. Scand.* **24**, 3416 (1970).
10. DINH-NGUYÊN, NG., RAAL A. and STENHAGEN, E. *Chem. Scr.* **2**, 171 (1972).
11. DINH-NGUYÊN, NG. and RAAL, A. *Chem. Scr.* **10**, 173 (1976).
12. DINH-NGUYÊN, NG. and STENHAGEN, E. *Swed. Pat.* 358,875; *Br. Pat.* 1,103,607; *Can. Pat.* 950,514; *Fr. Pat.* 1,466–088; *W. Ger Pat.* D 49,213.
13. DINH-NGUYÊN, NG. and STENHAGEN, E. *Acta Chem. Scand.* **20**, 1423 (1966).
14. DINH-NGUYÊN, NG. and VINCENT, J. *Mycopathologia* **58** (3), 137 (1976).
15. HOWE, I. and WILLIAMS, D. H. *Chem. Commun.* **1967**, 733 (1967).
16. HOWE, I. and WILLIAMS, D. H. *J. Chem. Soc.* **1968**, 202 (1968).
17. McCLOSKEY, J. In: *Methods in Enzymology*, Vol. 35, part B, pp. 340–348, Ed. J. M. LOWENSTEIN, Academic Press, New York and London (1975).
18. RYHAGE, R. and STENHAGEN, E. *Ark. Kemi* **13**, 523 (1959).
19. RYHAGE, R. and STENHAGEN, E. *Ark. Kemi* **14**, 497 (1959).
20. RYHAGE, R. and STENHAGEN, E. In: *Mass Spectrometry of Organic Ions*, Chapter 9, Ed. F. W. McLAFFERTY, Academic Press, New York (1963).
21. SPITELLER, M. and SPITELLER, G. *Massenspektrensammlung von Lösungsmitteln, Verunreinigungen, Säulenbelegmaterialien und einfachen aliphatischen Verbindungen*, Springer-Verlag, Wien and New York (1973).

Prog. Chem. Fats other Lipids. Vol. 16, pp. 207–230. Pergamon Press, 1978. Printed in Great Britain

MASS-SPECTROMETRIC SEQUENCE STUDIES OF LIPID-LINKED OLIGOSACCHARIDES BLOOD-GROUP FUCOLIPIDS, GANGLIOSIDES AND RELATED CELL-SURFACE RECEPTORS

KARL-ANDERS KARLSSON

Department of Medical Biochemistry, University of Göteborg, Fack, S-400 33 Göteborg 33, Sweden

CONTENTS

I. INTRODUCTION 207

II. CHOICE OF DERIVATIVES AND TECHNIQUE OF ANALYSIS 208

III. MASS SPECTRA 209
  A. Cytolipin S 209
  B. The receptor for cholera toxin 214
  C. Blood-group fucolipids 221

IV. CONCLUSION 224

ACKNOWLEDGEMENTS 229

REFERENCES 229

I. INTRODUCTION

Only 20 years have passed since one of the first successful applications of mass spectrometry for the structure analysis of more complex organic molecules was done by Stenhagen and collaborators (see the interesting historical survey by Einar Stenhagen).[33] The rapid development of this technique to become a central tool in organic and biochemical analysis is illustrated by the recent (July, 1976) landing on Mars of the spacecraft *Viking*, equipped with a gas chromatograph–mass spectrometer to test modern ideas on the origin of life.[2] Also on this combined method of separation and analysis, Einar and Stina Stenhagen made a pioneering contribution[32] (see also comments on apparatus in ref. 24).

Although Einar and Stina Stenhagen had their positions at Uppsala University and later in Göteborg, the earlier studies on mass spectrometry were done at the Karolinska Institute in Stockholm, where Einar Hammarsten had a mass spectrometric laboratory for isotope work, which was run by Ragnar Ryhage. It was not until 1968 that a commercial high resolution instrument was installed at our Department in Göteborg. This aroused my specific interest at that time because we had found a sulphatide (galactosylceramide 3-sulphate) related in its tissue appearance with Na^+ transport[14] and needed a specific detection of this glycolipid on a microscale. The first results[9,19] from analyzing some derivatives of sphingolipids then stimulated a more serious investigation of complex glycosphingolipids, which is our major research object at present.[14]

The analysis of an oligomeric compound, the *N*-trifluoroacetyl methyl ester of a dipeptide, was first reported in 1958 by Andersson,[1] a collaborator of Stenhagen. Some years later Lederer and coworkers gave a considerable stimulation to the work on sequencing of peptides by their mass spectrometric study of fortuitin, a peptidolipid isolated from a mycobacterium.[3] In addition to the recognition of the molecular weight (at m/e 1359), it was possible to deduce the entire sequence of the nine amino acids and identify the fatty acids. Present mass spectrometric techniques for the sequence analysis of oligopeptide derivatives (for a review, see ref. 4) are based on a direct inlet fractional vaporizaion,[20,36] or on highly developed gas chromatography–mass spectrometry.[22]

A similar approach for biologically important saccharides has not been presented, although small saccharides are within range of analysis (for a review, see ref. 23). Such a method is badly needed as a sensitive supplement to chemical degradative methods, which often create problems of interpretation due to various chemical changes of the monomers. Ideally, the mass spectrum should contain readily recognizable and intensity-balanced peaks corresponding to the molecular ion and to fragment ions derived from cleavage at the linkage of monomers (glycosidic oxygen). Although one would expect less success in this respect for saccharide than for peptide, due to a higher mass, more functional groups, different kinds of linkage (position and configuration) and the hetero-cyclic structure of the monomers, we have been able to get a reliable carbohydrate sequence analysis of sphingosine-linked oligosaccharides with as many as nine sugars. In addition, the lipid components are easily interpretable.

The glycosphingolipids are mostly cell surface receptors and antigens present in very low concentrations. This creates a particular need for a specific detection on a micro-scale. The examples for illustration in the present paper have been selected from our own project and from cooperative efforts with other research groups.

II. CHOICE OF DERIVATIVES AND TECHNIQUE OF ANALYSIS

A more detailed discussion on derivatives and sample handling has been presented elsewhere.[13] The preferred derivatives are outlined in Fig. 1. Although permethylation is more laborious than many other types of derivatizations, it has the advantage of giving a minimal mass increase. Also, this derivative produces relatively abundant primary ions and fewer rearrangements compared with other derivatives. With acetyl derivatives, pyrolysis appears if there are more than three sugars of the glycolipid, and although trimethylsilyl ethers may give important information on lipophilic components and indicate some positions of binding between sugars,[5] they are not particularly suitable for the sequencing of sugars, as shown in a comparative study on a tumor ganglioside.[17]

The reduction of methylated glycolipid with $LiAlH_4$, not used before in sugar analysis, removes amide oxygens to increase the volatility and improve stability upon electron impact. Also, a change in mass by 14 units is evidence for a reduced amide. In addition, the produced amines of amino sugars, by protonation, create a stability of vicinal glycosidic bonds upon hydrolysis,[13] presenting important information in degradative studies (to be published).

The mass spectrometer used is model MS 902 (AEI, Ltd.) with electron ionization and a separate probe heater. The trap current was 0.5 mA and the accelerator voltage 8 kV. High-molecular-weight fragments were brought into focus by decreasing the accelerator voltage. The mass numbers of individual spectra were obtained by counting by hand, which seems reliable for this type of derivatives, and are therefore nominal masses.

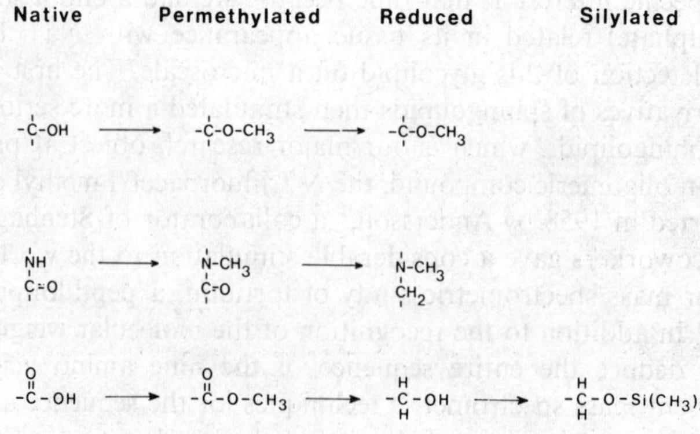

Native	Permethylated	Reduced	Silylated

$$-\overset{|}{\underset{|}{C}}-OH \longrightarrow -\overset{|}{\underset{|}{C}}-O-CH_3 \longrightarrow -\overset{|}{\underset{|}{C}}-O-CH_3$$

$$\underset{\overset{|}{C}=O}{\overset{NH}{|}} \longrightarrow \underset{\overset{|}{C}=O}{\overset{N-CH_3}{|}} \longrightarrow \underset{\overset{|}{CH_2}}{\overset{N-CH_3}{|}}$$

$$-\overset{O}{\overset{||}{C}}-OH \longrightarrow -\overset{O}{\overset{||}{C}}-O-CH_3 \longrightarrow -\overset{H}{\underset{H}{C}}-OH \longrightarrow -\overset{H}{\underset{H}{C}}-O-Si(CH_3)_3$$

FIG. 1. Outline of the three derivatives used for mass spectrometry of glycosphingolipids.

Although the detailed carbohydrate structures are known in several cases, the formulae for interpretation presented above the spectra are most often given in a schematic way, corresponding to the available mass spectrometric data.

III. MASS SPECTRA

Using the derivatives discussed above, we have now obtained highly informative spectra of all major types of glycosphingolipids, with varying number and sequence of hexose, fucose, hexosamine and the two types of sialic acid (N-acetyl- and N-glycolylneuraminic acid).[7,10–18,26–30] Based on this information, it has been possible to detect several previously unknown substances, even in mixtures with other glycolipids.[13,28]

A. *Cytolipin S*

M. M. Rapport and coworkers (New York) have recently isolated in small amounts a novel surface antigen from rat spleen (unpublished), designated cytolipin S and shown to be a glycosphingolipid with different chemical and immunochemical properties from cytolipin F, K and R characterized before in the same laboratory. It has, however, many characteristics in common with the glycolipid obtained by removal of sialic acid from the ganglioside shown in Fig. 6, an asialoganglioside or Galβ1 → 3GalNAcβ1 → 4Galβ1 → 4Glcβ1 → 1 Ceramide. Chromatographic properties, sugar components, and complement fixation ability are all very similar for the two glycolipids. However, in hemagglutination-inhibition studies, cytolipin S is considerably more effective than asialoganglioside, using antibodies raised against rat erythrocytes. The question was whether a dissimilarity in structure of the carbohydrate part might explain this clear difference in immunologic properties, and as a next step of the investigation, 200 μg of each glycolipid were sent to us for mass spectrometric analysis.

Spectra of methylated and methylated plus reduced derivatives of both glycolipids are shown in Figs. 2–5, and the two samples seem to be identical except for a difference

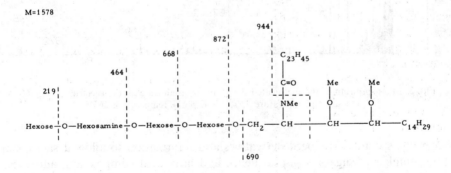

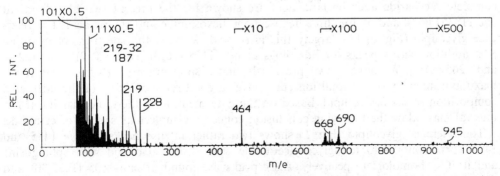

FIG. 2. Mass spectrum of the methylated derivative of cytolipin S. Electron energy 60 eV, ion source temperature 340°C, and probe temperature 260°C.

in ceramide components. The characteristic features of glycolipids in general is that spectra of methylated derivatives provide information on the sequence of the first few sugars from the nonreducing end, and show ceramide ions. An identical sequence in the two glycolipids (Figs. 2 and 3) is shown as hexose–hexosamine–hexose by the ions at m/e 187, 219, 464 and 668. Some evidence for an additional hexose is provided by the rearrangement ions at m/e 945 (944 + 1). Although the intensity scale was changed here to allow a comparison of sugar peaks, the ceramide regions differ for the two samples. Cytolipin S (Fig. 2) has clear peaks for nervonic acid and phytosphingosine, at m/e 690 and 658 (690-32), while the reference glycolipid (Fig. 3) lacks these peaks and shows ions at m/e 544 and 572, corresponding to stearic acid combined with sphingosine and its C_{20} homolog, respectively. Relative amounts of ceramide components may be supplemented from spectra of reduced derivatives.

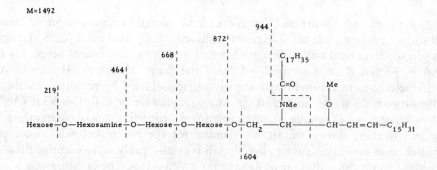

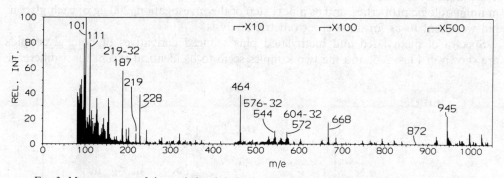

FIG. 3. Mass spectrum of the methylated derivative of asialoganglioside. Electron energy 60 eV, ion source temperature 330°C, and probe temperature 265°C.

Generally, spectra of reduced derivatives add information to allow a safe conclusion on total number of sugars, sugar sequence and fatty acid composition, and these data are finally supported by molecular weight ions. Very abundant ions containing the complete saccharide and the fatty acid are shown for the present two glycolipids at m/e 1183 (Figs. 4 and 5), for three hexoses, one hexosamine and stearic acid. The reference glycolipid (Fig. 5) has largely this fatty acid, whereas the spectrum of cytolipin S in addition shows peaks for homologs at m/e 1155 (16:0), 1211 (20:0), 1239 (22:0) and 1265 (24:1). We have shown previously that a spectrum properly recorded at the maximum intensity of the total ion curve gives a good representation of the fatty acid composition of the native lipid. Based on the data mentioned, the molecular ion (M-1) interval may allow the following conclusion concerning molecular species of ceramide.

The reference glycolipid (Fig. 5) shows two rather intense peaks at m/e 1435 and 1463, corresponding to tetraglycosylceramide with stearic acid and with sphingosine and its C_{20} homolog, respectively. Also, peaks are found at m/e 1405 (1435-30) and 1433 (1463-30). This conclusion is not only supported by the data presented above but also by the specific rearrangement ions found for the two bases at 338 and 366,

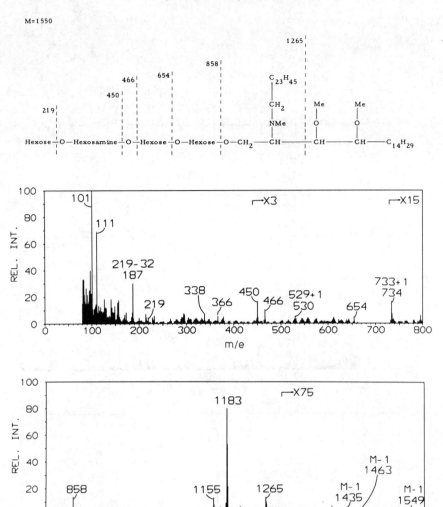

FIG. 4. Mass spectrum of the methylated and reduced derivative of cytolipin S. Electron energy 36 eV, ion source temperature 280°C, and probe temperature 240°C.

respectively. Evidence for these two molecular species in a similar relative concentration is found also for cytolipin S (Fig. 4). However, in addition, the ions at m/e 1549 and 1519 (1549-30) indicate the molecular species with nervonic acid and phytosphingosine. The absence of ions at m/e 1545 and 1515 (from a species with nervonic acid and the C_{20} homolog of sphingosine) and the low-abundant ion at m/e 1467 and 1437 (from a species with stearic acid and phytosphingosine) may be taken as evidence for a preferential combination in cytolipin S of dihydroxy base with shorter-chain fatty acid, and of trihydroxy base with longer-chain fatty acid, which has been observed for some glycosphingolipids before.[15]

The sequence of sugars is established as shown in the formulae of Figs. 4 and 5 by the ions at m/e 187, 219, 450, 466, 654 and 858. An isomeric, branched saccharide chain is not likely, due to the ions at m/e 530 and 734, being rearrangement ions with the fatty acid and one or two hexoses (compare indications below the formula of Fig. 5).

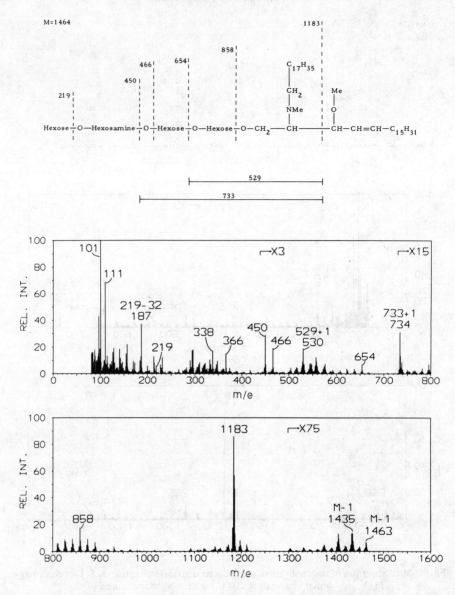

FIG. 5. Mass spectrum of the methylated and reduced derivative of asialoganglioside. Electron energy 36 eV, ion source temperature 290°C, and probe temperature 270°C.

The difference in agglutination inhibition for the two glycolipids cannot be definitely explained at the present stage of structural characterization. Usually the mass spectrometer cannot distinguish different positions or configurations of binding between monosaccharides. However, a $1 \rightarrow 3$ rather than a $1 \rightarrow 4$ binding of galactose to N-acetylgalactosamine of cytolipin S, as known for the reference, is likely as m/e 182 is absent in the spectra of methylated derivatives (Figs. 2 and 3). This ion is derived from a hexosamine with a $1 \rightarrow 4$ linked hexose (see below). Although a small difference in carbohydrate structure cannot yet be excluded, the identical chromatographic behaviour of both native and methylated samples, and the mass spectral data, speak against this. The different properties of the two glycolipids in agglutination inhibition, but not concerning complement fixation, could instead be due to the ceramide differences, affecting micelle formation. The association of ceramide with auxiliary lipid should be stronger with phytosphingosine present than sphingosine, due to one more hydroxyl group for lateral hydrogen bonding with lecithin (see the recent model for ceramide function in the plasma membrane).[14] If this is correct, the influence of ceramide structure on micelle

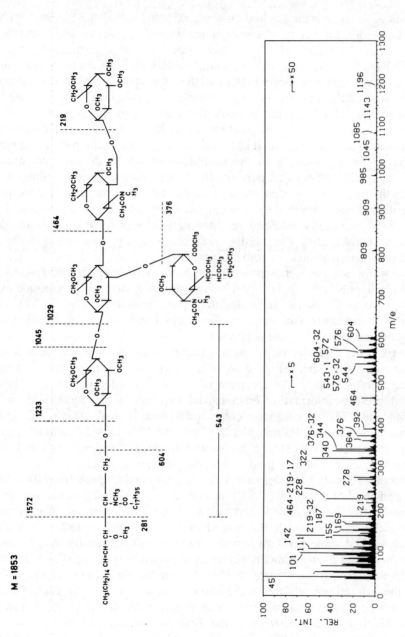

FIG. 6. Mass spectrum of the methylated derivative of a ganglioside, the receptor for cholera toxin. Electron energy 22.5 eV, ion source temperature 290°C, and probe temperature 280°C.

formation should be more critical in the agglutination inhibition experiment with only a 4-fold excess of auxiliary lipid (lecithin), than in the complement fixation test with a 100-fold excess (cholesterol–lecithin, 2:1).

B. *The Receptor for Cholera Toxin*

Some years ago, van Heyningen and co-workers[34] discovered the binding of cholera toxin to gangliosides, the group of glycosphingolipids that contain sialic acid. Several groups have later demonstrated a very high specificity for a particular ganglioside, Galβ1 → 3GalNAcβ1 → 4(NeuNAcα2 → 2)Galβ1 → 4Glcβ1 → 1 Ceramide (see formula of Fig. 6). Evidence has also been presented that the *in vivo* action of the toxin is mediated by this ganglioside and not by protein-bound receptor.[8] The possible therapeutic consequences of these findings make the study of mechanism of toxin action and of specificity of toxin binding of great relevance. H. Wiegandt and co-corkers (Marburg, W. Germany) have made several chemical modifications of the part of the receptor molecule that is most proximate to the membrane matrix, that is the ceramide and the glucose, to be able to tag this part specifically (e.g. fluorescence labelling) without interfering with toxin binding capacity. The covalent binding of modified receptor to large molecules or glass beads may be used therapeutically to adsorb ingested toxin molecules. These chemically modified receptor molecules may be difficult to identify conclusively by conventional degradative methods. It was in this phase of the work that we did mass spectrometry on some substances.

For gangliosides, we have chosen a third type of derivative, in addition to those for nonacid glycolipids (see Fig. 1). Upon reduction, the methylated ganglioside exposes a hydroxyl group at C_1 of the sialic acid (compare formula of Fig. 7). This results in a low yield of fragments which contain the reduced sialic acid. By silylation of this alcohol, a remarkable stabilization is effected.

Mass spectra of the three derivatives of intact receptor ganglioside have been discussed before[17] and will be used here (Figs. 6–8) for a comparison with spectra of chemically modified receptor molecules. The spectrum of the methylated ganglioside is shown in Fig. 6 and it presents information on a partial sequence of the saccharide (m/e 187, 219, 464, 344 and 376), on ceramide components (stearic acid at m/e 322 and 340, and sphingosine and its C_{20} homolog at m/e 364 and 392), and on ceramide (m/e 544 and 576 for the species with stearic acid and sphingosine, and m/e 572 and 604 for the species with stearic acid and the C_{20} homolog of sphingosine).

After reduction (Fig. 7), there appears a rather low-intensity peak from the complete saccharide and the fatty acid (m/e 1502), in contrast to the situation for nonacid glycosphingolipids (see Figs. 4 and 5, and below). Apparently, there is an easy losss of the reduced sialic acid to provide the very abundant ions at m/e 1169. Ions indicating the terminal reduced sialic acid are almost absent (m/e 334). On the other hand, the mass spectrum obtained after trimethylsilylation of the hydroxy group (Fig. 8) has clearly recognizable peaks for the complete saccharide and the fatty acid (at m/e 1574), as well as for the reduced and silylated sialic acid (m/e 374 and 406). The correct sequence and branching point are obtained from both reduced derivatives (Figs. 7 and 8) by ions at m/e 530 (see explanation below the formulae), showing that the hexose next to ceramide is monosubstituted. A disubstitution of the second hexose from ceramide is shown by the rearrangement ions at m/e 720. A monosubstitution should have given a shift to 734 (compare Figs. 4 and 5). This type of rearrangement ions are important for the identification of sialic acid binding in more complex gangliosides.[11] In collaboration with N. F. Avrova and co-workers, Leningrad, we have by this technique analyzed gangliosides of lamprey brain with three sialic acids bound in different ways (unpublished results). Ceramide ions are identical for both reduced derivatives (Figs. 7 and 8) and appear at m/e 296 and 310 (stearic acid), 388 and 366 (the two homologs of long-chain bases) and 548 and 576 (the combinations of stearic acid and long-chain base).

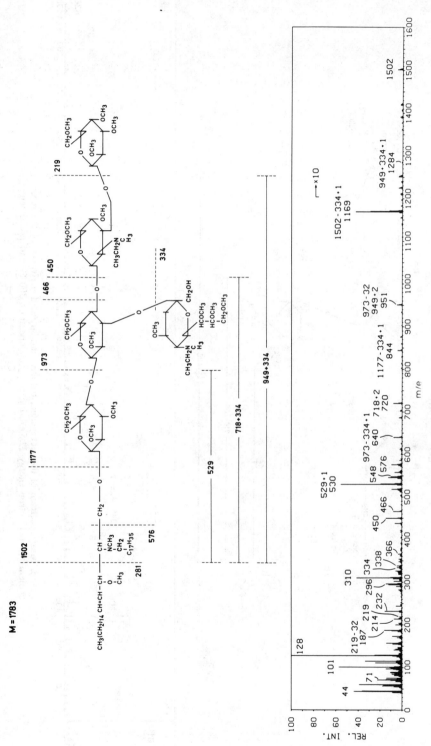

FIG. 7. Mass spectrum of the methylated and reduced derivative of the receptor for cholera toxin. Electron energy 22.5 eV, ion source temperature 270°C, and probe temperature 295°C.

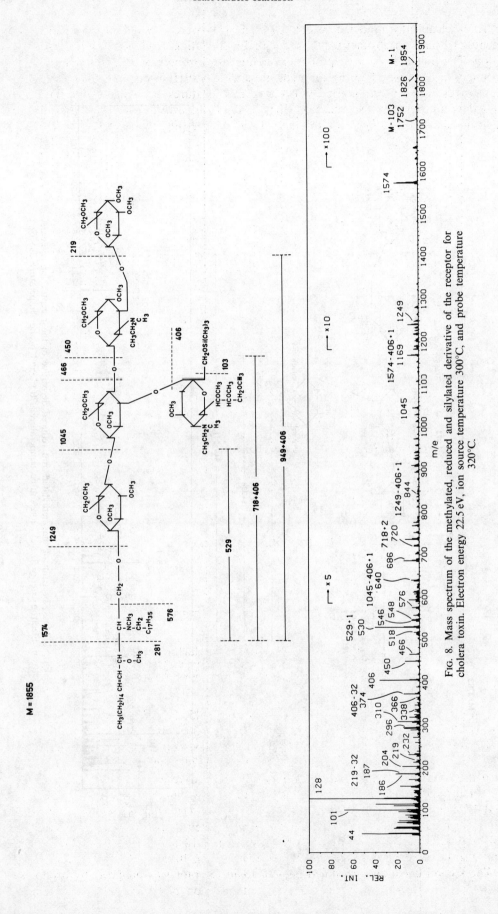

FIG. 8. Mass spectrum of the methylated, reduced and silylated derivative of the receptor for cholera toxin. Electron energy 22.5 eV, ion source temperature 300°C, and probe temperature 320°C.

Three chemically modified receptor gangliosides were checked for their structural identity by mass spectrometry after sequential methylation, reduction and trimethylsilylation. The first modified receptor was obtained by ozonolysis at the double bond of sphingosine (compare Fig. 6) and reduction with NaBH$_4$. Mass spectra of the three derivatives of this product (Figs. 9–11) have most features in common with the corresponding spectra of intact receptor (Figs. 6–8). Thus, evidence for a retained carbohydrate structure is shown for the methylated sample (Fig. 9 may be compared with

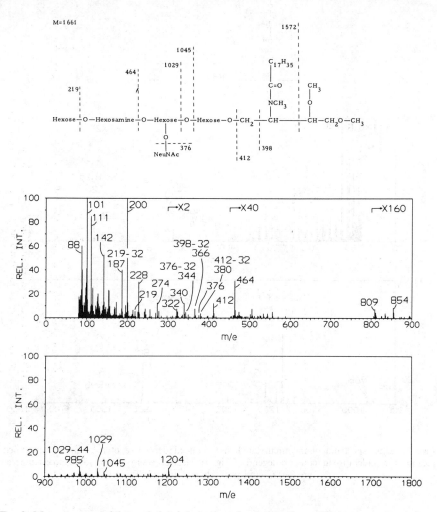

FIG. 9. Mass spectrum of the methylated derivative of a chemically modified receptor for cholera toxin (obtained by ozonolysis and NaBH$_4$ reduction; compare the formula of Fig. 6). Electron energy 32 eV, ion source temperature 310°C, and probe temperature 245°C.

Fig. 6) at m/e 187, 219, 344, 376, 464, 1029 and 1045; for the reduced sample (Fig. 10 may be compared with Fig. 7) at m/e 187, 219, 450, 466, 530, 720, 951, 1169 and 1502; and for the silylated derivative (Fig. 11 may be compared with Fig. 8) at m/e 187, 219, 374, 406, 450, 466, 530, 720, 1169 and 1574. The specific ions for stearic acid are found for both intact receptor and modified receptor, at m/e 322 and 340 for methylated derivatives (Figs. 6 and 9), and at m/e 296 and 310 (but see also 1502 and 1574) for reduced derivatives (Figs. 7, 8, 10 and 11). Finally, conclusive evidence for the modified ceramide is shown for the methylated derivative (Fig. 9) at m/e 380, 398 and 412, and for the reduced derivatives (Figs. 10 and 11) at m/e 384 and 398 (but also by molecular weight ions at m/e 1590 of Fig. 10 and 1662 of Fig. 11).

The second chemically modified cholera toxin receptor was obtained by ozonolysis and alkaline degradation, followed by reductamination of the reducing glucose with

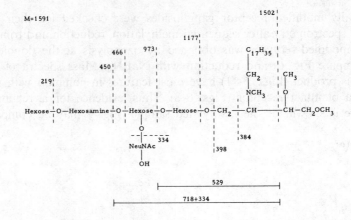

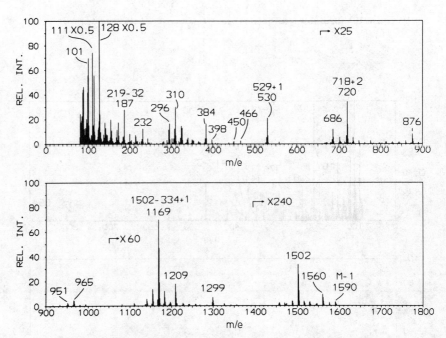

Fig. 10. Mass spectrum of the methylated and reduced derivative of a chemically modified receptor for cholera toxin (compare legend of Fig. 9). Electron energy 52 eV, ion source temperature 320°C, and probe temperature 240°C.

stearylamine. The expected substitution at carbon one of glucose is as follows (compare Fig. 12):

$$-CH_2-NH-C_{18}H_{37}.$$

Upon methylation, a quaternary amine is produced:

$$-CH_2-N_+(CH_3)_2C_{18}H_{37}.$$

With heating the sample in the ion source, this amine is expected to yield a tertiary amine by loss of one of the alkyl groups (see formulae of Fig. 12). The mass spectrum of the methylated derivative (Fig. 12) looks somewhat complex, partially due to two modes of alkyl chain loss, CH_3 or $C_{18}H_{37}$. Also, a small amount of a byproduct from the synthesis reaction (a dialkylated amine as shown by the formula at top, left) is indicated by molecular weight ions at m/e 1768 and 1738 (1768-30). Nevertheless, mass spectra of the usual type of derivatives (Figs. 12–14) show clearly recognizable peaks diagnostic for the expected unmodified terminal tetrasaccharide (compare the discussion above). Although there are several as yet unidentified rearrangement ions indicated in the spectra, the alkylaminated glucose part is easily detected. A loss of one of the methyl groups from the quaternary amine will give the product shown in the top right formula (M_2) of Fig. 12, producing ions from the modified glucose at m/e 296 and 502, and also molecular weight ions at m/e 1472, 1500 and 1530. Loss of the stearyl chain from the quaternary amine will give the species M_3 with molecular weight ions at m/e 1234, 1262 and 1292. A final species, M_4, is due to a loss of sialic acid from M_3 and is shown at m/e 887 and 917 (M_4-1).

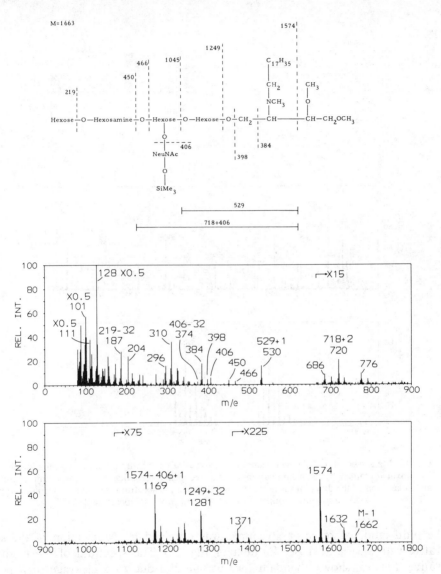

FIG. 11. Mass spectrum of the methylated, reduced and silylated derivative of a chemically modified receptor for cholera toxin (compare legend of Fig. 9). Electron energy 50 eV, ion source temperature 320°C and probe temperature 200°C.

The reduced derivatives of this modified receptor also give strong peaks for the N-alkylated glucose, at m/e 296 and 502 (Figs. 13 and 14), and supporting molecular weight ions at m/e 1444 and 1474 (Fig. 13) and 1516 and 1546 (Fig. 14). Rather abundant rearrangement ions are produced by a loss of 292 mass units and are indicated at m/e 1183 (Fig. 13) and 1255 (Fig. 14). The loss of reduced sialic acid is shown by ions at m/e 850 and 1141 (Fig. 13) and 850 (Fig. 14). (The peak at m/e 850 of Fig. 13 may also be derived from a second molecular species, M_2, by loss of 292 mass units, see top formula).

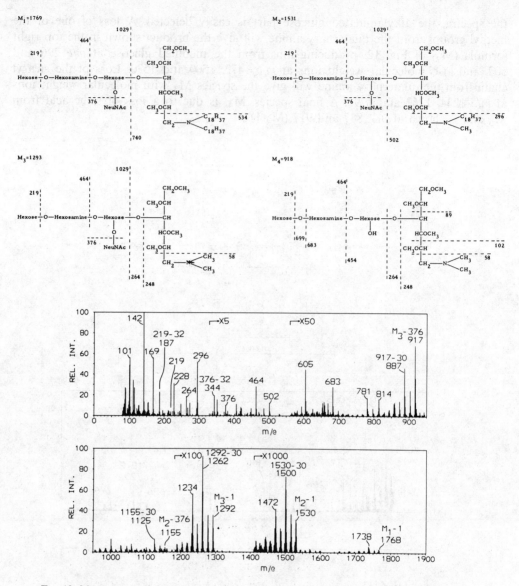

FIG. 12. Mass spectrum of the methylated derivative of a chemically modified receptor for cholera toxin (obtained by reductamination of the free pentasaccharide with stearylamine). For
an explanation of the different formulae presented above the spectrum, see text. Electron energy
44 eV, ion source temperature 300°C, and probe temperature 240°C.

The third modified receptor glycolipid was obtained by *N*-acylation of the second
derivative with stearic acid (see formula of Fig. 15). The spectrum of the methylated
derivative (Fig. 15) shows abundant ions at *m/e* 592 due to a cleavage in *β* position
to the nitrogen of the modified glucose. Upon reduction of the amide, the cleavage
appears in *α* position to give the very intense peak at *m/e* 534 (Figs. 16 and 17). The
intact sugar is evidenced by molecular weight ions. As for the second modified receptor,
there is a sequential loss in the reduced derivatives (Figs. 16 and 17) of 292 mass
units and the sialic acid.

Direct inlet mass spectrometry thus conclusively showed the identity of the modified
receptor molecules, a result not easily available by conventional degradative methods,
particularly on a micro scale. This is of importance for validating experiments on the
mechanism of cholera toxin action. Binding properties of different derivatives to cholera
toxin are discussed elsewhere,[35] and a recent experiment on lymphocytes utilized the
second modified receptor discussed above, after *N*-acylation with the fluorescing
DANSYL reagent to visualize the receptor on the living cell.[25]

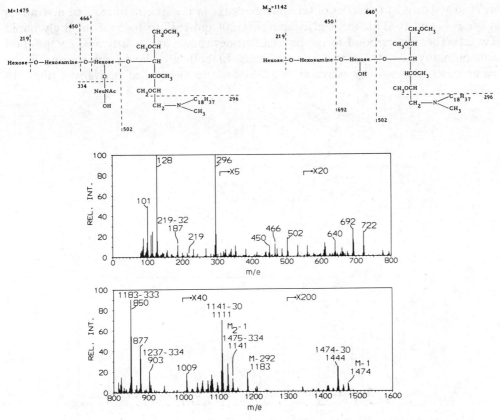

FIG. 13. Mass spectrum of the methylated and reduced derivative of a chemically modified receptor for cholera toxin (compare legend of Fig. 12). Electron energy 40 eV, ion source temperature 300°C, and probe temperature 250°C.

C. Blood-group Fucolipids

Although the saccharide determinants of the ABH and Lewis blood groups may be rather abundant in secretions (as glycoproteins in epithelial secretions like saliva and ovarial cyst fluid, or as free oligosaccharides of milk), the membrane-bound determinants of primarily glycosphingolipid nature are normally present in very small amounts. We have shown that mass spectrometry affords a sensitive and specific detection of these substances, which are thought to be involved in self and not self-discrimination processes between cells. In cooperation with J. M. McKibbin's group (Birmingham, Alabama), we have analyzed a number of blood-group active glycolipids isolated from human and dog small intestine.[13,26-30] We have also subjected a series of glycolipids from erythrocytes[12-14] of different species to mass spectrometry using the two types of derivatives (see Fig. 1). The results demonstrate in all cases an unambiguous identification of the saccharide making up the immunodeterminant. I will restrict the illustration here to three glycolipid fractions that we have isolated from human A_1 and B erythrocytes.

Mass spectra of the major A-active and the major B-active glycolipids are reproduced in Figs. 18–21. The two samples produce very similar spectra except for the specific mass shifts due to a terminal galactose in the B and a terminal N-acetylgalactosamine in the A glycolipid. These shifts are 41 mass units for methylated derivatives (Figs. 18 and 19) and 27 mass units for reduced derivatives (Figs. 20 and 21). These spectra represent in a strict sense the first conclusive overall saccharide sequence of these cell surface antigens, although the immunological determinants (terminal trisaccharides) are known from extensive studies of secreted glycoproteins. The type and ratio of sugars are clearly shown by the series of very intense peaks in the spectra of reduced derivatives at m/e 1644 and 1672 for the A glycolipid (Fig. 20) and 1617 to 1645 for the B glycolipid (Fig. 21). These correspond to the hexasaccharides shown and a fatty acid composition with primarily C_{22}, C_{23} and C_{24} fatty acids. In both glycolipids, sphingosine is the major base, as shown by peaks at m/e 364 for the methylated derivatives (Figs. 18

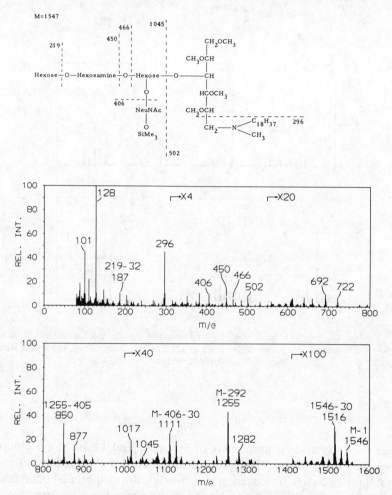

FIG. 14. Mass spectrum of the methylated, reduced and silylated derivative of a chemically modified receptor for cholera toxin (compare legend of Fig. 12). Electron energy 40 eV, ion source temperature 300°C, and probe temperature 260°C.

and 19) and 338 for the reduced derivatives (Figs. 20 and 21). Thus, the two glycolipids are hexaglycosylceramides with sphingosine and C_{22} to C_{24} fatty acids as major ceramide components. The sugar ratios are hexose:hexosamine:fucose, 3:2:1 and 4:1:1 for A and B glycolipid, respectively. Furthermore, relatively intense sequence ions confirm conclusively the sequences proposed. In addition, we have shown that the very intense m/e 182 is derived from a hexosamine substituted in position 4 but not in position 3. Therefore, this peak is direct evidence for type 2 saccharide chains (Gal$\beta1 \rightarrow$ 4GlcNAc) in these human erythrocyte glycolipids, which is supported by degradative data. Blood-group A active hexaglycosylceramide of dog small intestine is of this type, but the same glycolipid of human small intestine has type 1 saccharide chain (Gal$\beta1 \rightarrow$ 3GlcNAc), as shown by mass spectrometry[26] and by analysis after degradation.[31]

The apparent specificity in this type of fingerprinting should allow the specific detection of separate saccharide species in a mixture, and we have recently identified a novel

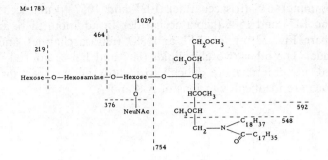

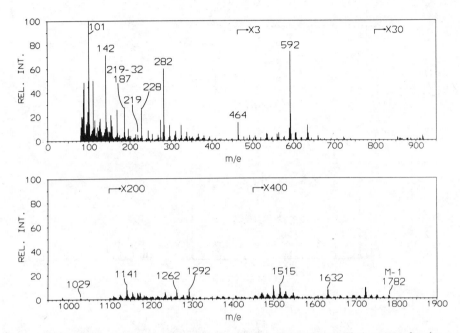

FIG. 15. Mass spectrum of the methylated derivative of a chemically modified receptor for cholera toxin (obtained by reductamination of the free pentasaccharide with stearylamine followed by N-acylation with stearic acid). Electron energy 44 eV, ion source temperature 300°C, and probe temperature 250°C.

heptaglycosylceramide in this way,[28] and have discovered several other glycolipids, primarily of human erythrocytes (in preparation). I will finally illustrate the analysis of a complex, slow-moving glycolipid fraction obtained from blood-group A_1 erythrocytes. After these spectra were recorded, characterizations have been done by further separation and degradations giving evidence for the presence of three different A-active glycolipids of the structures shown in Fig. 22, in addition to the glycolipids noted below. Mass spectra of the original mixture are shown in Figs. 23 and 24.

The spectrum of the reduced mixture (Fig. 24), which, as shown for the glycolipids discussed above, may inform on the exact number of sugars in the glycolipid, has a series of intense peaks at m/e 2253 to 2281, similar in profile to the saccharide-fatty acid ions of the two hexaglycosylceramides (Figs. 20 and 21). This corresponds to a nonasaccharide and C_{22} to C_{24} normal fatty acids, with the sugar ratio hexose:hexosamine:fucose, 4:3:2. Furthermore, there are ions indicating the presence of at least three more types of saccharide in small amounts combined with the same fatty acids. The peak at m/e 2107 corresponds to A-8 (see Fig. 22, hexose:hexosamine:fucose, 4:3:1), that at m/e 2050 corresponds to Le^b-8 (hexose:hexosamine:fucose, 4:2:2), and that at m/e 1876 to H-7 (hexose:hexosamine:fucose, 4:2:1).

The spectrum of the methylated mixture (Fig. 23) shows the following probable sequence ions for the nonasaccharide at m/e 157 and 189 (terminal fucose), 228 and

260 (terminal hexosamine), 638 (trisaccharide), 1025 and 1057 (pentasaccharide), 1261 (hexasaccharide), and 1474 and 1506 (heptasaccharide). In addition, there are sequence ions for A-8 (compare Fig. 22) at m/e 851 and 883 (tetrasaccharide), and 1055 and 1087 (pentasaccharide). The other two glycolipids (H-7 and Le^b-8) noticed in the spectrum of the reduced mixture are more difficult to identify in this spectrum as several of their sequence ions are identical with ions of A-8 and A-9.

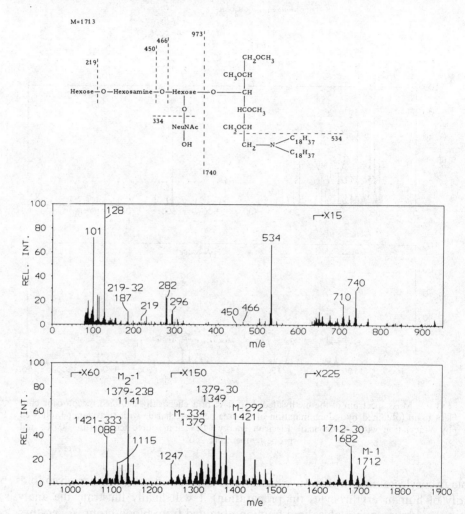

FIG. 16. Mass spectrum of the methylated and reduced derivative of a chemically modified receptor for cholera toxin (compare legend of Fig. 15). Electron energy 40 eV, ion source temperature 300°C, and probe temperature 250°C.

These two spectra (Figs. 23 and 24) provide conclusive evidence for the presence in the human erythrocyte of a nonaglycosylceramide of the composition and sequence shown in the formulae above the spectra (that sphingosine is the major base is shown by m/e 364 in Fig. 23 and 338 in Fig. 24). Besides several surface-antigenic implications, this is of particular interest since it shows the largest glycolipid, and also the largest cell surface saccharide, identified conclusively so far. Furthermore, these two spectra represent the largest biomolecules analyzed by mass spectrometry and providing basic information on overall structure including molecular weight.

IV. CONCLUSION

As illustrated in this paper, mass spectrometry is able to identify conclusively a lipid-linked nonasaccharide, which is the largest biomolecule identified so far with this tech-

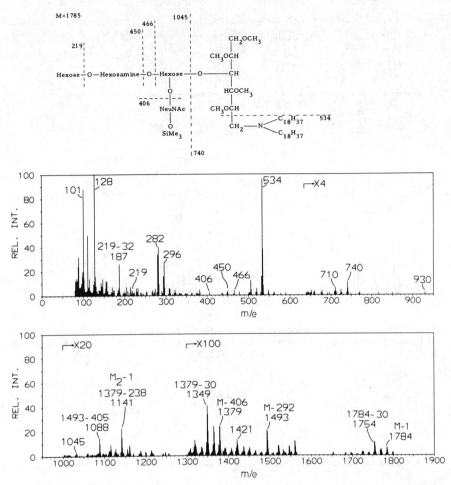

FIG. 17. Mass spectrum of the methylated, reduced and silylated derivative of a chemically modified receptor for cholera toxin (compare legend of Fig. 15). Electron energy 40 eV, ion source temperature 300°C, and probe temperature 220°C.

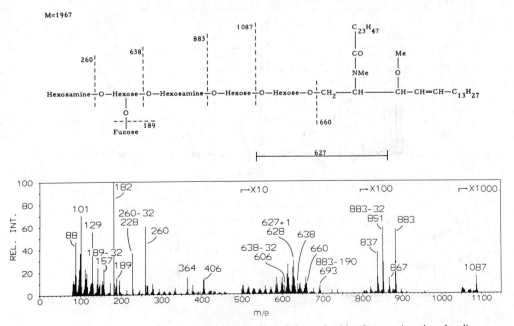

FIG. 18. Mass spectrum of the methylated derivative of the major blood-group A active glycolipid of human A_1 erythrocytes. Electron energy 40 eV, ion source temperature 290°C, and probe temperature 330°C.

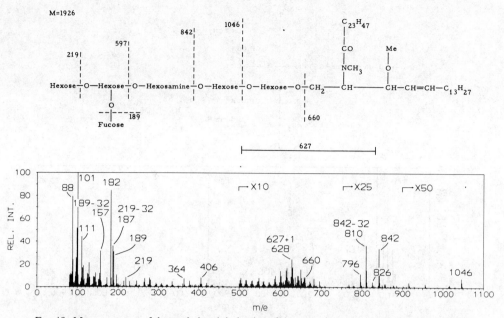

FIG. 19. Mass spectrum of the methylated derivative of the major blood-group B active glycolipid of human B erythrocytes. Electron energy 60 eV, ion source temperature 305°C, and probe temperature 295°C.

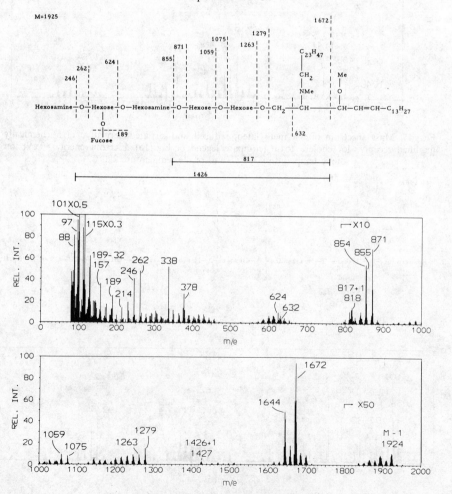

FIG. 20. Mass spectrum of the methylated and reduced derivative of the major blood-group A active glycolipid of human A_1 erythrocytes. Electron energy 56 eV, ion source temperature 305°C, and probe temperature 295°C.

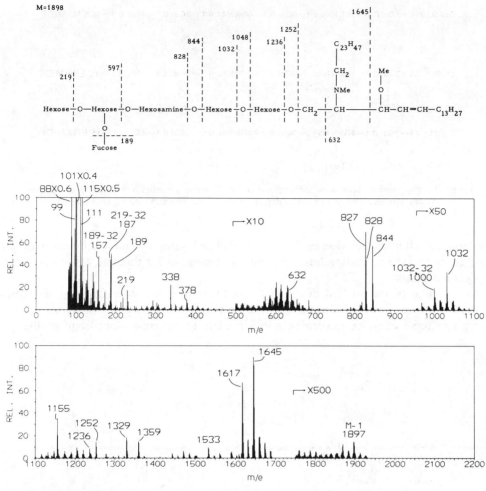

FIG. 21. Mass spectrum of the methylated and reduced derivative of the major blood-group
B active glycolipid of human B erythrocytes. Electron energy 65 eV, ion source temperature
300°C, and probe temperature 290°C.

nique. This achievement is not the result of recent technical development or the design
of new chemical reagents, for both the commercial apparatus used (MS 902) and the
derivatizations (methylation, reduction and trimethylsilylation) have been in use for more
than 10 years. Rather, it is the application of specific combinations of reagents and
the critical handling of samples that made this success. An equally good illustration
to this judgement is the sequential derivatizations needed for the optimal separation
and analysis of peptides, as recently documented by Nau and co-workers.[22] A further
advantage of the derivatives used by us is a continuous appearance, although somewhere
of low intensity, of peaks throughout the whole spectrum (in contrast to many trimethylsilyl derivatives), allowing a safe counting up of spectra. Still a problem is the lack
of a reliable mass marker in these high mass regions (1500–3000), although we plan
to apply fluorinated glycolipid.

Glycosphingolipids comprise a complex group of membrane bound substances, including sparsely existing cell surface antigens and toxin receptors as those discussed in
this paper. Recent data indicate that ganglioside-like compounds may be receptors also
for thyroid-stimulating hormone[21] and for interferon,[6] an antiviral agent. Mass spectrometry will certainly be of help in the detection and identification of such substances
on a microscale. A supplement in more detailed studies is a partial chemical degradation
and the analysis of the degradation mixture by temperature-programmed vaporization
in the ion source. Also, use of deuterated reagents may provide additional information,

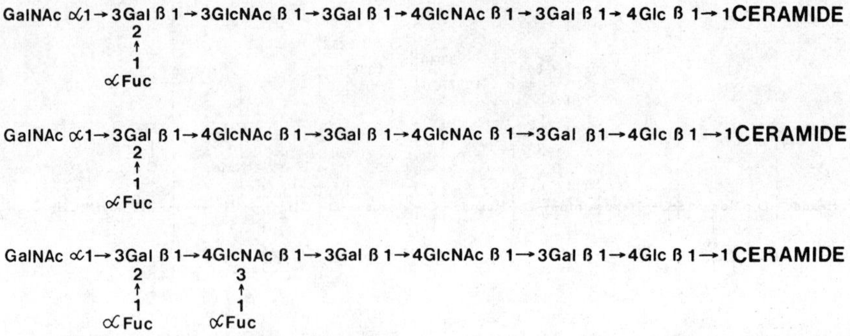

GalNAc α1→3Gal β 1→3GlcNAc β 1→3Gal β 1→4GlcNAc β 1→3Gal β 1→4Glc β 1→1CERAMIDE
2
↑
1
αFuc
GalNAc α1→3Gal β 1→4GlcNAc β 1→3Gal β 1→4GlcNAc β 1→3Gal β1→4Glc β 1→1CERAMIDE
2
↑
1
αFuc
GalNAc α1→3Gal β 1→4GlcNAc β 1→3Gal β 1→4GlcNAc β 1→3Gal β 1→4Glc β 1→1CERAMIDE
2
↑
1
αFuc
3
↑
1
αFuc

FIG. 22. Proposed structures of three blood-group A active glycolipids isolated from human A_1 erythrocytes and analyzed in mixture by mass spectrometry, see Figs. 23 and 24.

for instance after partial degradation of methylated samples and methylation of unmasked hydroxyls, indicating linkage points. Experiments on this line will be discussed in detail elsewhere.

At the time of the sudden death of Einar Stenhagen, we were finishing the study of the Forssman glycolipid hapten.[15] Stenhagen was fascinated by the highly informative spectra of these complex molecules, and his visits to us grew more frequent the last

Hexosamine–O–Hexose–O–Hexosamine–O–Hexose–O–Hexosamine–O–Hexose–O–Hexose–O–CH2–CH–CH–CH=CH–C13H27
Fucose
Fucose
C23H47
CO
NMe
Me
O
260
638
1057
1261
1506
189
189
660
627
REL. INT.
m/e
101
88
189-32
157
189
182
228
260
364
406
627+1
628
638
660
695
723
883-32
851
883
X10
X300
1025
1055
1057
1087
1087+44
1131
1261
1506-206
1300
1506-190
1316
1316+44
1360
1506-32
1474
1506

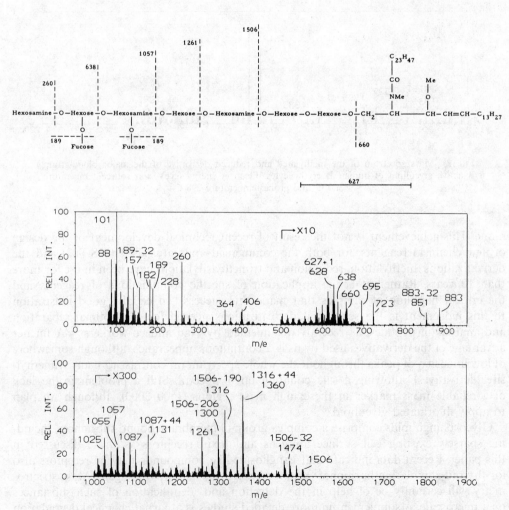

FIG. 23. Mass spectrum of the methylated derivative of a glycolipid fraction of human A_1 erythrocytes. The formula corresponds to the major molecular species identified (see Fig. 22). Electron energy 40 eV, ion source temperature 330°C, and probe temperature 325°C.

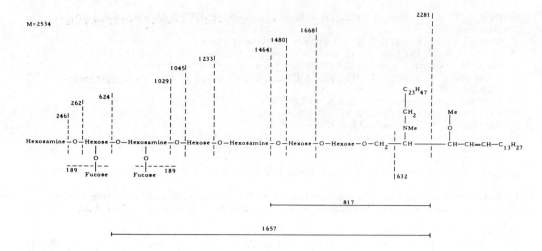

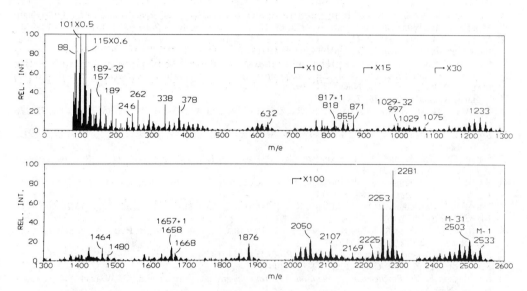

FIG. 24. Mass spectrum of the methylated and reduced derivative of a glycolipid fraction of human A_1 erythrocytes. The formula corresponds to the major molecular species identified (see Fig. 22). Electron energy 58 eV, ion source temperature 350°C, and probe temperature 320°C.

few months before his death. Although we were never a specific part of their project, Einar and Stina Stenhagen, with their exceptional human and scientific qualities, gave us the essential inspiration to initiate and continue our mass spectrometric studies on glycolipids.

Acknowledgements—The author is indebted to all people who have worked on this project, especially Ally Hansson, Karin Nilson, Irmin Pascher, Weston Pimlott and Bo E. Samuelsson. It is of interest that Nilson, our most qualified technician, started her lipid work on surface balance studies with Einar and Stina Stenhagen at Uppsala University in 1940. Pimlott came to Göteborg in 1968, engaged by AEI Ltd, to install the MS 902, but remained with Stenhagen to participate in construction work on mass spectrometers. He is now in charge of the MS 902 and other instruments within our project. A grant in support of this project was obtained from the Swedish Medical Research Council (No. 03X-3967).

REFERENCES

1. ANDERSSON, C.-O. *Acta Chem. Scand.* **12**, 1353 (1958).
2. ANDERSON, D. M., BIEMANN, K., ORGEL, L. E., ORO, J., OWEN, T., SCHULMAN, G. P., TOULMIN, P. and UREY, H. C. *Icarus*, 111–138 (1972).
3. BARBER, M., JOLLES, P., VILKAS, E. and LEDERER, E. *Biochem. Biophys. Res. Commun.* **18**, 469–473 (1965).

4. BIEMANN, K. In: *Biochemical Applications of Mass Spectrometry*, p. 405. Ed. G. WALLER, Wiley Interscience, New York (1972).
5. DAWSON, G. and SWEELEY, C. C. *J. Lipid Res.* **12**, 56–64 (1971).
6. FRIEDMAN, R. M. and KOHN, L. D. *Biochem. Biophys. Res. Commun.* **70**, 1078–1084 (1976).
7. HOLM, M., PASCHER, I. and SAMUELSSON, B. E. *Biomed. Mass Spectrom.* (in press).
8. HOLMGREN, J., LÖNNROTH, I., MÅNSSON, J.-E. and SVENNERHOLM, L. *Proc. Nat. Acad. Sci. USA* **72**, 2520–2524 (1975).
9. KARLSSON, K.-A. *Chem. Phys. Lipids* **5**, 6–43 (1970).
10. KARLSSON, K.-A. *FEBS Lett.* **32**, 317–320 (1973).
11. KARLSSON, K.-A. *Biochemistry* **13**, 3642–3647 (1974).
12. KARLSSON, K.-A. In: *Ganglioside Function, Adv. Exp. Med. Biol.* **71**, 15–25 (1976).
13. KARLSSON, K.-A. In: *Glycolipid Methodology*, p. 97, Ed. L. WITTING, American Oil Chemists' Society Press, Champaign, Illinois (1976).
14. KARLSSON, K.-A. *Nobel Symposia*, Nobel Foundation Series, Plenum Publishing Company, London, No. 34, p. 275 (1976).
15. KARLSSON, K.-A., LEFFLER, H. and SAMUELSSON, B. E. *J. Biol. Chem.* **249**, 4819–4823 (1974).
16. KARLSSON, K.-A., PASCHER, I., PIMLOTT, W. and SAMUELSSON, B. E. *Biomed. Mass Spectrom.* **1**, 49–56 (1974).
17. KARLSSON, K.-A., PASCHER, I. and SAMUELSSON, B. E. *Chem. Phys. Lipids* **12**, 271–286 (1974).
18. KARLSSON, K.-A., SAMUELSSON, B. E., SCHERSTÉN, T. and WAHLQVIST, L. *Biochim. biophys. Acta* **337**, 349–355 (1974).
19. KARLSSON, K.-A., SAMUELSSON, B. E. and STEEN, G. O. *Biochem. Biophys. Res. Commun.* **37**, 22–27 (1969).
20. MORRIS, H. R., WILLIAMS, D. H., MIDWINTER, G. G. and HARTLEY, B. S. *Biochem. J.* **141**, 701–713 (1974).
21. MULLIN, B. R., FISHMAN, P. H., LEE, G., ALOJ, S. M., LEDLEY, F. D., WINAND, R. J., KOHN, L. D. and BRADY, R. O. *Proc. Nat. Acad. USA* **73**, 842–846 (1976).
22. NAU, H., FÖRSTER, H.-J., KELLEY, J. A. and BIEMANN, K. *Biomed. Mass Spectrom.* **2**, 326–339 (1975).
23. RADFORD, T. and DEJONGH, D. C. In: *Biochemical Applications of Mass Spectrometry*, p. 313. Ed. G. WALLER, Wiley Interscience, New York (1972).
24. RYHAGE, R. *Anal. Chem.* **36**, 759–764 (1964).
25. SEDLACEK, H. H., STÄRK, J., SEILER, F. R., ZIEGLER, W. and WIEGANDT, H. *FEBS Lett.* **61**, 272–276 (1976).
26. SMITH, E. L., McKIBBIN, J. M., KARLSSON, K.-A., PASCHER, I. and SAMUELSSON, B. E. *Biochemistry* **14**, 2120–2124 (1975).
27. SMITH, E. L., McKIBBIN, J. M., KARLSSON, K.-A., PASCHER, I. and SAMUELSSON, B. E. *Biochim. biophys. Acta* **388**, 171–179 (1975).
28. SMITH, E. L., McKIBBIN, J. M., BREIMER, M. E., KARLSSON, K.-A., PASCHER, I. and SAMUELSSON, B. E. *Biochim. biophys. Acta* **398**, 84–91 (1975).
29. SMITH, E. L., McKIBBIN, J. M., KARLSSON, K.-A., PASCHER, I., SAMUELSSON, B. E. and LI, S.-C. *Biochemistry* **14**, 3370–3376 (1975).
30. SMITH, E. L., McKIBBIN, J. M., KARLSSON, K.-A., PASCHER, I., SAMUELSSON, B. E., LI, Y.-T. and LI, S.-C. *J. Biol. Chem.* **250**, 6059–6064 (1975).
31. SPENCER, W. A. and McKIBBIN, J. M. *Fed. Proc.* **35**, 1444 (1976).
32. STENHAGEN, E. *Z. Anal. Chem.* **205**, 109–124 (1964).
33. STENHAGEN, E. In: *Biochemical Applications of Mass Spectrometry*, p. 11, Ed. G. WALLER, Wiley Interscience, New York (1972).
34. VAN HEYNINGEN, W. E., CARPENTER, C. C. J., PIERCE, N. F. and GREENOUGH III, W. B. *J. Infect. Dis.* **124**, 415–418 (1971).
35. WIEGANDT, H., ZIEGLER, W., STÄRK, J., RONNEBERGER, H. J., ZILG, H., KARLSSON, K.-A. and SAMUELSSON, B. E. *Hoppe-Seyler's Z. Physiol. Chem.* **357**, 1637–1646 (1976).
36. WIPF, H.-K., IRVING, P., McCAMISH, M., VENKATARAGHAVAN, R. and McLAFFERTY, F. W. *J. Am. Chem. Soc.* **95**, 3369–3375 (1973).

Prog. Chem. Fats other Lipids. Vol. 16, pp. 231–255. Pergamon Press, 1978. Printed in Great Britain

ACYCLIC DITERPENE ALCOHOLS: OCCURRENCE AND SYNTHESIS OF GERANYLCITRONELLOL, PHYTOL AND GERANYLGERANIOL

Lars Ahlquist,* Gunnar Bergström† and Conny Liljenberg‡

Department for Structural Chemistry, Institute of Medical Biochemistry, Göteberg University, S-400 33 Göteborg, Sweden. †Department for Ecological Chemistry, Institute of Medical Biochemistry, Göteborg University, S-400 33 Göteborg, Sweden. ‡Department of Plant Physiology, Institute of Botany, Göteborg University, S-400 33, Göteborg, Sweden

CONTENTS

I. Introduction — 231
II. Synthesis of Optical Isomers of all-*trans*-Geranylcitronellol — 232
    A. Synthetic procedure — 233
    B. Characterization of the products — 234
III. Natural Geranylcitronellol — 234
    A. Comparison between synthetic isomers and natural geranylcitronellol — 234
    B. Occurrence in nature — 235
IV. Synthesis of Optical and Geometrical Isomers of Phytol — 240
    A. Synthetic procedure — 242
    B. Characterization of the products — 242
V. Natural Phytol — 245
    A. Comparison between synthetic isomers and natural phytol — 245
    B. Occurrence in nature — 247
VI. Chlorophyll in the Photosynthetic Membrane — 248
VII. The Role of Phytol and Geranylgeraniol in the Biosynthesis of Chlorophyll *a* — 250
    A. The esterification of chlorophyllide *a* — 251
    B. The pools of free phytol and phytol bound as acyl esters during greening — 253
Acknowledgements — 253
References — 253

I. INTRODUCTION

Diterpenes are quite widespread in nature, especially in plants. The metabolism of plant terpenoids, including diterpenes, has been reviewed by Waller,[107] by Nicholas[74] and by Oehlschlager and Ourisson *et al.*[75] *Geranylgeraniol*, in the form of its pyrophosphate ester, is a central building block in the biosynthesis of diterpenoids.[79] The alcohol was found in free form in tomato fruit,[104] and is one of the major wood extractions of *Cedrela toona.*[73a] Its role in the synthesis of phytol has been studied by one of us,[68] and we have found it to be present in volatile secretions of several insects. We have extracted geranylgeraniol from linseed oil through molecular distillation. *Geranyl-linalol* has been found in free form in jasmin oil,[64] and in the oleoresin of Norwegian spruce, *Picea abies*[59a] and has been implicated as a precursor of some diterpenes. The dihydroderivative of geranylgeraniol, *geranylcitronellol*, was identified as a major component of volatile marking secretions of male bumblebees.[62] In connection with the analytical work on these secretions, the two enantiomers of geranylcitronellol were prepared synthetically.[1] The aldehydes corresponding to geranylgeraniol and geranyl-citronellol, *geranylgeranial* and *geranylcitronellal*, were found to be present in volatile secretions from an ant, *Lasius carniolicus.*

Phytol is a hexahydroderivative of geranylgeraniol and is found in nature mainly esterified to chlorophyll. In connection with work on sesqui- and diterpenes from insects[1,2] and work on isoprenoid compounds in ancient sediments,[27] where phytol is transformed to several compounds, the question came up in the laboratory of the late Professors Einar and Stina Stenhagen about the structure and characteristics of phytol and its isomers. Strangely enough, a strict total synthesis of this very important natural product did not seem to have been undertaken. The diterpene alcohol phytol has two chiral centers and one double bond. There are, therefore, eight theoretically possible isomers. According to Weedon and coworkers,[19] the isomer of phytol which occurs in green plants is 3, 7R, 11R, 15-tetramethyl-2-*trans*-hexadecen-1-ol. A stereospecific synthesis has now been completed in this laboratory.[3] Four of the synthetic isomers have been compared with natural phytol. The synthesis has confirmed the structure given by Burrell et al.[19] *Isophytol* bears the same relation to phytol as geranyllinalol does to geranylgeraniol. It has been found in the extract of jasmin.[64] Phytanic and related acids, with different degrees of unsaturation, are found in ancient sediments and in living organisms as transformation products of phytol, see Lough.[70] The basis for this review is our work on acyclic diterpene alcohols from the aspects of occurrence in nature (especially in plants and insects), stereospecific synthesis and their role in the biosynthesis of chlorophyll *a*.

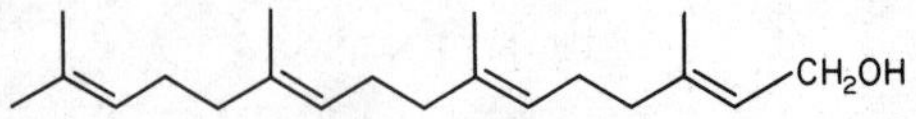

Scheme I. All-*trans*-geranylgeraniol

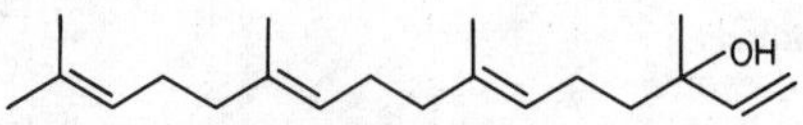

Scheme 2. All-*trans*-geranyllinalol

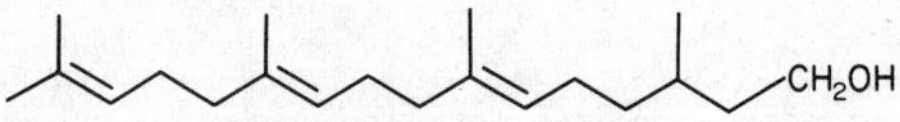

Scheme 3. All-*trans*-geranylcitronellol

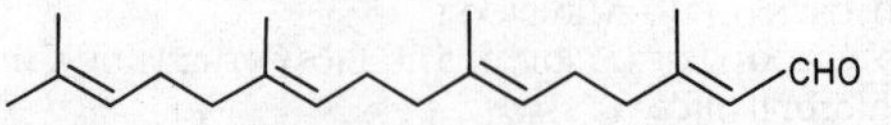

Scheme 4. All-*trans*-geranylgeranial

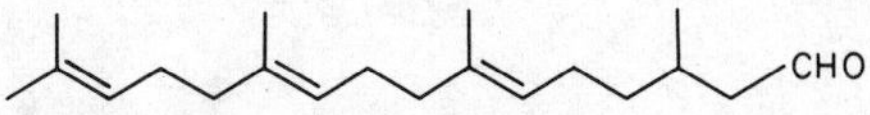

Scheme 5. All-*trans*-geranylcitronellal

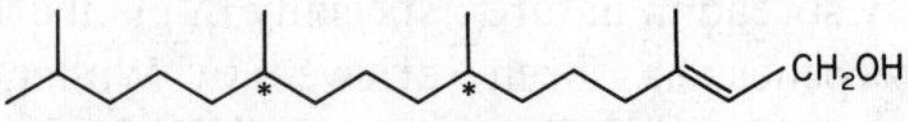

Scheme 6. Assymetric centers of phytol

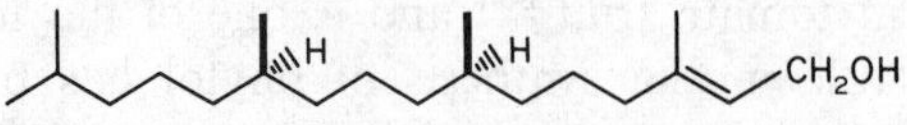

Scheme 7. Absolute configuration of natural phytol

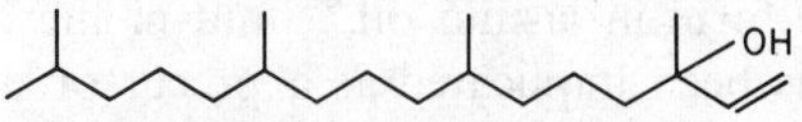

Scheme 8. Isophytol

II. SYNTHESIS OF OPTICAL ISOMERS OF ALL-*TRANS*-GERANYLCITRONELLOL

The route followed is similar to that used in the previously described synthesis of the corresponding sesquiterpene terrestrol from geraniol.[2] The present synthesis, outlined in Fig. 1, starts from farnesol. Barnard and Bateman[7] attempted the synthesis of homo-geranic acid via nitrile synthesis from geranyl chloride. The yield of the nitrile was extremely poor, however, and they therefore devised an alternative route to the nitrile via the oximinomalonate. We have found, however, that geranyl nitrile and farnesyl nitrile may be readily prepared from the chlorides by means of potassium cyanide in dimethylsulfoxide,[40,90] at a low temperature. The preparation of geranyl chloride has

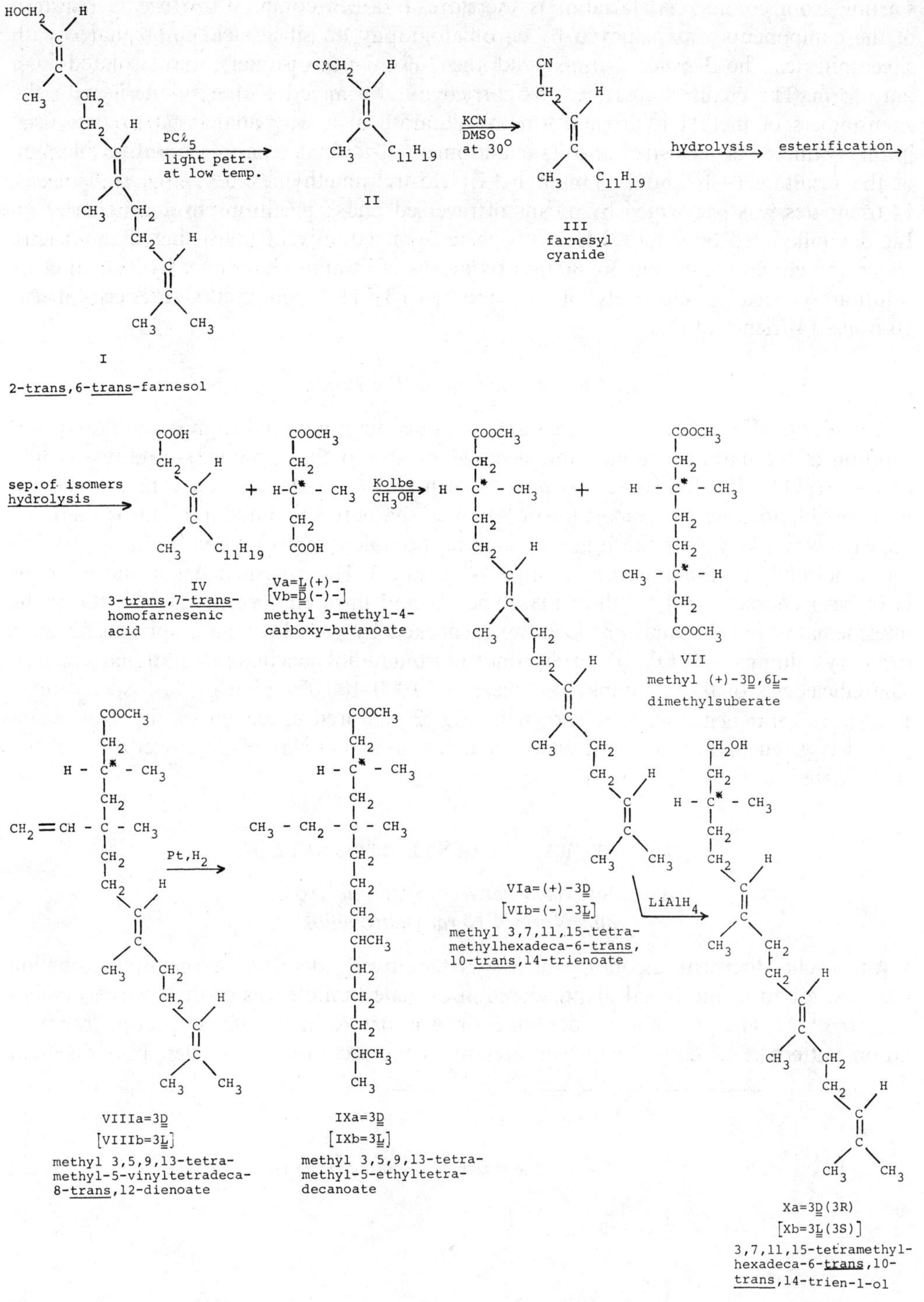

FIG. 1. The synthesis of the optical isomers of all-*trans*-geranylcitronellol. Component VIII is a side product produced during the electrolysis. Its structure is deduced from the mass spectrum of the hydrogenated compound IX.

been studied by Lääts and Teng[63] who obtained the best results using phosphorus pentachloride in light petroleum (b.p. 40–60°C) at a temperature below 0°C.

A. *Synthetic Procedure*

Commercial farnesol (Firmenich) is a mixture of the 2-*trans*, 6-*trans*- and the 2-*cis*, 6-*trans*-isomers[8] together with other minor components. Homofarnesenic acid prepared starting from commercial farnesol is therefore a rather complex mixture. Separation of the components was achieved by chromatography on silicic acid impregnated with silver nitrate. The 3-*trans*, 7-*trans*- and the 3-*cis*, 7-*trans*-isomers were isolated in a pure form. The chain elongation was carried out by mixed Kolbe synthesis with the enantiomers of methyl hydrogen 3-methylglutarate in a way analogous to that used in the syntheses of terrestrol and its enantiomer.[2] The final separation and purification of the resulting (+)- and (−)-methyl 3,7,11,15-tretramethylhexadeca-6-*trans*, 10-*trans*, 14-trienoates was performed by means of reversed phase partition chromatography on Lipidex (alkylated Sephadex LH 20)[28] using a mixed solvent consisting of methanol, water and ethylene chloride. Reduction by means of lithium aluminium hydride in ether solution yielded, respectively, (+)- and (−)-3,7,11,15-tetramethylhexadeca-6-*trans*, 10-*trans*, 14-trien-1-ol (X).

B. *Characterization of the Product*

In the case of terrestrol, it was possible to get enough material to measure the optical rotation of the natural product and it could be shown that its configuration was that of (−)-3L,7,11-trimethyldodeca-6-*trans*, 10-dien-1-ol (X). Also, in the present case, it was possible to observe a small levorotation of the natural compound, and it therefore appears very likely that the higher "isoprene homologue" from *Bombus terrestris* has the same configuration at carbon atom 3 as terrestrol. The intermediates in the synthesis have been characterized by their mass spectra and infrared spectra. The purity of the intermediates and the final products were checked by gas chromatography using glass capillary columns with OV-101 (a Silicone) and butandiol succinate as stationary phases. The efficiencies of the columns used were ∼50,000–100,000 plates. Mass spectrum of D-all-*trans*-geranylcitronellol is shown in Fig. 2. Infrared spectrum of the same compound is given in Fig. 3. Some data on the D- and L-forms of geranylcitronellol are summarized in Table 1.

III. NATURAL GERANYLCITRONELLOL

A. *Comparison between Synthetic Isomers and Natural Geranylcitronellol*

An acyclic diterpene alcohol, which was tentatively identified as geranylcitronellol, was first found in the labial gland secretion of male bumblebees of the species *Bombus terrestris*.[13,62] The preliminary identification was based on the mass spectral fragmentation pattern. This component was present in the secretion in smaller amounts than

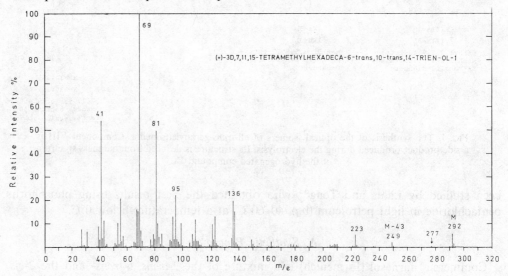

FIG. 2. Mass spectrum of synthetic D-geranylcitronellol.

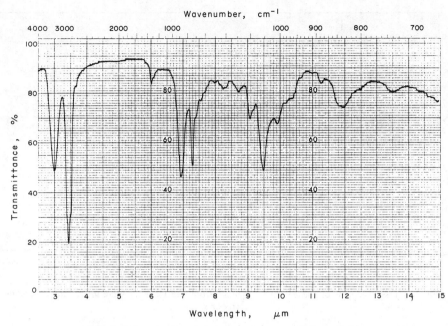

FIG. 3. Infrared spectrum of synthetic D-geranylcitronellol.

the major component, which was identified[2,98] as a dihydrofarnesol: $(-)$-3L,7,11-trimeth-yldodeca-6-*trans*, 10-dien-1-ol, named terrestrol. The diterpene would then be a higher "isoprene homologue" to this sesquiterpene. After completion of the synthesis of geranyl-citronellol, we were in a position to compare mass spectra, capillary gas chromato-graphic retention values and optical rotation of the synthetic and natural compound. Figure 4 shows the mass spectrum of the diterpene alcohol from *B. terrestris*. This should be compared with the mass spectrum of the synthetic compound given in Fig. 2. Capillary gas chromatogram with admixture of the synthetic compound to the natural secretion in approximately equimolecular amounts, see Figs. 5 and 6, gave a fully hom-ogenous diterpene alcohol peak. A small levorotation could be observed of the natural compound. With these data taken into account, the natural diterpene was identical with $(-)$-3L,7,11,15-tetramethylhexadeca-6-*trans*,10-*trans*,14-trien-1-ol (geranylcitronel-lol).

B. *Occurrence in Nature*

A diterpene alcohol, which is very probably the same compound, was identified from two other bumblebee species, viz. *Bombus hypnorum* (minor component) and *Psithyrus rupestris* (main component).[13,103] Again, the identifications were based on mass spectra and capillary gas chromatographic retention values. All-*trans*-geranylcitronellol has also been found to be the main component in the labial gland secretion of male *Bombus*

TABLE 1. Some Data on D- and L- All-*trans*-geranylcitronellol

	n_D^t α_D^t	d_4^t $|\alpha|_D^t$
D-all-*trans*-geranylcitronellol ($=(+)$-3D,7,11,15-tetramethyl hexadeca-6-*trans*,10-*trans*, 14-trien-1-ol)	1.4830(t = 25°) +1.59°(t = 22°)	0.870(t = 24°) +3.66°(t = 22°)
L-all-*trans*-geranylcitronellol ($=(-)$-3L,7,11,15-tetramethyl hexadeca-6-*trans*,10-*trans*, 14-trien-1-ol)	1.483(t = 25°) −1.57°(t = 23°)	0.874(t = 23°) −3.60°(t = 23°)

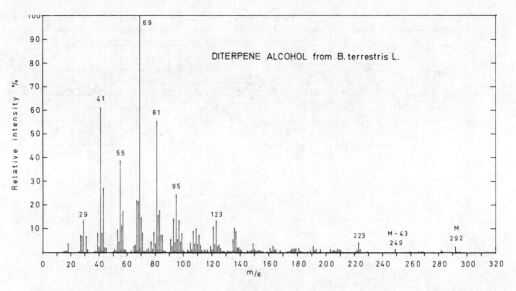

FIG. 4. Mass spectrum of the diterpene alcohol from the marking secretion of male *Bombus terrestris.*

lapponicus, blonde form. This form is also called *lapponicus*, as differentiated from the dark form called *scandinavicus*. The latter form does not contain any diterpene. Instead, the main component of the labial gland secretion in this form is a hexadecenyl acetate. In fact, the two forms are distinctly different in the composition of their secretions.[16] Typical capillary gas chromatograms of the volatile secretion from labial glands of *B. lapponicus lapponicus* (blonde) and *B. lapponicus scandinavicus* (dark) are shown in Figs. 7 and 8, respectively. The component indicated by an arrow in Fig. 7 corresponds

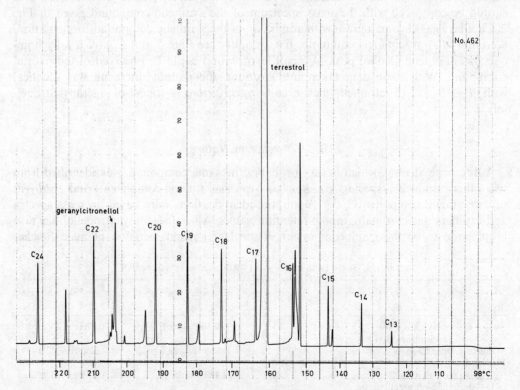

FIG. 5. Capillary gas chromatograms of the volatile material from heads of *Bombus terrestris.* Straight chain saturated hydrocarbons added as reference compounds. Rate of temperature programming 4°C per min.

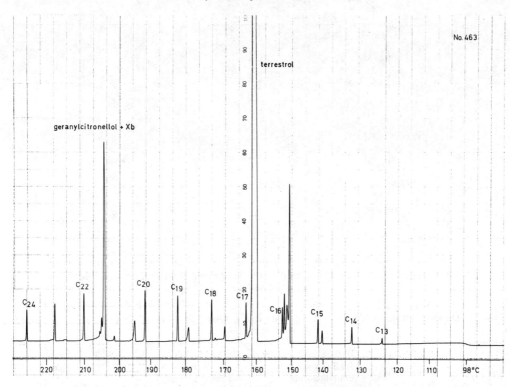

FIG. 6. Capillary gas chromatogram of the volatile marking secretion from *B. terrestris* with equimolecular amounts of synthetic geranylcitronellol (Xb) added.

to all-*trans*-geranylcitronellol. The secretions are employed for marking and help to differentiate between the two forms. They could therefore perhaps be better classified as two (sympatric) species. The recognition of the two chemically different forms was possible through chemical analysis of secretions from single individuals. To achieve this, precolumn tubes were used, in which volatile material from labial glands or whole heads could be driven off almost quantitatively.[12,99] The amount of secretion from one individual is in the order of 1 mg. In at least five other bumblebee species, all-*trans*-geranylgeraniol has been identified in the male marking secretion. In one of these, it is the main component, see Table 2. The corresponding acetate, all-*trans*-geranylgeranyl acetate, has been identified in at least six species, in three of them as main component.

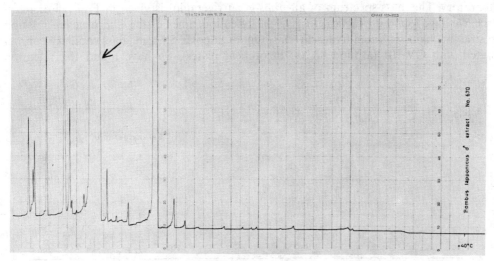

FIG. 7. Capillary gas chromatogram of the volatile material, blonde form, from *B. lapponicus* male, labial gland. The main component, indicated by an arrow, was identified as all-*trans*-geranylcitronellol.

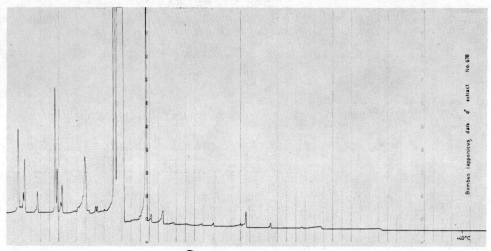

Fig. 8. Capillary gas chromatogram of the volatile material from *B. lapponicus*, dark form, male, labial gland.

TABLE 2. Diterpene Alcohols Identified in the Male Labial Gland Secretion of Bumblebees of the Genera *Bombus* and *Psithyrus*

Species	all-*trans*-geranylgeraniol	all-*trans*-geranylcitronellol	all-*trans*-geranylgeranyl acetate
B. hortorum	x		
B. subterranus	x		
B. soroensis			x
B. terrestris		x	
B. lucorum, "dark"	x		x·
B. cullumannus	x		x
B. pratorum	x		x
B. hypnorum		x	x
B. lapponicus, "blonde"		x	
Ps. rupestris		x	x

x indicates a relatively large amount, often the main component, of the respective secretion.

Geranylgeraniol has been extracted and purified for reference purposes by us from linseed oil according to Fedeli *et al.*[33] It has also been prepared synthetically in this laboratory. The mass spectrum of all-*trans*-geranylgeraniol is given in Fig. 9. For reference, we have found it practical to give the retention times of the diterpenes in relation to straight chain saturated hydrocarbons. On an unpolar Silicone glass capillary column coated with OV-101 (around 80,000 theoretical plates), we have found the retention

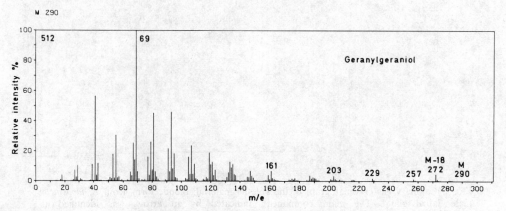

Fig. 9. Mass spectrum of all-*trans*-geranylgeraniol.

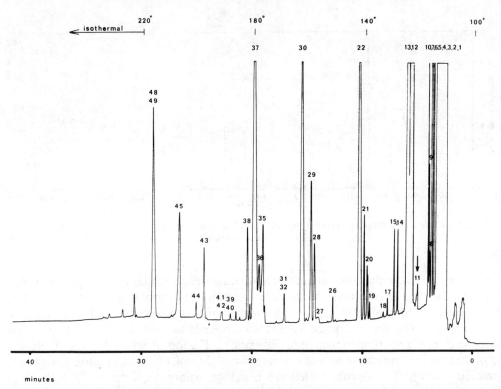

FIG. 10. Capillary gas chromatogram showing the volatile components of the Dufour gland secretion from worker ants of the species *Formica nigricans*. Components 45 and 48/49 correspond to geranylgeraniol and geranylgeranyl acetate, respectively.

values for all-*trans*-geranylcitronellol, all-*trans*-geranylgeraniol and all-*trans*-geranylgeranyl acetate to be 2133, 2164 and 2272, respectively (± 1). This refers to heneicosane = 2100, docosane = 2200, tricosane = 2300, etc. All-*trans*-geranylgeraniol and the corresponding acetate have also been identified[15] in three species of ants (*Formica nigricans*, *F. rufa* and *F. polyctena*, see Table 3. The compounds are present in the complex volatile secretion from the so-called Dufour gland in the abdomen of worker ants. Figure 10 shows an example (*F. nigricans*) of the complexity of these secretions. The main components are hydrocarbons, hendecane being the largest one. Peaks 45 and 48/49 correspond to geranylgeraniol and its acetate, respectively. In a volatile secretion from heads of worker ants of the species *Lasius carniolicus*, probably emanating from the mandibular gland, two diterpene aldehydes were found, see Table 3. They have been identified as all-*trans*-geranylcitronellal (main component) and all-*trans*-geranylgeranial.[14] The mass spectra of these two compounds are given in Figs. 11 and 12. They bear definite resemblances to the corresponding alcohols but do possess also some other characteristic fragments.

Oxygenated terpenes, such as mono- sesqui- and diterpenes, have been found to be quite common in volatile secretions from Hymenoptera (ants, wasps and bees). Although the exact biological functions of these diterpenes in bumblebees and ants are not yet

TABLE 3. Diterpene Alcohol, Acetate and Aldehydes Identified in the Dufour Gland (D) and Mandibular (m) Gland Secretion of Worker Ants. F = *Formica*, L = *Lasius*

Species	all-*trans*-geranylgeraniol	all-*trans*-geranyl geranyl acetate	all-*trans*-geranylgeranial	all-*trans*-geranylcitronellal
F. nigricans	x	x		
F. rufa	x	x		
F. polyctena	x	x		
L. carniolicus			x	x

x indicates a relatively large component.

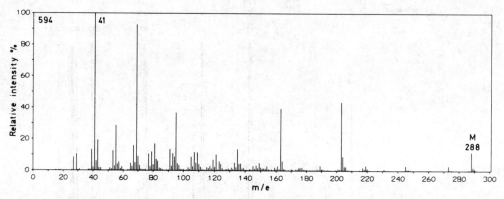

Fig. 11. Mass spectrum of geranylgeranial, identified from the cephalic volatile secretion of
Lasius carniolicus, worker ants.

known, we do know that they are active as volatile signals which release specific behavior. The function of the male bumblebee secretion has been interpreted as marking and species isolation and the secretion in worker ants is thought to function as recognition marks in the alarm–defence system. Compounds of such low volatility as these diterpenes can also function as trail substances. The occurrence of these diterpenes is of interest in connection with the reported inability of insects to cyclize acyclic isoprenoid precursors to sterols, which are therefore dietary requirements.[22] Kullenberg has reported field experiments[61a] in which geranylcitronellol exercised a rather strong attractive influence on the male *Psithyrus rupestris* during its daily route flying. It provoked inspection approaches and descents. Geranylgeraniol exercised some attraction on *P. rupestris* and *Bombus hortorum* males.

IV. SYNTHESIS OF OPTICAL AND GEOMETRICAL
ISOMERS OF PHYTOL

When we had encountered both geranylgeraniol and geranylcitronellol in several species of bumblebees and these compounds had been prepared synthetically, it became desirable to prepare also all the eight possible isomers of phytol. The reason for undertaking these syntheses was thus threefold: The need for a strict total synthesis of them was obvious and various chromatographic, spectroscopic and other data on them should be of interest for studies on phytol as a part of chlorophyll. Secondly, they should be of great value as reference compounds both in the biological and in the chemical analysis of bumblebee marking secretions and also in studies on other insect volatile

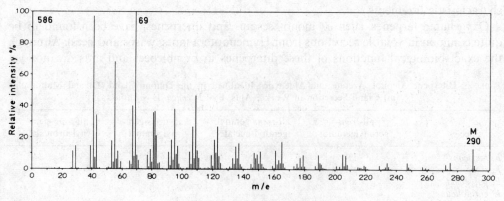

Fig. 12. Mass spectrum of geranylcitronellal, identified from the cephalic volatile secretion of
Lasius carniolicus, worker ants.

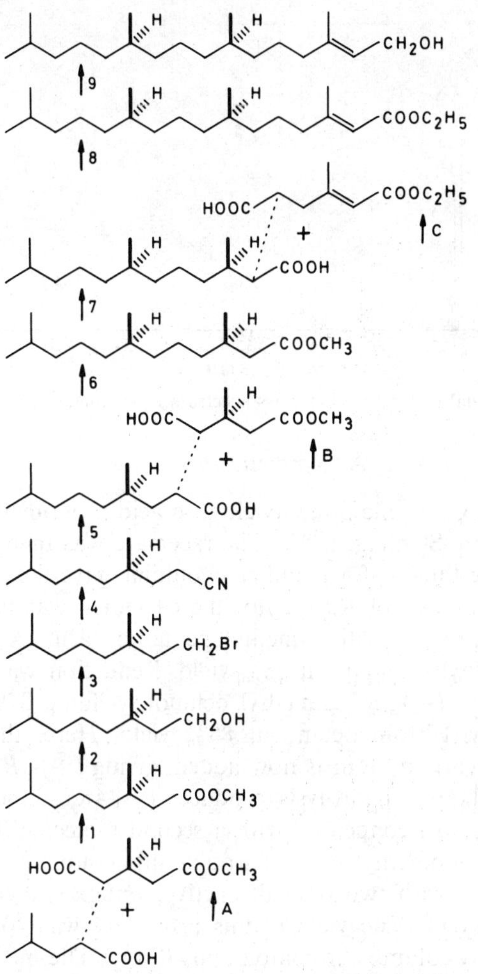

FIG. 13. The synthesis of the optical and geometrical isomers of phytol.

signals, Thirdly, it would indicate a way to a total synthesis of phytanic acid associated with Refsum's disease.

The synthetic procedure, designed by Stina and Einar Stenhagen, is summarized in Fig. 13. A detailed account of the synthesis is in preparation.[3] The procedure which proceeds in nine steps includes three condensations in the form of Kolbe electrolyses. The product of each step was purified by distillation and characterized by chromatographic (mainly capillary gas chromatography) and spectroscopic (mass spectrometry (MS), infrared spectroscopy (IR) and nuclear magnetic resonance spectroscopy (NMR)) methods as well as boiling point, refractive index and optical rotation. The products of the second and third electrolyses and the second reduction (step 9) were purified by column chromatography. For simplicity, only the synthesis of R,R-phytol is followed here. By substituting R(−)-3-methyl-4-methoxycarbonyl butanoic acid for the corresponding S(+) acid, L,L-phytol is produced.

Of the eight possible isomers, four have now been obtained in a very high degree of purity. Trans-R,R, cis-R,R, trans-SS and cis-SS were obtained in respective amounts: 921 mg, 549 mg, 633 mg and 382 mg. These amounts enabled us to characterize the compounds in several ways, including optical rotatory dispersion (ORD) studies. The synthesis of trans- and cis 7R, 11S-phytol has been carried through in part. The synthesis of trans- and cis forms of 7S, 11R-phytol is not yet begun.

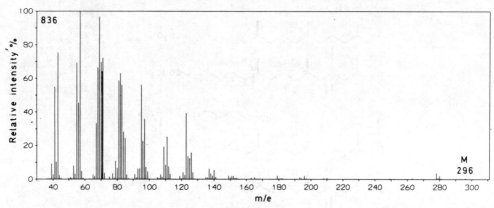

FIG. 14. Partial-high-resolution mass spectrum of synthetic *R,R-trans*-phytol.

A. *Synthetic Procedure*

In a preparatory step, racemic 3-methylglutaric acid monomethyl ester was prepared as described by Ställberg-Stenhagen.[96,97] The racemate was then resolved by using two alkaloids. Quinine gave the (−)-form and cinchonidine gave the (+)-form. The optically active half ester (in synthesis of *R,R*-phytol the (+)-form was used) was then coupled in the first Kolbe electrolysis with 4-methyl pentanoic acid (A in Fig. 13) producing (+)-3*R*,7-dimethyl methyloctanoate in 58% yield. Reduction was carried out with the use of LiAlH$_4$ and gave (+)-3*R*,7-dimethyl octanol, yielding 87%. Bromination (HBr) gave (−)-3*R*,7-dimethyl-1-bromooctane in 88% yield. Here, the sign of the optical rotation changes. One carbon atom is now added, giving (+)-3*R*,7-dimethyl-1-cyanooctane, yield 80%. An alkaline hydrolysis produces (−)-4*R*,8-dimethyl-nonanoic acid in 89% yield. The latter is now condensed with a second molecule (B in Fig. 13) of (+)-3-methyl-glutaric acid monomethyl ester. Through this coupling, a fifteen carbon atom sesquiterpenoid skeleton with two optically active centers is created, (+)-3*R*,7*R*,11-trimethyl-methyl dodecanoate. The yield of this procedure was 40%. Purification of the product was achieved by column chromatography (SiO$_2$). The methyl ester is hydrolyzed in alkali to the corresponding acid, (+)-3*R*,7*R*,11-trimethyldodecanoic acid in 90% yield. This C$_{15}$-skeleton is now reacted through a third Kolbe electrolysis with a mixture of *trans*- and *cis* 4-methyl-5-carboxy-ethylpent-4-enoic acid (C in Fig. 13). The latter compound was prepared according to Ställberg.[95] This reaction gave the *trans*- and *cis* isomers in the approximate proportions 2:1. The product then is a mixture of (−)-ethyl-3,7*R*,11*R*,15-tetramethyl-2-*trans*-hexadecenoate and (+)-ethyl-3,7*R*,11*R*,15-tetramethyl-2-*cis*-hexadecenoate. They could be separated by column chromatography in two steps using SiO$_2$ and then Lipidex as column packings. The latter separates the *trans*- and *cis* forms readily. The separated *trans*- and *cis* forms of the ethyl ester were reduced to the corresponding alcohols by LiAlH$_4$, followed by another column chromatography on Lipidex. In this way, *trans*- and *cis R,R*,-phytol were prepared.

B. *Characterization of the Products*

The final products, *trans*- and *cis R,R*- and *S,S*-phytol, were chromatographically pure as demonstrated by column chromatography on SiO$_2$ and Lipidex and by capillary gas chromatography on polar and unpolar phases. Spectroscopically, the isomers were studied by IR, MS, NMR and ORD techniques. Partial high resolution MS and IR spectra for *trans*- and *cis R,R*,-phytol are given in Figs. 14, 15, 16 and 17, respectively. Refractive indices are given in Table 4 and a summary of the ORD data in Table 5.

The mass spectra were obtained by an AEI 902 mass spectrometer in such a way that fragments containing the oxygen atom of the hydroxyl group were resolved in the low mass region. As shown by Figs. 14 and 15, there is one major oxygen containing fragment in this region, viz. $m/e = 71$. It corresponds to a combination of $[C_5H_{11}]^+$

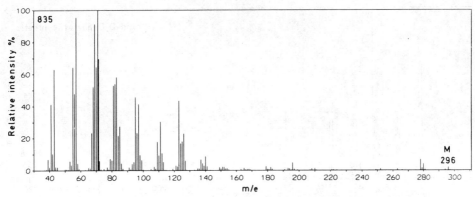

FIG. 15. Partial-high-resolution mass spectrum of synthetic *R,R-cis*-phytol.

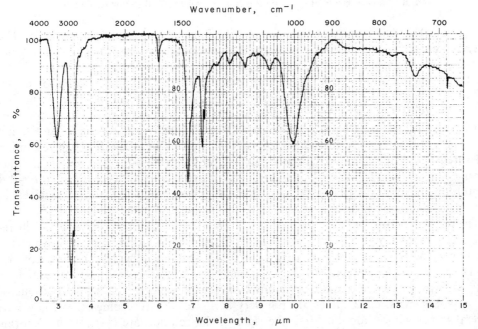

FIG. 16. Infrared spectrum of synthetic *R,R-cis* phytol.

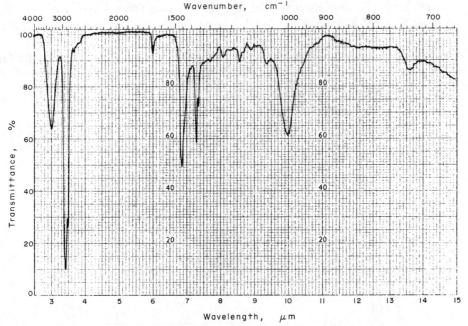

FIG. 17. Infrared spectrum of synthetic *R,R-trans* phytol.

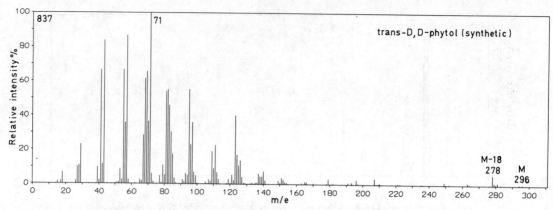

FIG. 18. Mass spectrum of synthetic *R,R-trans* phytol obtained at low resolution.

and $[C_4H_7O]^+$. The more exact masses of these fragments are 71.086 and 71.048. The oxygen-containing ion is formed by cleavage of the bond between carbon atoms 3 and 4. These two fragments are represented separately in the mass spectra in Figs. 14 and 15. Differences between the fragmentation patterns of *trans-* and *cis* forms of phytol are small, but a characteristic one is precisely in the proportion between the abundance of ions $[C_5H_{11}]^+$ and $[C_4H_7O]^+$. For *trans-* and *cis R,R*-phytol, we found the proportions to be 72:64 and 69:100, respectively. Other characteristic differences are found in the proportions between these fragments and $m/e = 69$ and $m/e = 57$. High resolution measurements of the mass of the molecular ion fully agreed with the theoretical value for the elemental composition $C_{20}H_{40}O$. A low-resolution mass spectrum of *trans-R,R* phytol, obtained with a LKB-2091 gas chromatograph–mass spectrometer, is given in Fig. 18 for comparison. In this spectrum, $m/e = 71$ is the dominant peak.

The optical rotations of the four isomers of phytol discussed here is small and therefore inaccurate as obtained with an ordinary polarimeter. An attempt to measure the rotation of *trans-S,S* phytol gave $|\alpha|_D$ of $+0.38°$. This was one reason for recording ORD spectra of the isomers. In Table 5, molecular rotations [M] are given at four different wavelengths. It should be pointed out that, owing to the small rotations of the phytol isomers, these values are not very accurate. The recordings show, however, several things without doubt. First, the signs of the rotation are $(-)$ for the *R*-forms and $(+)$ for the *S*-forms. Secondly, the ORD-curves of the *trans-* and *cis* pairs are near mirror images to one another. Thirdly, the *cis* pair show larger rotation than the *trans* pair. The ORD spectra of phytol isomers will be studied under more optimal conditions and the results given in a forthcoming report.

Techniques for the quantitative determination of phytol are, of course, important in the analytical work. Shimizu *et al.*[85] used a thin-layer chromatographic method and

TABLE 4

	n_D^{25}
trans-R,R phytol	1.4610
cis-R,R phytol	1.4618
trans-S,S phytol	1.4609
cis-S,S phytol	1.4618

TABLE 5

| | $[M]^{27}$ | | | |
	trans-R	*cis-R*	*trans-S*	*cis-S*
400	-10	-33	—	$+61$
350	-20	-55	$+35$	$+81$
300	-41	-122	$+70$	$+162$
250	-81	-332	—	$+446$

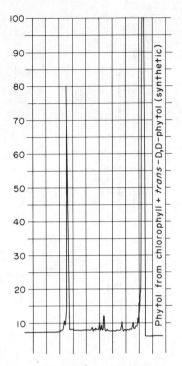

FIG. 19. Gas chromatogram—capillary column—of equimolecular mixture of natural phytol
and synthetic *R,R-trans* phytol.

Ellsworth and Perkins[32] determined phytol gas chromatographically through its degradation products on a Silicone SF-96 phase. With the help of gas chromatography on
glass column and Hyprose SP-80 as stationary phase,[69] a practical limit of sensitivity
of about 0.02 μg phytol could be attained. Our experiences from the synthetic work
as well as from analyses of both phytol and other acyclic diterpene alcohols, show
(see Sections II.A and IV.A above) that, for larger amounts of material, liquid chromatography on Lipidex is a good method. For small amounts, we have likewise found
that gas chromatography using glass capillary columns coated with Silicone OV-101
as stationary phase gives both good sensitivity and separation. With the mass spectrometer as detector, it should be possible to reach below the nanogram level. This
was attained on an instrument which gave a relatively intense peak at $m/e = 71$.

V. NATURAL PHYTOL

A. *Comparison between Synthetic Isomers and Natural Phytol*

Naturally occurring phytol was prepared by hydrolysis of chlorophyll from barley.
Mixed capillary gas chromatogram by use of a 24 m column coated with Silicone
OV-101 with equimolecular amounts of natural phytol and *trans-R,R*-phytol produced
one fully homogenous peak. This gas chromatogram is shown in Fig. 19. For comparison, Fig. 20 shows a chromatogram of pure *trans-R,R*-phytol run at the same conditions.
In Fig. 21, a mass spectrum of natural phytol is shown. This should be compared
with the one on synthetic phytol in Fig. 18.

Weedon and coworkers cited several earlier papers[37,47,48,56,58,71,82,92,93,109] dealing
with phytol in a report in 1966.[19] Much work has also been done by Swiss[53,54,73]
and Japanese[83,84,85] groups. Weedon *et al.* summed up the situation at the commence-
ment of their studies by saying: "Neither the geometrical nor the optical configuration
of phytol was known despite numerous "syntheses" of the alcohol and related compounds". These authors then describe their own synthetic and NMR work, of which

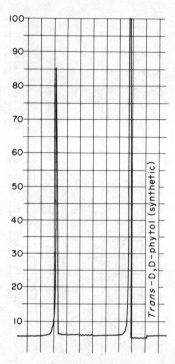

FIG. 20. Gas chromatogram—capillary column—of synthetic *R,R-trans* phytol in double the amount in Fig. 19.

brief accounts had been given previously.[20] The main problem in a strict total synthesis of geometrically and optically defined phytol seems to have been to get pure *trans*- and *cis* isomers and at the same time to have guaranteed absolute configurations at carbon atoms 7 and 11. Most workers from Fischer[37,38] on, who followed the pioneering work on phytol by Willställer,[113,114,115] have therefore inferred structural information from the C_{18}-methyl ketone which is obtainable directly from phytol. Weedon *et al.* also point out that the optical rotation of phytol from chlorophyll is very small, but that of the derived C_{18}-ketone and of the phytadienes produced on dehydration, make it clear that the natural alcohol is optically active.[57,59] By stereochemically controlled syntheses of the C_{18}-ketone, Weedon *et al.* concluded that phytol is 3,7*R*,11*R*,15-tetra-methylhexadec-*trans*-2-en-1-ol. Additional support for this structure of phytol has come

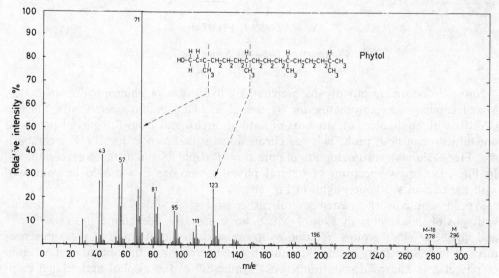

FIG. 21. Mass spectrum of natural phytol.

from optical rotatory dispersion studies.[24] The work now reported on the strict total synthesis of phytol isomers also confirms this structure. Bentley[11] outlined the evidence for the configurations assigned to compounds of major biological significance. He referred the earlier work of Crabbe,[24] Burrell,[19] Jackman[54] on the stereochemistry of phytol.

B. *Occurrence in Nature*

The main source of phytol is chlorophyll. This light-absorbing pigment which can be found in all photosynthesizing organisms has a porphyrin unit to which the phytol molecule is linked by an ester bond. Studies on the chemical nature and properties of the chlorophylls were initiated already in the nineteenth century.[51,52] Chlorophyll *a* was originally thought to exert only one function in the photosynthetic process, that of light absorption. When it became possible to measure photosynthesis in isolated systems (chloroplasts) it was found that energy transfer to chemical systems only occurred when the chlorophyll molecule was bound to a protein and when this complex was associated with membranes. Variations in proteins have later been shown to give the complexes different absorption maxima. Chlorophyll *a* has also been shown to have different chemical functions as the "antenna chlorophylls" and the "reaction center chlorophylls". It is reasonable to assume direct relations between the structures of the chlorophyll molecules and their different functions.

According to a hypothesis by Granick,[45] the biosynthetic pathway of chlorophyll reflects the biogenesis of the molecule during the evolutionary process. This would mean that the original photosynthetic pigments were probably water soluble and that the esterification occurred at the time of evolution when the compartmentalization of the "cell" by membranes took place. It must be obvious that a hydrophobic tail on the chlorophyll makes it possible for the molecule to associate with the lipids in membranes and also to keep the porphyrin part at a certain orientation in the membrane. The building up of these membranes goes hand in hand with the synthesis of the chlorophyll pigments. The pools of the esterifying prenols, free and activated, together with the esterification reaction between the alcohol and the porphyrin unit of the chlorophyll precursors have been studied.

Beside the biochemical and plant physiological interest for the absolute structure of phytol, it also has medical implications. Thus, the so-called Refsum's disease is associated with the accumulation of phytanic acid, a derivative of phytol. Lough has recently published a review[70] on the chemistry and biochemistry of phytanic and related acids. This includes treatment of the catabolism of phytanic acid and the role of phytanic acid in Refsum's disease. References to findings of isoprenoid acids in recent and ancient geolocial sediments are also discussed. The fate of phytol in animals has been studied by Baxter *et al.*[9,10,101]

The last step of the biosynthesis of chlorophyll *a* is the addition of the fatty alcohol phytol to the propionic acid residue of pyrrole ring IV of the porphyrin nucleus. This implies a thorough modification of the physicochemical properties of the pigment molecule. Several of the earlier porphyrin intermediates in the biosynthesis are water soluble and also the immediate precursors of chlorophyll *a*, protochlorophyllide and its reduced product chlorophyllide *a*, are rather hydrophilic. An esterification of chlorophyllide *a* with phytol increases the hydrophobic nature of the pigments without changing the absorbing properties in the visible wavelength region. As these photosynthetic pigments are bound to membrane systems, the thylakoids of the chloroplasts, the degree of esterification and also the nature of the alcohols involved must be important factors for the photosynthetic function of the chloroplasts. The hydrophobic phytol tail of the chlorophyll makes it possible for the molecule to associate either with the lipids or with the hydrophobic amino acid residues of the proteins in the membranes.

There are very few reports in the literature concerning the existence of free phytol in plants (Liljenberg,[66] Watts and Kekwick[108]). Phytol is not accumulated in dark grown leaves and seems to be intimately coupled to the light-induced chlorophyll biosynthesis when dark grown leaves are illuminated (Liljenberg[66]). Lederer and coworkers[26] have isolated phytol, isophytol and geranylgeraniol from an extract of jasmin. Sims and Pettus[88] have isolated free cis- and trans-phytol from the red alga, *Gracilaria andersoniana*. This is the first isolation of cis-phytol from a natural source. It is present only in the free form. Phytol has been found in esterified form both in a brown and in a blue–green alga as shown by de Souza and Nes.[94] Phytanic acid, together with lower isoprenoid acids, are found in sediments. Two pathways have been proposed and shown to occur. Boon et al.[18] discuss degradation via phytenic acid. It must be pointed out that through the whole plant kingdom with very few exceptions the chlorophylls are esterified to phytol. There seems to be an absolute demand that the esterifying alcohol be just phytol. The exceptions are chlorophylls and precursors of chlorophyll which contain geranylgeraniol, a possible precursor to phytol as esterifying alcohol (Wellburn et al.,[110] Liljenberg[67]).

Phytol in an activated form as phytylpyrophosphate has been reported from greening barley leaves. The phytylpyrophosphate was isolated from the etio-chloroplasts. The amount did not change very much during the first 8 hr of illumination (Liljenberg[68]). Except in greening leaves of higher plants, there are a few reports of phytyl esters in plant material. Steffens et al.[100] found phytol esterified to fatty acids in seeds of different species. Hexane extract of *Fatsia japonica* has been reported to contain phytyl palmitate and phytyl linoleate (Suga and Aoki[102]). Yellow leaves of *Acer platanoides* has been shown to contain phytyl linolenate (Csupor[25]). There are also observations that mosses contain acylesters of phytol (Gellerman et al.[41]). These fatty acids might probably be regarded as temporary acceptors for phytol either at the biosynthesis or at the breakdown of chlorophyll in the chloroplast.

VI. CHLOROPHYLL IN THE PHOTOSYNTHETIC MEMBRANE

A vast literature concerning the detailed composition and function of the different components of the thylakoids in the chloroplast exists, but very few experimental data and theoretical models can be found about the localization of the chlorophyll molecule in the membrane (see reviews by Kirk,[61] Thornber,[105] Anderson[4]). A model for the association of the phytol tail with the lipid bilayer in the membrane has been proposed by Rosenberg.[80] The mono- and digalactodiglycerides are the most prominent polar lipids of the thylakoid membrane which exists in such amounts that chlorophyll molecules can be accommodated. Besides, these galactodiglycerides of the thylakoid membranes have a very specific fatty acid composition. Bahl et al.[6] report up to 95% by weight linolenic acid of the monogalactosyldiglycerides. A rather stable lock-and-key interaction between the cis double bonds of the fatty acids in the galactodiglycerides and the protruding methyl groups of the terpenoid alcohol is proposed by Rosenberg.[80] Two chloroplasts from barley are shown in Fig. 22.

On the other hand, chlorophyll-protein aggregates have been isolated and characterized with the help of detergents. Different localizations of chlorophyll in the protein have been proposed (Thornber[105]). In one case, part of the photosynthesizing pigment-protein complex (in this case, a bacteriochlorophyll from *Chlorobium limicola*) was isolated without the use of detergents and the three-dimensional structure of the pigment-protein was determined by X-ray crystallography.[34,35] This is the first chlorophyll-containing protein for which the structure is known. The arrangement was deduced from a 2.8 Å resolution electron density map. The protein consists of three identical subunits, each containing a core of seven bacteriochlorophyll molecules. It has a molecular weight of about 150,000. Figures 23 and 24 are from Fenna and Matthews (*op. cit.*) and show, respectively, how phytol is attached to the porphyrin rings and how the seven bacteriochlorophyll molecules are arranged inside the protein. Although photosynthetic bacteria

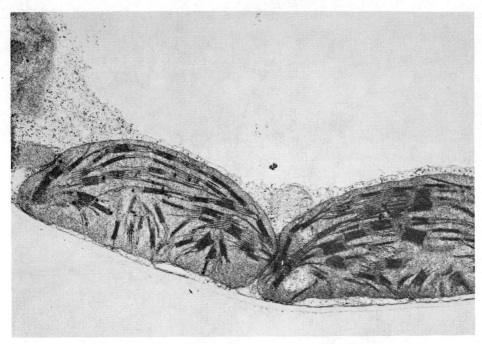

FIG. 22. Electron micrograph showing the ultrastructure of chloroplasts from barley seedlings. The two chloroplasts lie close to the cell wall, in the lower part of the picture, with a part of a vacuole visible in the upper part of the figure.

do not have chloroplasts, these results might provide evidences concerning the general principle of chlorophyll-protein association.

According to Fenna and Matthews, the phytyl chains in each subunit lie close together sandwiched between five of the porphyrin rings on one side and the other two rings and the beta sheet wall of the protein on the other. The hydrocarbon chains are, in most cases, in an extended conformation although in one instance (Bchl 1) the chain is bent into a complete U-shaped loop. The phytyl chains of chlorophylls 4, 5 and 6 lie parallel, forming a planar structure in close contact with the beta sheet wall of the protein.

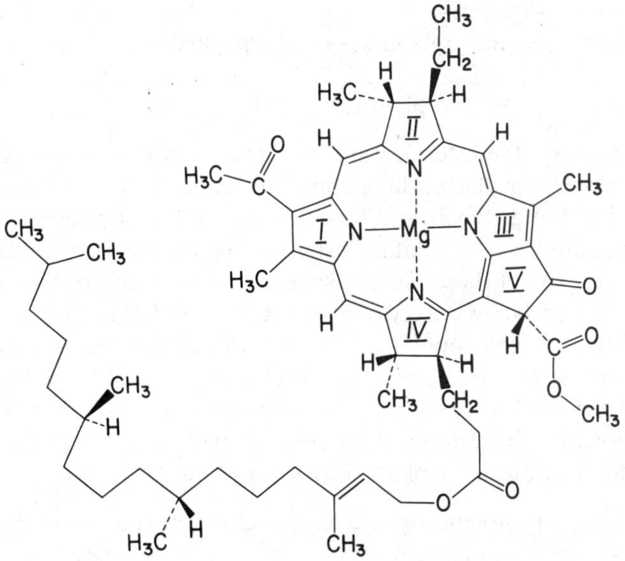

FIG. 23. Attachment of the phytyl chain to the porphyrin ring of bacteriochlorophyll *a* with the absolute configuration of the asymmetric carbon atoms indicated. Reproduced by kind permission of Dr. B. W. Matthews and Dr. R. E. Fenna, Institute of Molecular Biology, University of Oregon, Eugene, Oregon, USA.

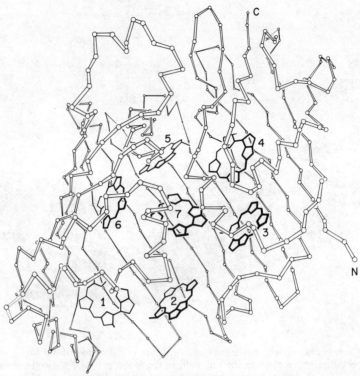

Fig. 24. Schematic diagram showing the arrangement of the polypeptide backbone and chlorophyll core of one subunit of the bacteriochlorophyll protein. Reproduced by kind permission of Dr. B. W. Matthews and Dr. R. E. Fenna, Institute of Molecular Biology, University of Oregon, Eugene, Oregon, USA.

VII. THE ROLE OF PHYTOL AND GERANYLGERANIOL IN THE BIOSYNTHESIS OF CHLOROPHYLL *a*

Biosynthetic pathways for the different protochlorophyll forms to chlorophyll a. The main pathway of chlorophyll *a* synthesis involves a photoreduction of Mg-vinylphaeo-porphyrin a_5 (protochlorophyllide) and an esterification of the reduced chlorophyllide with phytol [scheme (a) below]. Evidences for these biosynthetic steps are well documented.

(a) protochlorophyllide $\xrightarrow[2H]{hv}$ chlorophyllide $\longrightarrow$ chlorophyll *a*

  phytol

According to this pathway, the esterification thus takes place after the photoreduction to chlorophyllide *a* has taken place. The alcohol involved in this case is phytol (Goodwin,[44] Granick,[45] Kirk[60] and Rebeiz and Castelfranco[78]). It is, however, also well documented that nonirradiated plants contain esterified pigments to a certain extent. At the irradiation, at least part of these pigments will be photoreduced, thus forming some kind of chlorophyll *a* (see below). According to Rebeiz and Castelfranco,[77] these two pathways for formation of chlorophyll *a* can be summarized as follows [scheme (b) below]: From a common precursor, protochlorophyllide as well as prenylprotochlorophyllide can be formed. In the first case, the protochlorophyllide is photoreduced to chlorophyllide, whereafter the esterification (mainly with phytol) takes place. In the second case, the esterification takes place before the photoreduction.

protochlorophyllide $\xrightarrow[2H]{hv}$ chlorophyllide $\longrightarrow$ chlorophyll *a*

(b) common precursor   phytol

prenyl-protochlorophyllide $\xrightarrow[2H]{hv}$ chlorophyll *a*

This hypothesis for chlorophyll *a* formation is based on kinetic studies of ^{14}C-δ-amino-levulinic acid incorporation in the acid- and ester-form of the pigment and on the different cofactor requirements for biosynthesis of the two forms (Rebeiz et al.[76]).

According to a scheme proposed by Smith,[91] which is very similar to the second case in the scheme of Rebeiz and Castelfranco, the prenylprotochlorophyllide should actually be formed by an esterification of already existing protochlorophyllide [scheme (c)].

(c) protochlorophyllide ⟶ prenylprotochlorophyllide $\xrightarrow[2H]{h\nu}$ chlorophyll *a*

prenol

Rudolph and Bukatsch[81] proposed a specific enzyme, protochlorophyllase, for this reaction. According to Jones and Ellsworth,[55] such an enzyme is present in plants, but in a later paper by Ellsworth and Nowak,[30] it was shown that it was not possible to catalyze the esterification of protochlorophyllide with a prenol using a protochlorophyllase preparation from dark grown wheat leaves. The prenols used were phytol and farnesol. This fact thus raised the question whether the esterifying alcohol in this case could be some other than these two prenols. The chemical nature of this alcohol, probably a prenol, has been under discussion by several authors (Fisher and Bohn,[36] Fischer and Rüdiger[39] and Rebeiz and Castelfranco[78]). The protochlorophyll pigments found in the inner seed coats of pumpkin seeds do not photoconvert to chlorophyll *a*. It was recently shown by Ellsworth and Nowak[31] that these pigments were esterified with farnesol and a series of C_{20}-prenols from geranylgeraniol to the saturated phytanol including those with two and three double bonds. Also, the bacteriochlorophyll from several species of photosynthetizing bacteria have been shown to contain geranylgeraniol as esterifying alcohol.[112] Liljenberg[67] has shown that the above-mentioned esterifying alcohol of the protochlorophyll pigment in dark grown barley leaves is geranylgeraniol.

The content of the two forms of protochlorophyll pigments in dark grown plants and their photoreduction. The proportions of the acid- and ester-form of the protochlorophyll pigment differ between species (Godnev et al.,[42] Wolff and Price[116]) and is also age-dependent (Sironval et al.[89]). In a study of seventeen species, the ratio acid-to-ester-form varied between 0.14 and 9 (Godnev et al.[42]).

The literature treating the phototransformation of protochlorophyllide ester is confusing. Phototransformation is reported by Godnev et al.,[42] Rebeiz and Castelfranco[78] and Rudolph and Bukatsch.[81] In other reports, it is stated that no phototransformation could take place (Godnev et al.[43]) and, at least in one case, the protochlorophyllide ester is reported to remain unchanged after irradiation (Virgin[106]). It has been shown (Liljenberg[67]) that the protochlorophyllide ester of dark grown barley seedlings having geranylgeraniol as esterifying alcohol did, to a certain extent, photoconvert to the corresponding chlorophyll *a*. This would imply that, besides the main bulk of chlorophyll *a* having phytol as main esterifying alcohol, small amounts of chlorophyll *a* are esterified with geranylgeraniol. As photosystem I is the first system to start working at photosynthesis and this at very low chlorophyll concentrations, the question can be put whether the reaction center molecules are esterified with this alcohol.

A. *The Esterification of Chlorophyllide* a

It is generally believed that the esterification of chlorophyllide with phytol is catalyzed by the enzyme chlorophyllase (Holden,[50] Shimizu and Tamaki[86] and Böger[17]). As this enzyme catalyzes the hydrolysis of chlorophyll to chlorophyllide and free phytol, it has been regarded to function also in the synthetizing direction. The experimental support of this function is, however, rather doubtful. The esterification reaction in assays *in vitro* has been performed either in the presence of high concentrations of acetone

(40–60% aqueous acetone) or in the presence of detergents (Chiba et al.,[21] Wellburn,[111] Bacon and Holden[5] and Ellsworth[29]). The presence of small amounts of methylchlorophyllide in expanding wheat leaves and in passion flower leaves has also been taken as a proof that chlorophyllase can work as a transesterification enzyme changing methanol for phytol (Hines and Ellsworth[49]). Granick[45] has suggested that "the mechanism of phytylation is probably via phytol pyrophosphate rather than by the hydrolytic enzyme, chlorophyllase". Also, Goodwin[44] has suggested that, in analogy with other terpenoid reactions, phytylpyrophosphate would be the activated substrate at this reaction. The presence of phytylpyrophosphate in higher plants has been reported by Watts and Kekwick[108] who found both free phytol as well as phytylpyrophosphate in greening leaves of Phaseolus vulgaris. These analyses were, however, made on whole leaves. Therefore, nothing can be said about the contents of the plastids from these experiments. It has been shown (Liljenberg[68]) that phytylpyrophosphate is actually present in the etio-chloroplasts. This will thus be strong evidence that the esterification proceeds via a phosphorylated compound and that phytylpyrophosphate is the reactant during the esterification of chlorophyllide rather than free phytol. Phytylpyrophosphate may control the biosynthesis of the phytyl moiety by feedback inhibition.[46]

Changes in the rate of the esterification reaction induced by light. It has been shown[65] and later confirmed by Shlyk[87] that a pretreatment consisting of a short light impulse followed by 3 hr darkness given to dark grown seedlings will change the rate of the esterification of chlorophyllide. The same amount of chlorophyllide will, after such a pretreatment, be esterified in almost half the time compared with samples without pretreatment. The spectral dependence of this pretreatment effect has been investigated. The most effective wavelength region has been shown to be around 660–670 nm. This eliminates the possibility of protochlorophyllide being the light absorbing pigment at this reaction. In the blue region, part of the low effect could be attributed to carotenoids screening the active pigment. It is rather probable, however, that this effect is caused by light absorbed by phytochrome. The question then arises on which system the phytochrome is acting. It was shown by Liljenberg[66] that a pretreatment will not lead to a higher content of free phytol of the leaves. In other words, the higher rate of esterification is not due to a higher concentration of substrate. It is possible, however, that the phytochrome in the 730-nm-absorbing form activates or increases synthesis of the enzyme catalyzing the esterification.

Biosynthesis of esterifying prenols. Geranylgeraniol and phytol can be regarded as products from isopentenylpyrophosphate by enzyme systems in the etio-chloroplast. A less likely model for the building up of the prenyl chain of the chlorophyll molecule would be through a sequential addition of five carbon units to the porphyrin nucleus in analogy with prenyl chain additions to benzoquinone derivatives in animal tissues as postulated by Martinus.[72] Still another possibility would be that complete molecules of geranylgeranylpyrophosphate are first coupled to chlorophyllide and then hydrogenated to phytol.

Esterification of the chlorophyllide with phytylpyrophosphate or with free phytol must, however, be considered as the most likely way. This is also favored by the fact that free phytol and phytylpyrophosphate have been found in etio-chloroplasts. From results of *in vivo* labelling studies, Costes[23] has suggested that the immediate precursor of phytol is geranylgeraniol. Wellburn et al.[110] have studed the stereochemistry of phytol biosynthesis and summarized: "Thus, it appears that in the biosynthesis of phytol there is a stereospecific loss of hydrogen from the C-4 position of mevalonate, indicating that all the isoprenoid units are biogenetically *trans*". This suggests also that the enzymic hydrogenation to saturate the three ω-terminal isoprene residues does not involve removal of olefinic hydrogen atoms. The most likely explanation of these results is that phytol is formed by enzymic hydrogenation of all-*trans*-geranylgeranyl pyrophosphate.

However, the results do not preclude the possibility of hydrogenation occurring at earlier stages in the biosynthetic pathway.

B. *The Pools of Free Phytol and Phytol Bound as Acyl Esters during Greening*

With the gas chromatographic technique, it has been possible to detect free phytol in dark grown leaves (Liljenberg[66]). This means that a pool of free phytol is available for esterification of the chlorophyllide formed from the protochlorophyllide accumulated

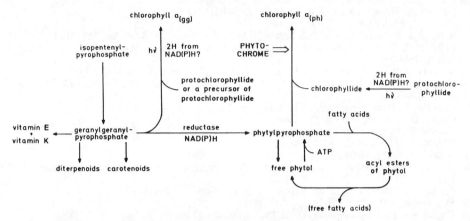

FIG. 25. Schematic and partially hypothetical chart showing the metabolic relationships between some components of the discussion in part VII.

in the dark. The amount of free phytol increases with age. Few data concerning free phytol in plants can be found in the literature (Watts and Kekwick[108]) as pointed out before. On the contrary, Fischer and Bohn[36] claim that such a pool does not exist. The pool of free phytol found in dark grown plants has been shown to change during irradiation (Liljenberg[66]). The first effect of irradiation is an immediate decrease which is interpreted as reflecting the esterification of chlorophyllide. The decrease is later followed by an increase during continuous irradiation. The concentration, however, never exceeded that of dark grown leaves. Determinations of esterified phytol (after 4 hr of irradiation) revealed the presence of more phytol than could be expected from the amount of chlorophyll pigments present. These results were explained by the presence of an acceptor to which this amount of phytol produced in excess was bound. It has been shown that this acceptor most likely is either saturated or monounsaturated fatty acids.[68] Dark grown seedlings after irradiation were thus shown to contain acyl esters of phytol. This discussion is summarized in Fig. 25.

Acknowledgements—This work is dedicated to the late Professors Einar and Stina Stenhagen who initiated it and completed most of the synthesis. Work reported here has been financially supported by the Swedish Natural Science Research Council. We thank Professor Hemming Virgin, Department of Plant Physiology, Institute of Botany, Göteborg University, Professor Torbjörn Norin, Department of Organic Chemistry, Royal Technical University, Stockholm, and Professor Sixten Abrahamsson, Department of Structural Chemistry, Institute of Medical Biochemistry, Göteborg University, for comments to the manuscript. Mrs. Ann-Britt Wassgren has done a substantial part of the preparative work on phytol. Mr. Max Lundgren has assisted in the late phase of this work. The help from Dr. Rolf Håkansson, Department of Organic Chemistry, Lund University, with obtaining ORD-data is gratefully acknowledged.

(*Received* 24 *March* 1977)

REFERENCES

1. AHLQUIST, L., OLSSON, B., STÅHL, A.-B. and STÄLLBERG-STENHAGEN, S., *Chem. Scr.* **1**, 237 (1971).
2. AHLQUIST, L. and STÄLLBERG-STENHAGEN, S., *Acta Chem. Scand.* **25**, 1685 (1971).
3. AHLQUIST, L. *et al.*, to be published in *Chem. Scr.* (1977).

4. ANDERSON, J. M., *Biochim. Biophys. Acta* **416,** 191 (1975).
5. BACON, M. F. and HOLDEN, M., *Phytochemistry* **9,** 115 (1970).
6. BAHL, J., FRANCKE, B. and MONÉGER, R., *Planta* **129,** 193 (1976).
7. BARNARD, D. and BATEMAN, L., *Proc. Chem. Soc.* 926 (1950).
8. BATES, R. B., GALE, D. M. and GRUNER, B. J., *J. Org. Chem.* **28,** 1086 (1963).
9. BAXTER, J. H. and MILNE, W. A., *Biochim. Biophys. Acta* **176,** 265 (1969).
10. BAXTER, J. H. and STEINBERG, D., *J. Lipid Res.* **8,** 615 (1967).
11. BENTLEY, R., *Molecular Asymmetry in Biology.* Vols. I and II, New York, London: Academic Press (1969, 1970).
12. BERGSTRÖM, G., *Chem. Scr.* **4,** 135 (1973).
13. BERGSTRÖM, G., KULLENBERG, B. and STÄLLBERG-STENHAGEN, S., *Chem. Scr.* **4,** 174 (1973).
14. BERGSTRÖM, G. and LÖFQVIST, J., *J. Insect Physiol.* **16,** 2353 (1970).
15. BERGSTRÖM, G. and LÖFQVIST, J., *J. Insect Physiol.* **19,** 877 (1973).
16. BERGSTRÖM, G. and SVENSSON, B. G., *Chem. Scr.* **4,** 231 (1973).
17. BÖGER, P., *Phytochemistry* **4,** 435 (1965).
18. BOON, J. J., RIJPSTRA, W. I. C., DE LEEUW, J. W. and SCHENCK, P. A., *Nature* **258,** 414 (1975).
19. BURRELL, J. W. K., GARWOOD, R. F., JACKMAN, L. M., OSKAY, E. and WEEDON, B. C. L., *J. Chem. Soc.* **C,** 2144 (1966).
20. BURRELL, J. W. K., JACKMAN, L. M. and WEEDON, B. C. L., *Proc. Chem. Soc.* 263 (1959).
21. CHIBA, Y., AIGA, I., IDEMORI, M., SATOH, Y., MATSUSHITA, K. and SASA, T., *Plant Cell Physiol. Tokyo* **8,** 623 (1967).
22. CLAYTON, R. B., *J. Lipid Res.* **5,** 3 (1964).
23. COSTES, C., *Phytochemistry* **5,** 311 (1966).
24. CRABBÉ, P., DJERASSI, C., EISENBRAUN, E. J. and LIU, S., *Proc. Chem. Soc.* 264 (1959).
25. CSUPOR, L., *Planta Med.* **19,** 37 (1970).
26. DEMOLE, E. and LEDERER, E., *Mém. Soc. Chim.* 1128 (1958).
27. EGLINTON, G., DOUGLAS, A. G., MAXWELL, J. R., RAMSEY, I. N. and STÄLLBERG-STENHAGEN, S., *Science* **153,** 1133 (1966).
28. ELLINGBOE, J., NYSTRÖM, E. and SJÖVALL, J., *J. Lipid Res.* **11,** 266 (1970).
29. ELLSWORTH, R. K., *Photosynthetica* **6,** 32 (1972).
30. ELLSWORTH, R. K. and NOWAK, C. A., *Photosynthetica* **7,** 246 (1973).
31. ELLSWORTH, R. K. and NOWAK, C. A., *Anal. Biochem.* **57,** 534 (1974).
32. ELLSWORTH, R. K. and PERKINS, H. J., *Anal. Biochem.* **17,** 521 (1966).
33. FEDELI, E., CAPELLA, P., CIRIMELE, M. and JACINI, G., *J. Lipid Res.* **7,** 437 (1966).
34. FENNA, R. E. and MATTHEWS, B. W., *Nature* **258,** 573 (1975).
35. FENNA, R. E. and MATTHEWS, B. W., *Brookhaven Symp. in Biol.*, No. 28 (1976).
36. FISCHER, F. G. and BOHN, H., *Liebigs Ann. Chem.* **611,** 224 (1958).
37. FISCHER, F. G. and LÖWENBERG, K., *Ann. Chem.* **475,** 183 (1929).
38. FISCHER, F. G., MÄRKL, G., HÖNEL, H. and RÜDIGER, W., *Ann. Chem. Bd.* **657,** 199 (1962).
39. FISCHER, F. G. and RÜDIGER, W., *Liebigs Ann. Chem.* **627,** 35 (1959).
40. FRIEDMAN, L. and SHECHTER, H., *J. Org. Chem.* **25,** 877 (1960).
41. GELLERMAN, J. L., ANDERSON, W. H. and SCHLENK, H., *Lipids* **10,** 656 (1975).
42. GODNEV, T. N., ROTFARB, R. M. and AKULOVICH, N. K., *Photochem. Photobiol.* **2,** 119 (1963).
43. GODNEV, T. N., GALAKTIONOV, S. G. and RASKIN, V. I., *Dokl. Akad. Nauk SSSR* **181,** 237 (1968).
44. GOODWIN, T. W. In: *Structure and Function of Chloroplasts* (GIBBS, M. ed.), p. 475, Berlin, Heidelberg, New York: Springer Verlag (1971).
45. GRANICK, S. In: *Biochemistry of Chloroplasts* p. 373 (GOODWIN, T. W. ed.) London, New York: Academic Press (1967).
46. GRAY, J. C. and KEKWICK, R. G. O., *Biochim. Biophys. Acta* **279,** 290 (1972).
47. GUNAR, V. I., GUSSEV, B. P. and NAZAROV, I. N., *Zh. Obshch. Khim.* **28,** 1444 (1958).
48. GUNAR, V. I. and ZAV'YALOV, S. I., *Dokl. Akad. Nauk. SSSR* **132,** 829 (1960).
49. HINES, G. D. and ELLSWORTH, R. K., *Plant Physiol.* **44,** 1742 (1969).
50. HOLDEN, M., *Biochem. J.* **78,** 359 (1961).
51. HOPPE-SEYLER, F., *Z. Phys. Chem.* **3,** 339 (1879).
52. HOPPE-SEYLER, F., *Z. Phys. Chem.* **4,** 193 (1880).
53. ISLER, O. and DOEBEL, K., *Helv. Chim. Acta* **27,** 225 (1954).
54. JACKMAN, L. M., RÜEGG, R., RYSER, G., VON PLANTA, C., GLOOR, U., MAYER, H., SCHUDEL, P., KOFLER, M. and ISLER, O., *Helv. Chim. Acta* **48,** 1332 (1965).
55. JONES, C. B. and ELLSWORTH, R. K., *Plant Physiol.* **44,** 1478 (1969).
56. KARRER, P. and BRETSCHER, E., *Helv. Chim. Acta* **26,** 1758 (1943).
57. KARRER, P., GEIGER, A., RENTSCHLER, H., ZBINDEN, E. and KUGLER, A., *Helv. Chim. Acta* **26,** 1741 (1943).
58. KARRER, P. and RINGIER, B. H., *Helv. Chim. Acta* **22,** 610 (1939).
59. KARRER, P., SIMON, H. and ZBINDEN, E., *Helv. Chim. Acta* **27,** 313 (1944).
59a. KIMLAND, B. and NORIN, T., *Acta Chem. Scand.* **21,** 825 (1967).
60. KIRK, J. T. O., *Ann. Rev. Plant Physiol.* **21,** 11 (1970).
61. KIRK, J. T. O., *Ann. Rev. Biochem.* **40,** 161 (1971).
61a. KULLENBERG, B., *Zoon, Suppl.* **1,** 31 (1973).
62. KULLENBERG, B., BERGSTRÖM, G. and STÄLLBERG-STENHAGEN, S., *Acta Chem. Scand.* **24,** 1481 (1970).
63. LÄÄTS, K. and TENG, S., *Izv. Akad. Nauk Est. SSR* **16,** 292 (1967).
64. LEDERER, E., *France Parfums* **3,** 28. *Chem. Abs.*, **54,** 14579 (1960).
65. LILJENBERG, C., *Physiol. Plant.* **19,** 848 (1966).
66. LILJENBERG, C., *Physiol. Plant.* **25,** 358 (1971).

67. LILJENBERG, C., *Physiol. Plant.* **32**, 208 (1974).
68. LILJENBERG, C., PhD thesis, Göteborg Univ., Göteborg, Sweden, p. 70 (1975).
69. LILJENBERG, C. and ODHAM, G., *Physiol. Plant.* **22**, 686 (1969).
70. LOUGH, A. K., *Prog. Chem. Fats other Lipids* **14**, 5 (1975).
71. LUKES, R. and ZOBACOVA, A., *Chem. Listy* **51**, 330 (1957).
72. MARTIUS, C. In: *Vitamins and Hormones*, p. 457. (HARRIS R. S. and WOOL, I. G. eds.) London, New York: Academic Press (1962).
73. MAYER, H., GLOOR, U., ISLER, O., RÜEGG, R. and WISS, O., *Helv. Chim. Acta* **47**, 221 (1964).
73a. NAGASAMPAGI, B. A., YANKOV, L. and SUKH, DER, *Tetrahedron Lett.* 189 (1967).
74. NICHOLAS, H. J. In: *Biogenesis of Natural Compounds*, 2nd edn., p. 829. (BERNFELD, P., ed.). Pergamon Press, (1967).
75. OEHLSCHLAGER, A. C. and OURISSON, G. In: *Terpenoids in Plants*, p. 83 (PRIDHAM, J. B., ed.) Academic Press (1967).
76. REBEIZ, C. A., YAGHI, M., ABOU-HAIDAR, M. and CASTELFRANCO, P. A., *Plant Physiol.* **46**, 57 (1970).
77. REBEIZ, C. A. and CASTELFRANCO, P. A., *Plant Physiol.* **47**, 24 (1971).
78. REBEIZ, C. A. and CASTELFRANCO, P. A., *Ann. Rev. Plant Physiol*, p. 129. (BRIGGS, W. R., GREEN P. B. and JONES, R. L. eds.) Palo Alto, California: Annual Reviews Inc. (1973).
79. RICHARDS, J. H. and HENDRICKSON, J. B. *The Biosynthesis of Steroids, Terpenes and Acetogenins*, Benjamin (1964).
80. ROSENBERG, A., *Science* **157**, 1191 (1967).
81. RUDOLPH, E. and BUKATSCH, F., *Planta* **69**, 124 (1966).
82. SARYCHEVA, I. K., VOROB'EVA, G. A., KUZNETSOVA, N. A. and PREOBRAZHENSKII, N. A., *Zh. Obsheh. Khim.* **28**, 647 (1958).
83. SATO, K., KURIHARA, Y. and ABE, S., *J. Org. Chem.* **28**, 45 (1963).
84. SATO, K., MIZUNO, S. and HIRAYAMA, M., *J. Org. Chem.* **32**, 177 (1967).
85. SHIMIZU, S., FUKUSHIMA, H. and TAMAKI, E., *Phytochemistry* **3**, 641 (1964).
86. SHIMIZU, S. and TAMAKI, E., *Bot. Mag.* **75**, 462 (1962).
87. SHLYK, A. A., RUDOI, A. B. and VEZITSKII, A. Y., *Photosynthetica* **4**, 68 (1970).
88. SIMS, J. J. and PETTUS, J. A., *Phytochemistry* **15**, 1076 (1976).
89. SIRONVAL, C., MICHEL-WOLWERTZ, M. R. and MADSEN, A., *Biochim. Biophys. Acta* **94**, 344 (1965).
90. SMILEY, R. A. and ARNOLD, C., *J. Org. Chem.* **25**, 257 (1960).
91. SMITH, J. H. C. In: *Comparative Biochemistry of Photoreactive Systems*, (ALLEN, M. B. ed.) New York, London: Academic Press (1960).
92. SMITH, L. I. and ROUALT, G. F., *J. Am. Chem. Soc.* **65**, 745 (1943).
93. SMITH, L. I. and SPRUNG, J. A., *J. Am. Chem. Soc.* **65**, 1276 (1943).
94. DE SOUZA, N. J. and NES, W. R., *Phytochemistry* **8**, 819 (1969).
95. STÄLLBERG, G., *Acta Chem. Scand.* **11**, 1430 (1957).
96. STÄLLBERG-STENHAGEN, S., *Ark. Kemi, Mineral. Geol.* **23**, 1 (1946).
97. STÄLLBERG-STENHAGEN, S., *Ark. Kemi, Mineral. Geol.* **25**, 1 (1947).
98. STÄLLBERG-STENHAGEN, S., *Acta Chem. Scand.* **24**, 358 (1970).
99. STÄLLBERG-STENHAGEN, S., *Chem. Scr.* **2**, 97 (1972).
100. STEFFENS, D., BLOS, I., SCHOCH, S. and RUDIGER, W., *Planta* **130**, 151 (1976).
101. STEINBERG, D., AVIGAN, J., MIZE, C. E., BAXTER, J. H., CAMMERMEYER, J., FALES, H. M. and HIGHET, P. F., *J. Lipid Res.* **7**, 684 (1966).
102. SUGA, T. and AOKI, T., *Phytochemistry* **13**, 1623 (1974).
103. SVENSSON, B. G. and BERGSTRÖM, G., *Insectes Soc.*, **24**, 213 (1977).
104. THOMPSON, G. A., PURCELL, A. E. and BONNER, J., *J. Plant Physiol.* **35**, 678 (1960).
105. THORNBER, J. P., *Ann. Rev. Plant Physiol.* 127 (1975).
106. VIRGIN, H. I., *Physiol. Plant.* **13**, 155 (1960).
107. WALLER, G. R., *Prog. Chem. Fats other Lipids* **10**, 151 (1969).
108. WATTS, R. B. and KEKWICK, R. G. O., *Arch. Biochem. Biophys.* **160**, 469 (1974).
109. WEICHET, J., HODRAVA, J. and KVITA, V., *Chem. Listy* **51**, 568 (1957).
110. WELLBURN, A. R., STONE, K. J. and HEMMING, F. W., *Biochem. J.* **100**, 23c (1966).
111. WELLBURN, A. R., *Phytochemistry* **9**, 2311 (1970).
112. WELLBURN, A. R., *Biochem. Physiol. Pflanz.* **169**, 265 (1976).
113. WILLSTÄTTER, R. and HOCHEDER, F., *Liebigs Ann. Chem.* **354**, 205 (1907).
114. WILLSTÄTTER, R., HOCHEDER, F. and HUG, E., *Liebigs Ann. Chem.* **371**, 1 (1909).
115. WILLSTÄTTER, R. and OPPÉ, A., *Ebenda* **378**, 1 (1910).
116. WOLFF, J. B. and PRICE, L., *Arch. Biochem. Biophys.* **72**, 293 (1957).

Prog. Chem. Fats other Lipids. Vol. 16, pp. 257–278. Pergamon Press, 1978. Printed in Great Britain

FUNCTIONAL ASPECTS OF ODONTOCETE HEAD OIL LIPIDS WITH SPECIAL REFERENCE TO PILOT WHALE HEAD OIL*

JONAS BLOMBERG

Department of Medical Biochemistry, University of Göteborg, Sweden†

CONTENTS

I. INTRODUCTION 257
  A. The toothed whales (*Odontoceti*) 257
  B. The pilot whale (*Globicephala melaena melaena*) 258
  C. Historical perspectives 258

II. LIPID STRUCTURE 259
  A. Composition of head oil lipids 259
    1. Lipid class pattern 259
      (a) Pilot whale 259
      (b) Other odontocetes 260
    2. Fatty acids 261
      (a) Pilot whale 261
      (b) Other odontocetes 261
    3. Fatty alcohols 263
    4. Hydrocarbons 264
    5. Waxes 264
    6. Triglycerides 265
      (a) Pilot whale 265
      (b) Other odontocetes 265
    7. Other lipid classes 266
  B. Compositional topography of melon lipids 266
    1. Relation to jaw and blubber fat 266
    2. Lipid content 267
    3. Lipid class 268
    4. Fatty acids 268

III. FUNCTIONAL ASPECTS 269
  A. The echolocation of odontocetes 269
    1. Emission of ultrasound 269
    2. Reception of ultrasound 270
  B. The possible role of lipids in sound processing 271
    1. The study of sonic properties of odontocete lipids 271
    2. Molecular structure and sound velocity 271
      (a) The Rao function 271
      (b) Schaaffs' function 272
      (c) Calculation of sound velocity 272
    3. Sound absorption 273
    4. Topographical distribution of sound velocity in melon tissue 273
    5. The odontocete melon as a sonic lens 273
  C. Other functional aspects 275
IV. CONCLUSIONS AND PERSPECTIVES 275
ACKNOWLEDGEMENTS 276
REFERENCES 276

I. INTRODUCTION

A. *The Toothed Whales* (*the* Odontoceti) (*Table 1*)

Because of their oceanic and submerged life, the odontocetes have not been subject to basic biological observation to the same extent as other mammals. Quite often human contact with these animals is confined to dying or dead animals either caught or

*This paper is part of a doctor's thesis from the Institutes of Medical Biochemistry and Virology, University of Göteborg, delivered by the author in May 1977.
†Present address: Division of Virology, Department of Medical Microbiology, University of Göteborg, Guldhedsgatan 10 B, 413 46 Göteborg, Sweden.

258   Jonas Blomberg

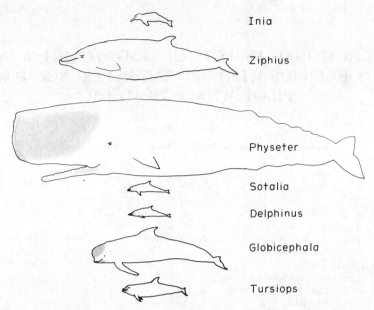

FIG. 1. Some important odontocete species. The animals have been drawn to approximate scale.
The smallest genera have been assumed to be 2 m long.

stranded. Consequently, we are sometimes in a situation where we can describe chemical
structures peculiar to these species in molecular detail, but we have only faint ideas
of their functional meaning to the animal. As will be detailed in the further discussion,
the odontocetes generally have two specialized fat depots in their head: a fat body
resident in the forehead often referred to as the melon and the mandibular fat body.
Their lipids constitute the head oils discussed in this review. Some important odontocete
species are shown in Fig. 1. The present review attempts to gather pertinent chemical
and physical information of odontocete head oil lipids, with special reference to the
pilot whale, and to draw some functional conclusions. Total coverage has not been
attempted. A review on the subject was made by Malins and Varanasi.[76]

B. *The pilot whale* (Globicephala melaena melaena)

The black ("blackfish") animal up to 9 m long, which has a characteristic fat-filled
bulge on its head ("melon"), seems to feed largely on squid and to follow their migrations. (See e.g. Spotte[107]). As squid rise from the depths at night, and become possible
to catch, a good nonoptical orientation and spotting system is a necessity for the pilot
whale. The pilot whales (genus *Globicephala*) are generally divided into several species,
but our ignorance is so profound that even this basic fact is disputed (Cf. note to
Table 1). *Globicephala melaena*, *Globicephala scammoni* and *Globicephala macrorhynca*
are generally accepted names for the Atlantic, East Pacific and West Pacific species,
which are discernible on the basis of fin length, color pattern and skull and tooth
anatomy. Pilot whales are distributed worldwide. They are, however, most prevalent
in the North Atlantic and Pacific Oceans. Nowadays they are caught mainly in Japan
and the Faroe Islands. The catches have declined steadily, from about 10,000 animals
per year in the fifties to 500–1000 animals per year in the seventies.[31]

C. *Historical Perspectives*

Chevreul[29] reported the occurrence of a volatile, strongly smelling acid in the saponified oil of a pilot whale. He named it *acide delphinique*. During the rest of the 19th
century, this acid was called *acide phocenique* (Chevreul[30]). Later this acid was shown

to be 3-methylbutanoic acid, and to be identical with isovaleric acid (André[7]). In the 1930s, Tsujimoto and Koyanagi[111] analyzed pilot whale head oil using the methods of their time, low pressure distillation and titration indices. The presence of isovaleroyl waxes and isovaleroyl triglycerides were deduced. Other early investigators that deserve mention are: Denisov[34] (*Delphinapterus*); Van Gaver[45] (Several species); Marcelet[77] (*Delphinus*); Hilditch and Lovern[56-58] (*Physeter*); Gill and Tucker[48] (*Tursiops*); Margaillan[78] (*Tursiops*); Lovern[73] (*Phocaena*); Toyama[109] (*Globicephala*); Williams and Maslow[128] (*Delphinapterus*); Kizevetter.[61]

At the beginning of the seventies, three different laboratories independently published detailed chemical information on the composition of dolphin head oil lipids.[17,115,116,126] The occurrence of isovaleric acid in wax and triglycerides was confirmed. The connection with the developing hypotheses of Wood[129] and Norris[87] on the role of melon fat in ultrasonic echolocation was recognized in all three laboratories.

Methodological developments during the fifties and sixties like thin layer chromatography (TLC), gas chromatography (GC) and gas chromatography coupled to mass spectrometry (GC–MS) were the probable causes of this simultaneous outburst of investigation. The first information on the lipids in the melon and mandible of Odontoceti obtained with these modern methods came from Varanasi and Malins[114,115] who found small amounts of dialkoxypentanes in the jaw oil from *Tursiops Truncatus* (bottlenose dolphin). The unusually prominent short-chain fatty acid content of the triglycerides of jaw and melon lipids in this animal was noted. At this time, the groups of Litchfield and Ackman which earlier had considerable experience of the GLC analysis of triglycerides and fatty acids, respectively, announced results on the occurrence of wax and triglycerides containing 1 and 2 moles of isovaleric acid in *Delphinapterus* (beluga) and *Globicephala* head fats.[68,126]

The studies in our own laboratory were initiated by Dr. Klaus Serck-Hanssen who, in 1962, observed that the watchmaker used a free-flowing oil to lubricate his wristwatch. As a keen lipidist, Dr. Serck-Hanssen took an infrared spectrum of the oil, and found some unusual features indicative of short and branching carbon chains. He traced the oil (Ezra Kelley Oil, Blackfish Head Oil) to William F. Nye & Co. of New Brunswick, who reported its origin from the pilot whale head. Under the supervision of Dr. Serck-Hanssen and Prof. Stina Stenhagen, the structural investigation of this oil was begun in 1968. The superb instrumentation and experience with lipid analysis in the laboratory was of great help.

II. LIPID STRUCTURE

A. *Composition of Head Oil Lipids*

1. *Lipid Class Pattern*

(a) *Pilot whale*. Since the time of Chevreu,[29] Tsujimoto and Koyanagi,[111] Blomberg[17,18] and Wedmid et al.[126,127] have analyzed the lipid composition of pilot whale head fat. Working with head oil from *Globicephalus sieboldii* Gray (= *Globicephala macrorhyncha*), Tsujimoto and Koyanagi[111] by means of titrations and vacuum distillation found that the oil to a large part consisted of triglycerides with one or two isovaleroyl moieties. A small part were wax esters with properties similar to cetyl isovalerate. Triisovalerin was not found. Blomberg[17,18] using silica gel column chromatography, could divide the head oil of *Globicephala melaena* into hydrocarbon (3 wt %), wax and cholesterol esters (9 wt %), triglycerides (87 wt %), and cholesterol and diglycerides (1 wt %). By another silica gel column chromatography, the triglycerides were further subdivided into non-isovaleroyl (XXX) (11 wt %), mono-isovaleroyl (XXV) (19 wt %), and diisovaleroyl (VXV) (57 wt %) species. Triisovalerin (VVV) was not found. The composition is clearly borne out by the TLC pattern of Fig. 2. A similar separation of phytanate

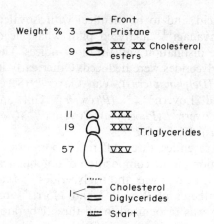

FIG. 2. TLC pattern of pilot whale head oil, modified from Blomberg.[18] Silica Gel H. Petroleum
ether: Diethyl ether : Acetic acid 80 : 20 : 1.

triglycerides was obtained by Karlsson et al.[62] After saponification, the oil yielded 10
wt % of lipid soluble unsaponifiables, consisting mainly of alcohols and hydrocarbon.

Wedmid et al.[126] and Wedmid et al.[127] studied the variations in lipid class content
in pilot whale melon tissue. These results will be discussed further in Section IIB. Wax
esters and triglycerides were separated, and analyzed for fatty acid and fatty alcohol
composition. The hydrogenated wax esters and triglycerides were also analyzed as such
by GC.

Variable amounts of VXV, XXV and XXX triglycerides (57–96, 4–25 and 0.1–18 mole
% of triglycerides, respectively) and wax esters with (XV) and without (XX) isovaleric
acid (60–96 and 3–40 mole % of wax esters, respectively) were found, depending on
the location in the melon tissue. This variability means that exact comparisons between
analyses of samples from different animals and different locations are impossible.

(b) *Other odontocetes.* Triglycerides, wax esters, diacyl glyceryl ethers, cholesterol, diglycerides and phospholipids have been identified. In a comparative survey, Litchfield
and Greenberg[69] found that odontocetes could be divided into four groups on the
basis of the composition of their head oil lipids (Fig. 3, Table 1). The tendency toward
accumulation of wax esters and triglycerides of low molecular weight in the melon
and jaw tissues as compared to blubber fat seems to be a general feature of odontocetes.[1,2,45,56–58,68,71,73,77,78,94,102,117,120] This feature may have functional meaning
because in lipids a lower molecular weight is associated with a lower sound velocity
(see below). By means of a quantitative infrared spectrometric method, the wax ester
content of melon and jaw oils was found to vary widely between species.[69,70] Values
from all odontocete families were given. The following variations were found: Platanistidae: 0–66 wt %, Physeteridae: 18–84 wt %, Ziphiidae: 19–36 wt %, Monodontidae:
0 wt %, Phocaenidae: 0–3 wt %, Stenidae: 22–49 wt %, Delphinidae: 3–45 wt %. The

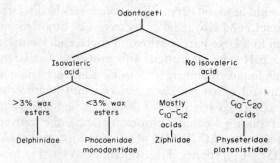

FIG. 3. Chemical taxonomy of odontocetes. Taken from Litchfield and Greenberg.[69]

TABLE 1. Taxonomy of Toothed Whales*

Order	Suborder	Superfamily	Family	Subfamily	Genus
Cetacea	Odontoceti	Platanistoidea	Platanistidae	Platanistinae	*Platanista*
				Iniinae	*Inia*
					Lipotes
				Stenodelphininae	*Stenodelphis*
		Physeteroidea	Ziphiidae		*Mesoplodon*
					Ziphius
					Tasmacetus
					Berardius
					Hyperoodon
			Physeteridae	Physeterinae	*Physeter*
				Kogiinae	*Kogia*
		(×) Monodontoidea	Monodontidae		*Delphinapterus*
					Monodon
		(×) Delphinoidea	Stenidae		*Steno*
					Sousa
					Sotalia
			Delphinidae	Delphininae	*Stenella*
					Delphinus
					Grampus
					Tursiops
					Lagenorhynchus
					Lagenodelphis
				Cephalorhynchinae	*Cephalorhynchus*
				Orcinae	*Orcinus*
					Pseudorca
					Orcaella
					Globicephala
					Feresa
				Lissodelphininae	*Lissodelphis*
			Phocaenidae		*Phocaena*
					Phocaenoides
					Neomeris

*(×) Denotes that these species have significant amounts of isovaleric acid in their acoustic tissues. Taxonomic authorities: Morris,[80] Ellerman and Morrison-Scott,[36] Simpson.[106] Fraser[42] uses the same taxonomy, although *Phocoena* and *Phocoenoides* are used instead of *Phocaena* and *Phocaenoides*. The genus *Globicephala* was specifically treated by Fraser,[40,41] Hershkovitz,[55] Walker *et al.*,[122] and Rice and Scheffer.[101]

rest of the weight was mostly made up of triglycerides. A small part of diacyl-glyceryl ethers was present in Platanistidae, Physeteridae and Ziphiidae.

2. Fatty Acids

(a) *Pilot whale.* Blomberg[17,18] and Wedmid *et al.*[127] analyzed the fatty acids from the whole oil and from triglyceride subclasses by means of GC or GC–MS. A series of normal saturated and monounsaturated, and isobranched saturated, C_5–C_{18} fatty acids were found (Table 2). A generally good agreement between the values of the investigators can be seen. Blomberg[17,18] found 2-methylbutyric acid, the anteiso analog of isovaleric acid. The trace amounts of anteiso acids may to some extent be dietary, because squid, the principal food source of pilot whales, contain iso and anteiso acids.[4]

(b) *Other odontocetes.* As mentioned above Litchfield and Greenberg[69] divided the odontocetes into four groups based on the lipid composition of their melon and jaw fats (Fig. 3). Stenidae, Delphinidae, Phocaenidae and Monodontidae have much isovalerate in the form of triglycerides. Stenidae and Delphinidae species have isovalerate also in the form of wax esters. Ziphiidae, Physeteridae and Platanistidae have only trace amounts of isovalerate lipids in these depots. Instead, C_{10}–C_{20} acids occur in the form of wax esters, with smaller amounts of triglycerides. Ziphiidae uniquely have much C_{10}–C_{12} acids, and in contrast to all other whales, their blubber fat is largely (> 94 wt %) composed of wax.[71]

TABLE 2. Comparative Melon Fatty Acid Patterns from Triglycerides of some Odontocetes

	Delphinidae			Stenidae	Monodontidae	Physeteridae
	Globicephala		Tursiops	Sotalia	Delphinapterus	Physeter
	1	2	3	4	5	6
				(Mole %)		
iso 5:0[7]	48	57	51	68	60	2
n, iso and anteiso 10–13 (:0 and :1)	3	1	5	1	6	27
iso 14:0	5	3	3	2	1	0.5
n 14:1	1	1	2	tr	0.8	13
n 14:0	7	5	4	3	5	17
iso 15:0	10	8	13	14	10	1
n 15:1	ND	0.1	ND	tr	0.3	ND
n 15:0	0.2	0.3	1	0.4	0.4	0.7
iso 16:0	3	3	2	3	1	0.2
n 16:1	13	10	7	2	8	13
n 16:0	3	3	4	4	4	12
n, iso 17 (:0 and :1)	0.7	0.8	0.9	0.6	0.6	1
iso 18:0	tr	tr	ND	tr	tr	0.3
n 18:1	6	7	4	0.9	2	9
n 18:0	0.7	0.1	tr	0.5	0.1	1
n 19–22 (:0 and :1–3)	0.2	1	ND	ND	0.3	3
	100.8	100.3	96.9	99.4	99.5	100.7

(1) The values for total oil have been recalculated to triglyceride values by subtraction of the isovaleric acid contributed by wax.[18] (2) Means of outer, center and inner parts of the melon are shown. Small amounts of anteiso 13:0, 15:0 and 17:0 were detected. Traces of 4,8,12 trimethyltridecanoic, pristanic and phytanic acids were present. 14:1 was ω7 and ω5. 16:1 was ω11, ω9, ω7 and ω5. 18:1 was ω7 and ω5. 18:2 ω6 and 18:3 ω3 were also detected. 7-methylhexadecenoic acid was present.[127] (3) n 7:0 and n 9:0 comprised 1.4 mole %. An unknown fatty acid in the C 18–20 region comprised 3.0 mole %.[117] (4) Small amounts of anteiso 11:0, 13:0, 15:0 and 17:0 were detected. 14:1 and 16:1 were ω9, ω7 and ω5. 18:1 was ω11, ω9, ω7 and ω5. Traces of 15:1 ω? and 17:1 ω8 were also detected. 7-methylhexadecenoic acid was present.[1] (5) Small amounts of anteiso 11:0, 13:0, 15:0 and 17:0 were detected. 4,8,12-trimethyltridecanoic, pristanic and phytanic acids were also detected in trace amounts.[68] (6) The values were obtained from the front of the junk which is rich in wax esters and correspondingly relatively poor in triglycerides. This was the only location where isovaleric acid was detected. Means of measurements from two whales are shown.[81] (7) Abbreviated nomenclature: no of carbon atoms: number of double bonds. n = normal. ND = not detected.

The similarity in the dolphin melon triglyceride fatty acid patterns can be seen in Table 2 where analyses of one Stenidae, two Delphinidae, one Monodontidae and one Physeteridae species are presented. Because of the known variations of fatty acid composition with topographical location, no exact comparison between different samples can be made (cf. Section II B). With the exception of *Physeter*, the similarities are nevertheless remarkable. Generally, isovalerate species also have longer, C_{13}–C_{16}, isobranched fatty acids in triglycerides and wax esters (cf Table 2). The dominating unsaturated fatty acid generally is 16:1 (Nomenclature: see Vol. IX, pp. 3–6.) Trace amounts of isoprenoid fatty acids, which are dietary,[3,6] and mono- and polyunsaturated fatty acids with unusual double bond locations have also been found (cf. notes of Table 2). Anteisobranched fatty acids are generally found in small amounts.[1,5,127] The dietarily contributed part of the different fatty acid types is unknown. However, iso-acids are rapidly synthesized in melon tissue[74] (cf. Ackman and Sipos[4]).

Several authors[1,5,17,18,79,84,105] described short-chain fatty acids other than 3-methylbutanoic (isovaleric) acid. 2-Methylpropanoic (isobutyric) acid was present in *Delphinus delphis*, *Stenella attenuata*, *Phocaenoides dalli*, *Tursiops truncatus* and *Sotalia fluviatilis* heads oils. The dextrorotatory 2-methylbutanoic (the anteiso isomer of isovaleric) acid was found in *Delphinus delphis*, *Tursiops truncatus*, *Sotalia fluviatilis* and *Globicephala melaena* (see above). It was probably the main contributor to the optical activity of the body oil from *Delphinus delphis*.[105]

These branched short-chain fatty acids are synthesized from coenzyme A thioester amino acid intermediates or by deamination of leucine (isovaleric acid), isoleucine (2-methylbutanoic acid) and valine (isobutyric acid).[59,65] Malins *et al.*[74] and Varanasi and Malins[119] made metabolic studies of the synthesis of lipids containing the isobranch. A slice from melon of *Tursiops gilli* rapidly converted radioactive leucine to isovaleric acid (and later to long-chain iso fatty acids) esterified in triglycerides. Wax esters were considerably slower to incorporate radioactivity, which came to reside mainly in the fatty acid portion. Although more data are needed it is tempting to postulate that the high and selective metabolic activity of isovalerate-containing triglycerides is part of an adaptatory mechanism for regulation of the sonic refractory power of the melon tissue either by variation of VXV content, or the use of isovalerate for generation of heat *in situ* (cf. Section B, 5).

Morris[81,82] found small amounts of isovaleric acid in the so-called "junk" of the sperm whale (*Physeter catodon*) which probably is phylogenetically related to the melon of other odontocetes (cf. Figs. 10 and 12). This and the occurence of minor amounts of C_{14}–C_{16} isobranched fatty acids in triglycerides and waxes from so-called "non-isovalerate" odontocetes[2,81,82] suggests that the capacity for the synthesis of isobranched fatty acids is widely distributed in odontocetes. In studies of phospholipids of the melon of *Tursiops gilli*, Varanasi and Malins[120] found no isovaleric acid, but some mole % of isobranched long-chain fatty acids. On the basis of this finding, they suggested that pathways involving phosphoglyceride intermediates may not be operative in the biosynthesis of the isovaleroyl triglycerides.

3. Fatty Alcohols

Blomberg[17,18] analyzed the alcohols of the unsaponifiables from pilot whales as acetates, trimethyl silyl ethers and hydrocarbons by GC and GC–MS. The hydrocarbons were made by a simple tosylation-reduction reaction. With the exception of iso 5:0 the general pattern from the fatty acids was qualitatively repeated (Tables 2 and 3). However, large quantitative differences were present. The pattern of alcohols from other odontocetes largely repeats the pattern of fatty acids. However, some quantitative differences were notable (Tables 2 and 3). Varanasi and Malins[119] found that the ^{14}C incorporation in isobranched wax ester alcohols from ^{14}C leucine proceeded after that of the wax ester acids. Isovaleric acid became radioactive before the long-chain acids. This suggested conversion of isobranched long-chain fatty acid to isobranched long-chain alcohol. The quantitative differences between acid and alcohol patterns show that the conversion of acids to alcohols is a discriminating process. As can be seen, the alcohol composition is similar in the Stenidae and Delphinidae species.

TABLE 3. Comparative Melon Fatty Alcohol Patterns from Wax of some Odontocetes

		Globicephala		Tursiops	Sotalia
		1	2	3	4
			(Mole %)		
iso	14:0	ND	0.3	tr	0.3
n	14:0	2	1	1	1
iso	15:0	20	22	24	17
n	15:0	3	1	3	2
iso	16:0	37	48	31	46
n	16:1	ND	1	2	2
n	16:0	20	15	24	21
iso	17:0	5	3	3	5
n	17:0	ND	2	ND	0.4
iso	18:0	ND	0.1	ND	0.2
n	18:1	9	6	6	2
n	18:0	2	0.5	2	0.6
		98	99.9	96	97.5

(1) The occurrence of a branched C_{19} alcohol was tentatively identified.[18] (2) Trace amounts of anteiso 15:0 and 16:0 were detected. The positions of the double bounds in 16:1 were $\omega9$ and $\omega7$, in 18:1 they were $\omega11$, $\omega9$ and $\omega5$. Means of outer, center and inner parts of the melon are shown.[127] (3) Iso and n 5:0 and 6:0, together about 4 mole %, were detected.[117] (4) Anteiso 15:0 and 17:0 were present in 0.7 and 1.3 mole %, respectively. 16:1 was $\omega7$ and 18:1 was $\omega11$ and $\omega9$.[1]

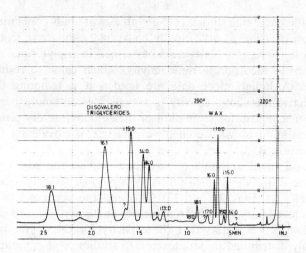

FIG. 4. GLC of pilot whale head oil. Wax esters and VXV triglycerides have been eluted. 1%
SE-30. Glass column, 2 m long. 3 mm inside diameter, 220–290° C, 10° C/min.[18]

4. *Hydrocarbons*

Blomberg[17,18] found hydrocarbons from the pilot whale which consisted almost entirely of pristane. It was identified by GC–MS (cf. Bendoraitis *et al.*[13] Hallgren and Larsson[52]). Squalene or phytane could not be detected. The pristane which is common in marine lipids is probably of dietary origin. Avigan and Blumer[8] traced pristane to zooplankton which synthesizes the bulk of marine pristane from phytol. A similar situation was found in herring oil by Lambertsen and Holman[64] and Hallgren and Larsson[52] where pristane also proved to be the dominant hydrocarbon. This is in contrast to shark liver oil which is largely composed of squalene.[110] The hydrocarbon content of the head oils of other odontocetes has been only sparsely investigated. In a review of lipid waxes, Warth[123] reported that spermaceti contained "0%" hydrocarbon. However, Blumer and Thomas[21] found "zamene", i.e. monounsaturated pristane analogs, in oil from sperm whale.

5. *Waxes*

The isovalerate waxes from pilot whale were subjected to analysis by GC and GC–MS[17,18] (Figs. 4 and 5). Each of the major wax peaks were analyzed by mass spectrometry, thus giving further structural information. The waxes yielded the expected fragmentation patterns with the main cleavages occurring around the ester bond, which proved their structure. As noted above, the Phocaenidae and Monodontidae families, and the genus Platanista[69,70] lack wax esters in their head oil. Varanasi and Malins[115] found isovaleroyl wax esters in *Tursiops gilli* jaw oil. A small amount (2.6 mole %)

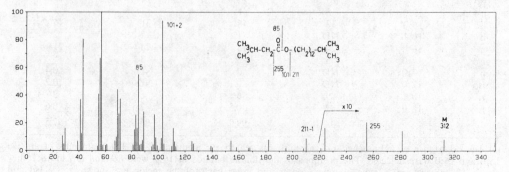

FIG. 5. Mass spectrum of isopentadecanyl isovalerate obtained on gas chromatography of pilot
whale head oil. Electron energy: 70 eV. Source temp.: 250°C.[18]

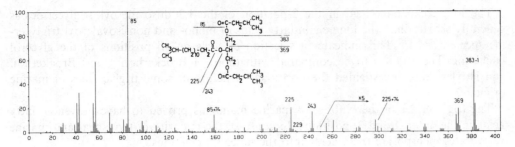

FIG. 6. Mass spectrum of diisovalero isopentadecanoin obtained on gas chromatography of pilot whale head oil. Electron energy: 70 eV. Source temp.: 250°C.[18]

of isoamyl wax esters were also present. Litchfield et al.[71] recently compared the average carbon numbers of melon and jaw wax esters from many different odontocete genera. The following variations were found: Platanistidae (Inia): 29.5, Ziphiidae: 28.2–30.2, Physeteridae: 29.7–29.8, Stenidae (Sotalia): 21.1 and Delphinidae: 21.1–23.0. Delphinidae and Ziphiidae were especially rich in branched chain structures.

In a review of waxes of marine animals, Nevenzel[86] reported a carbon no. range of 30–42 for wax esters from copepods and fish. Thus, odontocete head oil waxes are generally of lower molecular weight than other marine waxes.

6. Triglycerides

(a) *Pilot whale.* The individual VXV triglycerides (cf. Fig. 4) gave MS spectra that were easily interpretable[17,18] (cf. Barber et al.[10]). All had a comparatively small peak at m/e 229 (VV$^+$), which indicated that the isovaleric acid moieties were esterified at the 1 and 3 positions of glycerol (Fig. 6). The cleavage of a 1–2 or 2–3 bond in the glycerol backbone cannot give an ion with two isovaleroyls if they are situated symmetrically at the 1 and 3 positions. This was confirmed by the mass spectral analysis of a synthetic asymmetric dipentanomyristin (PPM), which had a prominent m/e 229 (PP$^+$) ion. The symmetric character of the VXV triglycerides was further confirmed by polarimetry.[17,18] This was possible because of the work of Schlenk Jr.[105] who made polarimetric studies on synthetical symmetric (VXV) and asymmetric (VVX) triglycerides. He found that an asymmetric diisovalero myristin (VVM) had an α_{21}^{D} of + 0.75° (10% in methanol). The value for pilot whale VXV triglycerides was + 0.2 ± 0.2°, rather close to zero, giving support to a symmetric structure.

(b) *Other odontocetes.* The GC–MS results[17,18] from *Globicephala* were later confirmed by Varanasi et al.,[112] who compared a synthetic asymmetric diisovalero myristin (VVM) with a mixture of VXV triglycerides prepared from *Tursiops truncatus*. The polarimetric findings[17,18] were later confirmed by Varanasi et al. (unpublished, 1972) who made similar measurements on VXV triglycerides of *Tursiops truncatus*. Recently, extensive PMR investigations including XXV and VXV triglycerides from thirteen different odontocete species by Wedmid and Litchfield,[124,125] further confirmed the esterification of the outer positions of glycerol by isovaleric acid.

Indirectly, this was also confirmed by the demonstration of lipase resistance of the VXV triglycerides by Malins and Varanasi.[15] Attempts to determine the triglyceride structure by means of pancreatic lipase failed because of the presence of the short and isobranched fatty acid at the 1 and 3 positions. This was probably due to steric hindrance by the iso-branching, as it was earlier shown that butyric acid is readily cleaved from an outer position.[33] Similar situations are known to occur with long-chain, highly unsaturated fatty acids in the same positions.[23] (cf. Desnuelle & Savary.[35]) Consequently, the 1,3 structure of the dolphin head oil VXV triglycerides was proven by polarimetry, lipase resistance, mass spectrometry and PMR.

The long-chain fatty acid in the 2-position of the 1,3 diisovaleroyl triglycerides is generally shorter than the long chain fatty acids of mono- and non-isovaleroyl triglyceride species.[17,18,127] This indicates a selection of acyls on all positions of the glycerol backbone. These data can be compared with those of Brockerhoff[25] and Brockerhoff and Hoyle[26] who investigated the positional structure of some triglycerides of marine origin.

The triglycerides of the blubber of marine mammals proved to have polyenoic fatty acids in the outer (1 and 3) positions. In contrast, the majority of terrestrial and marine vertebrates incorporate these acids into the inner (2) positions.

Both the bulky polyenoic C_{20}–C_{24} fatty acids and branched C_5 acids should be expected to keep the triglycerides fluid down to at least 0°C, which may be a prerequisite for fat which is in close contact with cold water (cf. Gilmore[49]). It may be significant that they both occupy the outer positions of the triglyceride.

Litchfield et al.[71] reported average carbon numbers for head oil triglycerides of odontocete genera. The results were similar to those reported above for waxes. The following variations were found: Platanistidae and Physeteridae: 42.0–50.0. Ziphiidae: 33.1–40.8. Monodontidae, Stenidae Delphinidae and Phocaenidae: 25.5–33.3. Thus, the isovalerate species have generally shorter acyl chains. In these species, isovaleric acid occurs in the form of XXV and VXV triglycerides. So far, no investigator has been able to find triisovalerin (VVV).[17,18,68,118]

The exact position on sn-glycerol of the isovaleric acid in the XXV triglycerides has so far not been determined. However, in a recent PMR study, Wedmid and Litchfield,[125] found that the isovaleroyl in the monosubstituted triglycerides of several genera was situated at the outer positions of glycerol, which is similar to the sn-3 position of butyric acid in butterfat.[24,95]

7. Other Lipid Classes

Blomberg[18] reported the occurrence of periodate-oxidizable diols in the unsaponifiable fraction of pilot whale oil and obtained mass spectral evidence of traces of other lipids of the neutral plasmalogen type at GC–MS of pilot whale head oil. At TLC, small spots moving just ahead of the XXX triglycerides further indicated that this type of lipid was present. Varanasi and Malins[114] reported the presence of dialkoxypentane diethers in the jaw oil of Tursiops truncatus. They were, however, present in only trace amounts. Diacyl glyceryl ethers analogous to the triglycerides are generally found in 1–2 wt % in non-isovalerate genera.[69,70] Varanasi and Malins,[116] however, reported trace amounts of chimyl and batyl alcohols in the unsaponifiables of jaw oil from the common porpoise (Phocaena Phocaena) which belongs to the isovalerate genera.

Tursiops gilli melon phospholipids were examined by Varanasi and Malins[120] who found up to 8 mole % of the acids to be C_{14}–C_{16} isobranched. They were present mainly in the 2-position. The same authors[121] also found nearly 40 mole % of C_5–C_{16} isobranched fatty acids in phosphatidyl choline from Delphinus delphis brain; 20 mole % of the acids were due to isovaleric acid. Phosphatidyl ethanolamine contained smaller amounts. Both ethanolamine and choline phosphatides were resistant to phospholipase A 2, analogous to the behavior of isovaleroyl triglycerides.[75] This astonishing finding implicates that isovaleric acid may have other functions than lowering sound velocity. Nitrogen solubility may be another variable that should be investigated.

B. Compositional Topography of Melon Lipids

1. Relation to Jaw and Blubber Fat

The differences between the head and blubber fats were early recognized. Van Gaver[45] and Margaillan[78] described differences in iodine, acid and Reichert-Meissl indices

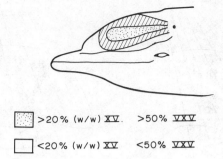

▨ >20% (w/w) XV.   >50% VXV

□ <20% (w/w) XV   <50% VXV

Fig. 7. Megaphonic arrangement of lipids in the melon of *Tursiops*. (Constructed from the data of Litchfield *et al.*[72])

between fat samples obtained from different parts of the dolphin body. The melon and jaw fats are generally very similar. This may be related to the common function of these fats in sound transmission. On the other hand, blubber fat is mostly strikingly different from the jaw and melon fats, having lipids of higher average molecular weight, and a higher level of unsaturation. In the isovalerate species, wax is much less common in the blubber than in the head fat.[1,2,68,70,71,102,117] The internal organs contain neutral lipids of the usual mammalian type.[73,85] As mentioned above, an interesting exception to this rule was found by Varanasi and Malins,[121] who demonstrated the presence of isovaleric acid and long-chain isobranched fatty acids in *Delphinus delphis* brain. Lesch *et al.*[66] did not report such acids from the brain of *Globicephala melaena*, but they may not have been looked for.

2. Lipid Content

Variations in lipid content have been recorded by Litchfield *et al.*[72] for *Tursiops truncatus* melon. In a detailed topographical investigation, the midsagittal central parts

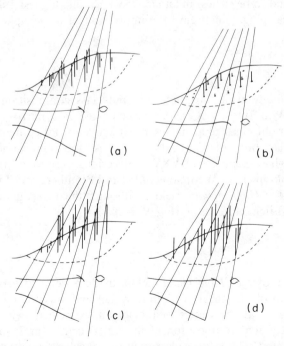

Fig. 8. Chemical and acoustical topography of *Tursiops* melon. Midsagittal distribution of (a) lipid content (wt % of tissue); (b) XV wax content (wt % of fat); (c) VXV triglyceride content (wt % of fat); (d) Sound velocity (water velocity at 23°C (1518 m/s)—tissue velocity at 23°C). The length of the spikes is directly proportional to the figures found by Litchfield *et al.*,[72] (a–c) and Norris and Harvey[92] (d). The angle of view is indicated by the auxiliary perspectivistic lines. The length of the spikes has been adjusted to conform with the perspective.

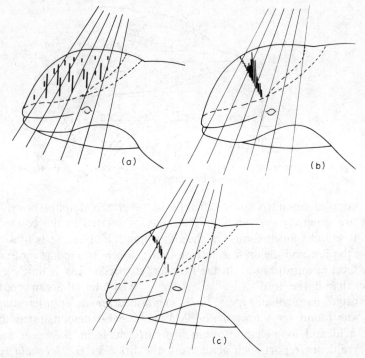

FIG. 9. Chemical and acoustical topography of *Globicephala* melon. Midsagittal distribution of (a) XV wax esters (wt % of fat); (b) XV wax esters (micromoles per microliter oil); (c) Sound velocity (water velocity at 23°C (1518 m/s)—tissue velocity at 23° C.) The length of the spikes is directly proportional to the values found by Wedmid et al.[127] (a), and Blomberg and Lindholm[19] (b and c). The angle of view is indicated by the perspectivistic lines. The length of the spikes has been adjusted to conform with the perspective.

of the melon were found to be richest in fat (75–95 wt %). The highest value was obtained close to the right nasal plug, which gave the impression of a megaphonic construction (Figs. 7, 8a).

3. *Lipid Class*

Litchfield et al.[72] found wide variations of lipid class composition in the *Tursiops* melon. (Figs. 7, 8b). Wax (mostly XV) was concentrated in a narrow zone within the melon, ending at the right nasal plug, as described above for fat content. VXV triglycerides exhibited a similar, but not identical, pattern. At the skin surface, ahead of the "megaphone", still high percentages of VXV triglycerides were found (Fig. 8c).

Wedmid et al.,[126] Blomberg,[17] Wedmid et al.[127] and Blomberg and Lindholm[19] found similar variations in the *Globicephala* melon, although the investigations were not as extensive as those of Litchfield et al.[72] (Fig. 9 a–e).

4. *Fatty Acids*

Wedmid et al.,[127] studying *Globicephala* melon, made a detailed investigation of the midsagittal distribution of unsaturation (iodine value) and wax ester content (Fig. 9a). Unsaturation was lowest at the center of the melon, coincident with a peak in wax ester content. The fatty acid composition of the inner, center, and outer parts of the melon was also presented. The fatty acids varied as expected from the results on lipid class and unsaturation. Similar results were obtained by Blomberg[17] and Blomberg and Lindholm,[19] who by means of radiotracer methods,[16] investigated the lipid distribution on a line perpendicularly from the apex into the melon (Fig. 9b, c). Morris[81,82] found fatty acid variations in the junk and spermaceti organ of *Physeter* (Fig. 10).

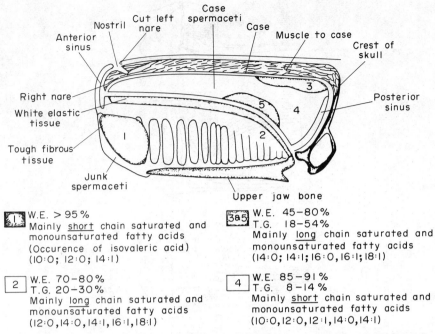

FIG. 10. Compositional lipid topography of the forehead of *Physeter*. Taken from Morris.[81]

III. FUNCTIONAL ASPECTS

A. *The Echolocation of Odontocetes*

1. *Emission of Ultrasound*

Several authors have recorded a beam of 20–200 kHz ultrasound emitted as click trains from some odontocetes (for reviews, see Evans[37] and Norris[87–90]). At the higher frequencies, the beam was sharp[91] (Fig. 11). Thus, some mechanism concentrating sound ahead of the animal must exist. Despite a decade of investigation, the exact location and mechanism of sound production is debated.[90] Diffractive mechanisms are important at low sound frequencies (e.g. 20 kHz) which have a wavelength (ca 7 cm) close to

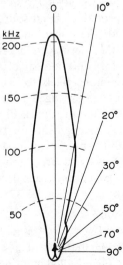

FIG. 11. Upper frequency of echolocation clicks of *Steno bredanensis* as a function of angular position around animal. Taken from Norris and Evans.[91]

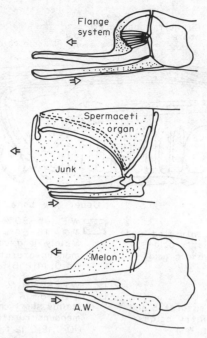

FIG. 12. Schematic diagram of some anatomical features from the heads of genera from three different odontocete families. (a) *Platanista* (Platanistidae). (b) *Physeter* (Physeteridae). (c) *Tursiops* (Delphinidae). The dotted areas denote lipid-rich compartments of the head. A.W. = Acoustic window. The arrows indicate emission and reception of ultrasound, respectively.

the dimensions of the melon (10–100 cm). Further proof of efficient echosounding was obtained by experiments on blindfolded animals. Busnel and Dziedzic[28] found that a blindfolded *Phocaena* could navigate past 0.4 mm thin wires. Evans and Powell[38] (1967) found that *Tursiops* was able to distinguish copper discs from exact replicas made from other metals. Wood,[129] Bel'kowitsch and Yablokov[12] and Norris[87] postulated that the fatty melon may collimate emitted sound much like an optical lens (cf. Bel'kowitsch and Nesterenko.[11] However, at least three principles for the emission of sound beams can be deduced from the odontocete head anatomy (Fig. 12). The flange system of *Platanista* (Platanistidae) have been suggested to act as baffles which direct sound ahead of the animal by a non-refractive mechanism[57] (Fig. 12a). Specialized maxillary flange systems are also prominent in Ziphiidae, although their influence on sound emission is difficult to assess.[87] The heat of *Physeter* (Physeteridae) exhibits a radically different design (Fig. 12b). The probable sound source, the "museau de singe" (monkey's snout), is situated at the front of the enormous forehead, which is composed of two lipid-containing compartments, the spermaceti organ and the "junk".[97,104] Norris and Harvey[92] proposed a reverberatory mechanism of sound production, by which sound is reflected on the posterior air sinus back to the "museau de singe". The acoustic tissues of *Tursiops* (Delphinidae) are arranged in a more lens-and-mirror-like manner (Fig. 12c). Thus, the superfamilies Platanistoidea, Physeteroidea and Delphinoidea probably utilize at least three different mechanisms for the emission of directive ultrasound. Although the function of these different systems is still obscure, it is probably significant that in Fig. 12 the (a) and (b) systems are "non-isovalerate", and the (c) system is "isovalerate". The connection between anatomical and chemical differentiation is, however, difficult to explain at present.

2. *Reception of Ultrasound*

Bullock *et al.*[27] made suggestive findings of a sound sensitive spot on the lateral side of the mandible. This has been called "the acoustic window" by Norris.[88,89] It is covered only by a thin bone lamella, over-and underneath which isovalerate rich lipids are concentrated.[102] Norris[88,89] suggested that sound reflected by objects ahead of the animal mainly enters through this window, and is then guided through the mandibular fat body straight to the inner ear, although the exact mode of transmission from fat to cochlea has not been demonstrated. The role of the auditory meatus is unclear, as is also the possible sound reception by the melon. The well developed neural sound processing apparatus of the odontocetes, eighth cranial nerve, inferior colliculi (Flanigan[39]) and temporal lobe cortex (Lilly[67]) illustrates the profound importance of the echolocation system to the odontoceti.

Summarizing, the odontocetes seem to be able to project a beam of ultrasound structured in space by frequency variation and structured in time as click trains, and to be able to receive selected parts of the reflected sound in the mandible, and possibly more generally through the melon, using also scanning movements of the head during swimming. The odontocete brain seems well adapted for analysis of the incoming complex acoustic information.

B. *The Possible Role of Lipids in Sound Processing*

1. *The Study of Sonic Properties of Odontocete Lipids*

As mentioned above, the narrow ultrasonic beam emitted by some and probably all odontocetes suggests a lens-like action of the melon (in Delphinoidea). Several authors have undertaken the study of acoustic properties of the melon and its constituent lipids.[17,19,20,93,113]

By analogy with optical lenses, positive acoustic lenses should be constructed from material having a low sound velocity relative to the surrounding medium. Also, the postulated wave-guide function of the jaw fat necessitates a low sound velocity of this fat, in order to secure complete reflection against the walls of the wave-guide. Thus, the sound velocity is important in the assessment of the function of lipids in sound processing.

2. *Molecular Structure and Sound Velocity*

Fluid lipids of short chain length tend to have a low sound velocity (see e.g. Hustad et al.[60]). Additionally, several investigators have noted a correlation between chain branching and decreased sound velocity.[14,22,46,47,103] Structural changes affect both adiabatic compressibility and density, which together determine the sound velocity.[99] In this review, two empirical relationships between molecular structure and sound velocity will be considered.

(a) *The Rao function.* This relationship[98] states that

$$u^{1/3} \cdot \frac{M}{d} = R, \qquad (1)$$

where u is the sound velocity in m/s, M the molecular weight, d the density in g/cm^3 and R a constant often referred to as the molar sound velocity, R is independent of temperature, and has additive properties, i.e. different structural details of a molecule can be assigned values that are part of R, which can be added or subtracted as the molecular structure—and thereby molecular volume—is changed. Once the assignments for these structural elements are known, a value of R can be calculated. Provided that an accurate estimate of the density is used, a predictive accuracy of sound velocity within 0.1–0.2% can be obtained at least for many hydrocarbons.[47] Structural complications like branching of a hydrocarbon chain have to be compensated for by means of empirical factors.[9,47,50,51]

The *Rao* function (1) has become widely used because of its good additive properties and temperature independence. Gouw and Vlugter[50,51] employed this relation in studies of triglycerides and methyl esters. In studies of natural triglyceride oils, they found that

$$R_{TG} = \Sigma(R_{Me}) - 391. \tag{2}$$

By subtracting 391 (corresponding to the 4 hydrogens less in the glyceride) from the sum of R values of the three methyl esters of the component fatty acids, the R value of the triglyceride could be calculated. Provided that the density of the oil was known, the sound velocity could be accurately predicted.

(b) *Schaaffs' function.* The formula constructed by Schaaffs[103] states that

$$u = 4450 \cdot d \cdot \frac{B}{M} - \frac{4450}{1 + B/\beta}, \tag{3}$$

u is the sound velocity in m/s, B and β are measures of the molecular volume. ($B = \Sigma_i$ $(zA)_i$; $\beta = \Sigma_i (z\,\alpha)_i)$. A and α are called inner and outer atomary volumes. B and β have additive properties, and can be calculated by addition of appropriate A and α values.[103] Provided that the density and molecular weight are known, a fairly accurate value of the sound velocity of an organic fluid can be calculated. The Schaaffs formula (3) and its atomary volumes have been calculated for 20°C and are not independent of temperature.

These two formulae can be applied in the correlation of chemical structure, sound velocity and echolocation in odontocetes.

(c) *Calculation of sound velocity.* The mathematical relationships (1), (2) and (3) can be employed to calculate approximate sound velocities of individual lipid species, lipid mixtures, and the influence of structural modifications on the sound velocity. The acoustic properties of pilot whale head oil lipids will be mainly discussed. The dominant species of lipids in this oil are, as stated above, among the triglycerides diisovaleroisopentadecanoin (VXV) (MW 484) and among the wax esters isopentadecanyl isovalerate (XV), (MW 312). The densities of these substances are not precisely known. A fair approximation can, however, be derived from the work of Varanasi et al.,[113] who studied the densities of mixtures of dolphin (*Stenella*) triglycerides and wax esters. They found a nonlinear relation between density and % wax in wax–triglyceride mixtures. The formulae (1) and (2) give values of 1390 m/s and 1354 m/s for densities of 0.95 and 0.86 g/cm³ for the VXV triglyceride and the XV wax ester, respectively. Velocities calculated by formula (3) are 1364 and 1325 m/s, respectively. The known triglyceride and wax composition of pilot whale head oil[18] can be used to compute the sound velocities of total triglyceride (0.93 g/cm³) and total wax (0.87 g/cm³). Using relations (1) and (2), velocities of 1382 and 1408 m/s, respectively, were obtained. Relation (3) gave 1364 and 1346 m/s, respectively.

Varanasi et al.[113] found a velocity decrement of 0.7% relative to the expected value for 10% *Stenella* wax in *Stenella* triglycerides. Applying this to pilot whale head oil, the sound velocity can be calculated to 1375 m/s (1) and (2), or 1362 m/s (3). The experimental value is 1387 m/s (20°C)[20] a fair agreement between observation and prediction.

The empirical formulae (1), (2) and (3), despite some inherent uncertainties, can be used to assess the role of iso-branching vs chain length in the lowering of sound velocity by head oil lipids. In formula (1) R is decreased by 3.5 by an iso branch compared to the straight chain isomer.[9,14,47,108] The density decrement for one iso branch in hydrocarbons and monoesters is about 0.6%. (Data from Geelen[47]). If this is approximately valid for triglycerides, a sound velocity decrement of roughly 3 (1) or 4–7 (3) m/s for one iso branch is obtained. From the data of Gouw and Vlugter[50] (cf. Blomberg and Nordby-Jensen[20]), an increment of 3.8 m/s/methylene in a series of simple triglycerides can be computed (20°C). Judging from these estimates, one iso branch would give about the same decrease of sound velocity as the deletion of one methylene group. Thus, it seems that in the head oil lipids a low sound velocity is easier to obtain by shortening chain length than by extensive isobranching. However, structural details like isobranching may participate in intermolecular interactions which are designed to minimize density, and thus further decrease the sound velocity. Such effects were noted by Varanasi et al.[113] Their physiological importance is, however, difficult to assess. The higher average chain lengths of lipids from non-isovalerate species may only mean that they utilize other sound-focusing mechanisms or that the dimensions of their sonic lens are large enough to collimate sound despite low refractivity of their head oil lipids.

Summarizing, then, it seems that the low average chain length of the head oil is the main cause of its low sound velocity. Isobranching further lowers the sound velocity somewhat.

3. Sound Absorption

An acoustic lens should have a low sound absorption. No studies at the frequencies used by the odontocete have been published, however. In their study of pilot whale head oil, Blomberg and Nordby-Jensen[20] found an absorption coefficient of $0.531 \pm 0.011 \, \text{cm}^{-1}$ at 9.92 MHz (20°C). As a comparison, corn oil had $0.718 \pm 0.015 \, \text{cm}^{-1}$ which is appreciably higher. By dividing the absorption coefficient by the square of the frequency, a value (α') which is constant over a wide frequency range is obtained.[53] This relationship should facilitate comparisons with other oils. However, measurements at the frequencies used by the odontocete are needed because relaxation can create non-linear sound absorption behavior. A low sound attenuation of whale blubber was also found by Reysenbach De Haan[100] and Purves.[96]

4. Topographical Distribution of Sound Velocity in Melon Tissue

Norris and Harvey[92] studied the distribution of sound velocity in the melon tissue of Tursiops truncatus. A core of low sound velocity led to the probable sound source, the right nasal plug (Fig. 8c). This distribution is similar to the distribution of isovalerate rich wax esters and triglycerides found by Litchfield et al.[72] in the same animal (Figs. 7 and 8). It is also in line with the predictions of the empirical relationships between lipid structure and sound velocity discussed in the preceding section, according to which a mixture of short-chain waxes and short-chain triglycerides should have a low sound velocity.

Blomberg and Lindholm[19] found a similar sound velocity gradient corresponding to variations in lipid composition in Globicephala melaena melon (Fig. 9c).

The region of high wax content coincided with a minimum of sound velocity. A core of low sound velocity should refract sound within the melon, and may thus be a collimating system for the production of directional ultrasound in the echolocation system of odontocetes. Whether the expected refraction is large enough to actually focus ultrasound is the subject of the subsequent section, where a tentative answer is derived from simple calculations.

5. The Odontocete Melon as a Sonic Lens

As we have seen, sonic data, lipid compositional data, anatomical data and physiological data favor a collimating role of the melon in the species of the Delphinoidea superfamily. A tentative sonic model of the odontocete melon could be constructed by combining the data that have been gathered. Only two genera have been studied enough to attempt such a construction, i.e. Globicephala and Tursiops. A sonic model could indicate whether the refraction of the melon could concentrate the sound into a beam as observed for many odontocetes (see Section II.A.1). Attempts to make tentative sonic models are presented in Figs. 14 (Tursiops) and 15 (Globicephala). The sound velocities

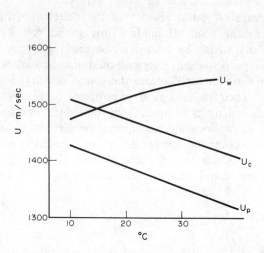

FIG. 13. The relationship between sound velocity (U) and temperature for the three liquids: sea water (U_w), corn oil (U_c) and pilot whale head oil (U_p). Taken from Blomberg and Nordby-Jensen.[20]

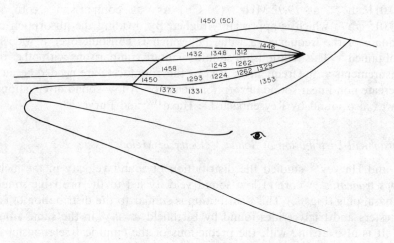

FIG. 14. Tentative model of sonic refraction in the melon of *Tursiops*, based on the data of Norris and Harvey.[93] Reflection off air sinuses, muscles and bone has not been considered.

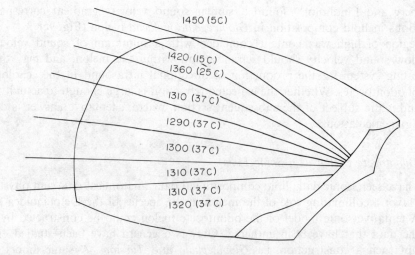

FIG. 15. Tentative model of sonic refraction in the melon of *Globicephala* based on the data of Blomberg and Lindholm[19] and inferences from lipid compositional data of Blomberg and Lindholm[19] and Wedmid *et al.*[127] Reflection off air sinuses, muscle and bone has not been considered.

in the *Globicephala* melon have been approximated by comparison with the velocities found by Norris and Harvey[92] and the lipid distribution found by Wedmid et al.[127] The one-dimensional velocity data of Blomberg and Lindholm[19] have also been incorporated.

A temperature gradient across the melon tissue has been surmised, and the velocity data of Norris and Harvey[92] and Blomberg and Lindholm[19] have been modified accordingly, using the relationship between sound velocity and temperature (a decrease of 3.5 m/s per °C) (Fig. 13). Although this relation was found for pilot whale head oil, it is typical of triglycerides, and is probably approximately valid also for most of the melon tissue, which contains 50–90% fat. In Figs. 14 and 15, wavefronts have first been calculated. Then, sound paths have been constructed. They are shown together with the calculated velocities. The emission of collimated sound beams can be deduced. However, the models contain many approximations. They are only intended for a discussion of possible directive mechanisms of sound projection. Only refractive mechanisms have been considered in this review. Diffraction and reflection may be as important but cannot be treated because of lack of data. At low frequencies (e.g. 20 kHz) the whole melon should act as a resonator.

Figures 14 and 15 indicate that most of the refraction is located within the melon tissue, and not at the melon/water interface. The sound velocity of water seldom exceeds that of melon tissue by more than 10%, i.e. "refractive indices" for sound tend to be 1.0–1.1 (Fig. 13). This is a low value for a lens, and it is probably only through the internal velocity gradient and the comparatively large dimensions that the melon may perform as a lens. The graded refraction within the melon also gives a minimum of reflective losses. These are known to give serious losses at interfaces between media of widely different velocities. An acoustic lens should experience a considerable influence from water temperature. A diving odontocete may experience a change in water temperature from 15°C to 5°C. This will drastically change the refraction of sound at the melon/water interface (Fig. 13). The odontocete must have a mechanism in order to cope with this, either by changes in melon temperature, or by changes in melon form.[20]

C. *Other Functional Aspects*

There have been suggestions of several alternative functions of the head oil lipids of odontocetes.

Clarke[32] suggested that the sperm whale head oil is used for control of buoyancy. Kooyman[63] and Fraser and Purves[43] suggested that the high nitrogen solubility of fat may remove nitrogen to the fat depots, and prevent diving sickness due to nitrogen bubbles in the central nervous system. Short-chain aliphatic acids like isovaleric acid are known to be sexual attractants—pheromones—of many mammals.[15,83] A trail of excreted isovaleric acid from a dolphin could conceivably be followed by taste for long distances by another dolphin. Gilmore[49] suggested that isovaleric acid-containing lipids may have an especially pronounced freeze-point depressing effect. These possibilities are not improbable. However, they have not received experimental attention and are consequently still only postulates.

IV. CONCLUSIONS AND PERSPECTIVES

The study of the unusual head oil lipids of odontocetes inevitably leads the investigator to fascinating interdisciplinary research with implications in such fields as taxonomy, biochemistry, acoustics and ecology. The aim of this review has been to unite, within the limits of existing data, the results of different investigators into a coherent scheme. The prime example has been the pilot whale which has been studied by several authors. To conclude, the head oils of odontocetes have a lipid composition different from that of blubber and internal body fat. The head oil lipids have a lower average molecular weight than those of the other fat depots. Furthermore, in most odontocete families, wax is more abundant in the head oils than in the other fat bodies.

Within the superfamily Delphinoidea isovaleric acid is a dominating fatty acid in melon and jaw oils. No significant amounts of this acid exist in other fats, with the exception of the brain. Isovaleric acid is mostly esterified in triglycerides, although sometimes also in wax esters. In triglycerides it occupies the outer (1 and 3) positions of glycerol. The head anatomy and topographical lipid distribution indicates a functional role of lipids in the emission of ultrasound for echolocation. On the basis of morphology and lipid composition at least three different sonic collimating mechanisms can be postulated. The low-molecular, isovalerate-laden triglycerides and waxes of the Delphinoidea species can be expected to transmit sound at a comparatively low velocity. In topographical studies of melon tissue, they seem to be concentrated in a central core of low sound velocity. On the basis of these findings, hypothetical sonic lens models of the melon can be constructed. The results of studies on pilot whales fit well into the general pattern. The experimental unavailability of odontocetes has been the major obstacle to a more profound understanding of the physiology of their echolocation system. Chemical studies on the tissues implicated have progressed much since 1970. However, there is still a long way before we fully understand the mechanisms of this system and the role of lipids therein although they surely can yield to determined investigators.

Hopefully, there will still be odontocetes to investigate in that future.

Acknowledgements—The author is indebted to Dr. Klaus Serck-Hanssen for support throughout this work. Some of the rare references were found by him. The late Professors Stina and Einar Stenhagen provided many valuable suggestions and comments. The generous help from the virological laboratory of Göteborg during the preparation of this manuscript is also acknowledged.

(*Received 25 February* 1977)

REFERENCES

1. ACKMAN, R. G., EATON, C. A., KINNEMAN, J. and LITCHFIELD, C. *Lipids* **10**, 44–49 (1975).
2. ACKMAN, R. G., EATON, C. A. and LITCHFIELD, C. *Lipids* **6**, 69–77 (1971).
3. ACKMAN, R. G. and HOOPER, S. N. *Comp. Biochem. Physiol.* **24**, 549–565 (1968).
4. ACKMAN, R. G. and SIPOS, J. C. *Comp. Biochem. Physiol.* **15**, 445–456 (1965).
5. ACKMAN, R. G., SIPOS, J. C., EATON, C. A., HILAMAN, B. L. and LITCHFIELD, C. *Lipids* **8**, 661–667 (1973).
6. ACKMAN, R. G., SIPOS, J. C. and TOCHER, C. S. *J. Fish. Res. Board Can.* **24**, 635–650 (1967).
7. ANDRÉ, E. *Comptes Rendus* **178**, 1188–1191 (1924).
8. AVIGAN, J. and BLUMER, M. *J. Lipid Res.* **9**, 350–352 (1968).
9. BACCAREDDA, M. and PINO, P. *Gazz. Chim. Ital.* **81**, 205–211 (1951).
10. BARBER, M., MERREN, T. O. and KELLY, W. *Tetrahedron Lett.* No. 18, 1063–1067 (1964).
11. BEL'KOWITSCH, W. M. and NESTERENKO, J. I. *Naturwiss. Rundsch.* **25**, 143–147 (1972).
12. BEL'KOWITSCH, W. M. and YABLOKOV, A. *Yuni Tekhnik* **3**, 76–79 (1963).
13. BENDORAITIS, J. G., BROWN, B. L. and HEPNER, L. S. *Anal. Chem.* **34**, 49–53 (1962).
14. BERGMANN, L. *Der Ultraschall.* 6th ed. Zürich: S. Hirzel Verlag (1954).
15. BERUTER, J., BEAUCHAMP, G. K. and MUETTERTIES, E. L. *Physiol. Zool.* **47**, 130 (1974).
16. BLOMBERG, J. Abstract No. 41, 11th World Congress, International Society for Fat Research, Göteborg (1972).
17. BLOMBERG, J. Abstract No 223, 11th World Congress, International Society for Fat Research, Göteborg (1972).
18. BLOMBERG, J. *Lipids* **9**, 461–470 (1974).
19. BLOMBERG, J. and LINDHOLM, L.-E. *Lipids* **11**, 153–156 (1976).
20. BLOMBERG, J. and NORDBY-JENSEN, B. *J. Acoust. Soc. Am.* **60**, 755–758 (1976).
21. BLUMER, M. and THOMAS, D. W. *Science* **148**, 370–371 (1965).
22. BOELHOUWER, C., VAN ELK, J. and WATERMAN, H. I. *Erdöl Kohle* **11**, 778–781 (1958).
23. BOTTINO, N. R., VANDENBURG, G. A. and REISER, R. *Lipids* **2**, 489–493 (1968).
24. BRECKENRIDGE, W. C. and KUKSIS, A. *J. Lipid Res.* **9**, 388–393 (1968).
25. BROCKERHOFF, H. *Comp. Biochem. Physiol.* **19**, 1–12 (1966).
26. BROCKERHOFF, H. and HOYLE, R. *J. Arch. Biochem. Biophys.* **102**, 452–455 (1963).
27. BULLOCK, T. H., GRINNELL, A. D., IKEZONO, E., KAMEDA, K., KATSUKI, Y., NOMOTO, M., SATO, O., SUGA, N. and YANAGISAWA, K. *Z. Vgl. Physiol.* **59**, 117–156 (1968).
28. BUSNEL, R. G. and DZIEDZIC, A. in: *Animal Sonar Systems: Biology and Bionics*, Busnel, R. G. ed., Lab. Physiol. Acoustique, Jouy-en-Josas-78, France (1967).
29. CHEVREUL, M. E. *Ann. Chim. Phys.* **7**, 264 (1817).
30. CHEVREUL, M. *Recherches sur les Corps Gras d'Origine Animale*, pp. 104 and 115, Paris: F. G. Levrault (1823).

31. COMMITTEE FOR WHALING STATISTICS. *International Whaling Statistics*, Vols 37, 47, 57, 65, 75, Oslo: Grøndal and Søn Trykkeri (1958, 1962, 1966, 1970, 1975).
32. CLARKE, M. R. *Nature* **228**, 873–874 (1970).
33. CLEMENT, G., CLEMENT, J. and BEZARD, J. *Biochem. Biophys. Res. Comm.* **8**, 238–242 (1962).
34. DENISOV, I. A. Khimicheski Sostav Zhirov Belukhi (Chemical Comp. of the Fat of the Beluga) (1835). ref. in: *Trudy Vniro* **3** (1936).
35. DESNUELLE, P. and SAVARY, P. *J. Lipid Res.* **4**, 369–381 (1963).
36. ELLERMAN, J. R. and MORRISON-SCOTT, T. C. S. Checklist of Palaearctic and Indian Mammals 1758–1946. British Museum, Nat. Hist. pp. 1–810, London (1951).
37. EVANS, W. E. *J. Acoust. Soc. Am.* **54**, 191–199 (1973).
38. EVANS, W. E. and POWELL, B. A. in: *Animal Sonar Systems: Biology and Bionics*, Busnel, R. G. ed., Lab. Physiol. Acoustique, Jouy-en-Josas-78 (1967).
39. FLANIGAN, N. J. in: *Mammals of the Sea*, Ridgeway, S. H., ed., pp. 215–246. Boston: Charles Thomas (1972).
40. FRASER, F. C. *At. Rep.* **1**, 49 (1950).
41. FRASER, F. C. *Annals and Magazine of Natural History Incl. Zool., Bot. and Geol.* Vol. IV. 12th Series, pp. 942–944. London: Taylor and Francis Ltd (1951).
42. FRASER, F. C. in: *Whales, Dolphins and Porpoises*, Norris K. K. ed., pp. 7–31. Berkeley and Los Angeles: Univ. of Calif. Press (1966).
43. FRASER, F. C. and PURVES, P. E. *Nature* **176**, 1221–1222 (1955).
44. GALLIANO, R. E., MORGANE, P. J., McFARLAND, W. L., NAGEL, E. L. and CATHERMAN, R. *Anat. Rec.* **155**, 325–338 (1966).
45. VAN GAVER, F. Contribution Zoologique à l'Etude des Huiles d'Animaux Marins. Thesis, Fac. Sci. Marseille (1923).
46. GEELEN, H., WATERMAN, H. I., WESTERDIJK, J. B. and KLAVER, R. F. *Riv. Combust.* **9**, 355–364 (1955).
47. GEELEN, H. Het Onterzoek van Koolwaterstoffen met behulp van de Voortplantingssnelheid van Ultrasonore Trillingen. Thesis, Delft Technical High School, Gravenhage, Holland: Excelsior (1956).
48. GILL, A. H. and TUCKER, O. M. *Oil Fat Ind.* **7**, 101–102 (1930).
49. GILMORE, R. M. *Oceans* 1, 9–20 (1969).
50. GOUW, T. H. and VLUGTER, J. C. *J. Am. Oil. Chem. Soc.* **44**, 524–526 (1964).
51. GOUW, T. H. and VLUGTER, J. C. *Fette Seifen Anstrichm.* **69**, 159–164 (1967).
52. HALLGREN, B. and LARSSON, S. *Acta Chem. Scand.* **17**, 543–545 (1963).
53. HEASELL, E. L. and LAMB, J. *Proc. Phys. Soc.* **69**, 869–877 (1956).
54. HERALD, E. S., BROWNELL, R. L. JR., FRYE, F. L., MORRIS, E. J., EVANS, W. E. and SCOTT, A. B. *Science* **166**, 1408–1410 (1969).
55. HERSHKOVITZ, P. *U.S. Natl. Mus. Bull.* **246**, 90 (1966).
56. HILDITCH, T. P. and LOVERN, J. A. *J. Soc. Chem. Ind.* **47**, 105–111 T (1928).
57. HILDITCH, T. P. and LOVERN, J. A. *J. Soc. Chem. Ind.* **48**, 359–368 T (1929).
58. HILDITCH, T. P. and LOVERN, J. A. *J. Soc. Chem. Ind.* **48**, 369–372 T (1929).
59. HORNING, M. G., MARTIN, D. B., KARMEN, A., VAGELOS, P. R. *J. Biol. Chem.* **236**, 699 (1961).
60. HUSTAD, G. O., RICHARDSON, T., WINDER, W. C. and DEAN, M. P. *Chem. Phys. Lipids* **7**, 61–74 (1971).
61. KIZEVETTER, I. U. *Zhiry Morskikh Mlekopitay Uschchikh* (Fats of Marine Mammals), Vladivostok (1953).
62. KARLSSON, K. A., NILSSON, K. and PASCHER, I. *Lipids* **3**, 389–390 (1968).
63. KOOYMAN, G. L. in: *The Biology of Marine Mammals*, Andersen, H. T. ed., pp. 65–94, New York: Assoc. Press (1969).
64. LAMBERTSEN, G. and HOLMAN, R. T. *Acta Chem. Scand.* **17**, 281–282 (1963).
65. LEDERER, E., *Biochem. J.* **93**, 449–468 (1964).
66. LESCH, P., NEUHAUS-MEIER, S. and BERNHARD, K. *Helv. Chim. Acta* **51**, 1655–1662 (1968).
67. LILLY, J. C. *Man and Dolphin*, Garden City, New York: Doubleday (1961).
68. LITCHFIELD, C., ACKMAN, R. G., SIPOS, J. C. and EATON, C. A. *Lipids* **6**, 674–681 (1971).
69. LITCHFIELD, C. and GREENBERG, A. J. *Comp. Biochem. Physiol.* **47 B**, 401–407 (1974).
70. LITCHFIELD, C., GREENBERG, A. J., CALDWELL, D. K., CALDWELL, M. C., SIPOS, J. C. and ACKMAN, R. G. *Comp. Biochem. Physiol.* **50 B**, 591–597 (1975).
71. LITCHFIELD, C., GREENBERG, A. J. and MEAD, J. G. *Cetology* No. 23, 1–10 (1976).
72. LITCHFIELD, C., KAROL, R. and GREENBERG, A. J. *Marine Biol.* **23**, 165–169 (1973).
73. LOVERN, J. A. *Biochem. J.* **28**, 394–402 (1934).
74. MALINS, D. C., ROBISCH, P. A. and VARANASI, U. *Biochem. Biophys. Res. Comm.* **48**, 314–319 (1972).
75. MALINS, D. C. and VARANASI, U. in: *Protides of the Biological Fluids*, Peeters, H. ed., **19**, 127–129. Oxford: Pergamon Press (1971).
76. MALINS, D. C. and VARANASI, U. in: *Biochemical and Biophysical Perspectives in Marine Biology*, Malins, D. C. and Sargent, J. R. eds. Vol. 2, pp. 237–290, New York: Academic Press (1975).
77. MARCELET, H. *Compt. Rend.* **182**, 1416–1417 (1926).
78. MARGAILLAN, L. Le Pouvoir Rotatoire des Corps Gras Naturels. Thesis, Fac. Sci Paris. Gauthier-Villars (1930).
79. MORII, H. and KANAZU, R., *Bull. Jap. Soc. Scient. Fish.* **38**, 599–605 (1972).
80. MORRIS, D. *The Mammals. A Guide to the Living Species*, pp. 237–245, London: Hodder and Stoughton (1965).
81. MORRIS, R. J. *Deep-Sea Res.* **20**, 911–916 (1973).
82. MORRIS, R. J. *Deep-Sea Res.* **22**, 483–489 (1975).
83. MYKYTOWYCZ, R. and GOODRICH, B. S. *J. Invest. Dermatol.* **62**, 124–131 (1974).
84. NAKAMURA, Y. and TSUJINO, I. *J. Agric. Chem. Soc. Jap.* **27**, 642–645 (1952).
85. NELSON, G. J. *Comp. Biochem. Physiol.* **46 B**, 257–268 (1973).
86. NEVENZEL, J. C. *Lipids* **5**, 308–319 (1970).

87. NORRIS, K. S. in: *Marine Bioacoustics*, Tavolga, W. N. ed., pp. 317–336, Oxford: Pergamon Press (1964).
88. NORRIS, K. S. in: *Evolution and Environment*, Drake, E. T. ed., pp. 297–324, New Haven, Connecticut: Yale Univ. Press. (1968).
89. NORRIS, K. S. in: *The Biology of Marine Mammals*, Anderson, H. T. ed., pp. 391–423, New York.: Acad. Press (1969).
90. NORRIS, K. S. in: *Biochemical and Biophysical Perspectives in Marine Biology*, Malins, D. C. and Sargent, J. R. eds, pp. 215–236, Vol 2, New York: Acad. Press (1975).
91. NORRIS, K. S. and EVANS, W. E. in: *Marine Bioacoustics*, Tavolga, W. N. ed., pp. 305–316, Vol. II, Oxford: Pergamon Press (1967).
92. NORRIS, K. S. and HARVEY, G. W. in: *Animal Orientation and Navigation*, Galler, S. R., Schmidt-Koenig, K., Jacob, S. and Belleville, R. E., eds, pp. 397–417, NASA, Washington D.C.: US Government Printing Office (1972).
93. NORRIS, K. S. and HARVEY, G. W. *J. Acoust. Soc. Am.* **56,** 659–664 (1974).
94. PATHAK, S. P., SUWAL, P. N. and AGARWAL, C. V. *Biochem. J.* **62,** 634–637 (1956).
95. PITAS, R. E., SAMPUGNA, J. and JENSEN, R. G. *J. Dairy Sci.* **50,** 1332-1336 (1967).
96. PURVES, P. E. in: *Whales, Dolphins and Porpoises*, Norris, K. S., ed., pp. 320–380, Berkeley and Los Angeles: Univ. of Calif. Press (1966).
97. RAVEN, H. C. and GREGORY, W. K. *Am. Mus. Novit.* 677, pp. 1–17 (1933).
98. RAO, M. R. *J. Chem. Phys.* **9,** 682–685 (1941).
99. RAO, M. R. *J. Chem. Phys.* **14,** 699 (1946).
100. REYSENBACH DE HAAN, F. W. *Acta Oto-Laryngol.* **134,** 1–114 (supplement) (1957).
101. RICE, D. W. and SCHEFFER, V. B. A List of Marine Mammals of the World. Special Scientific Report— Fisheries No. 579, Washington D.C.: US. Fish and Wildlife Service (1968).
102. ROBISCH, P. A., MALINS, D. C., BEST, R. and VARANASI, U. *Biochem. J.* **130,** 33–348 (1972).
103. SCHAAFFS, W. *Ergeb. Exakten Naturwiss.* **25,** 109–192 (1951).
104. SCHENKKAN, E. J. and PURVES, P. E. *Bijdr Dierk* **43,** 93–112 (1973).
105. SCHLENK, W., JR. *J. Am. Oil Chem. Soc.* **42,** 945–957 (1965).
106. SIMPSON, G. G. *Bull. Am. Mus. Nat. Hist.* **85,** 1–350 (1945). (See especially pp. 100–105, 215–216).
107. SPOTTE, S. *Anim. Kingdom* **79,** 21–25 (1976).
108. TAYLOR, W. J., PIGNOCCO, J. M. and ROSSINI, F. D. *J. Res. Nat. Bur. Standards* **34,** 413–416 (1945).
109. TOYAMA, Y. *J. Soc. Chem. Ind.* (Japan) (Supplemental Binding) **37,** 537 B (1934).
110. TSUJIMOTO, M. *Ind. Eng. Chem.* **8,** 889–896 (1916).
111. TSUJIMOTO, M. and KOYANAGI, H. *J. Soc. Chem. Ind.* (Japan) (Supplemental Binding) **42,** 272 B–274 B (1937).
112. VARANASI, U., EVERITT, M. and MALINS, D. C. *Int. J. Biochem.* **4,** 373–378 (1973).
113. VARANASI, U., FELDMAN, H. R. and MALINS, D. C. *Nature* **255,** 340–343 (1975).
114. VARANASI, U. and MALINS, D. C. *Science* **166,** 1158–1159 (1969).
115. VARANASI, U. and MALINS, D. C. *Biochemistry* **9,** 3269–3631 (1970).
116. VARANASI, U. and MALINS, D. C. *Biochemistry* **9,** 4576–4579 (1970).
117. VARANASI, U. and MALINS, D. C. *Biochim. Biophys. Acta* **231,** 415–418 (1971).
118. VARANASI, U. and MALINS, D. C. *Science* **176,** 926–927 (1972).
119. VARANASI, U. and MALINS, D. C. *Biochem. Soc. Trans.* **2,** 1277–1279 (1974).
120. VARANASI, U. and MALINS, D. C. *Biochim. Biophys. Acta* **348,** 55–62 (1974).
121. VARANASI, U. and MALINS, D. C. *Biochim. Biophys. Acta* **409,** 304–310 (1975).
122. WALKER, E. P., HAMLET, S. E., LANGE, K. I., DAVIS, M. A., UIBLE, H. E. and WRIGHT, P. F., *Mammals of the World*, Vol. 2, p. 1125, Second Edition, Baltimore, MD: The John Hopkins Press (1968).
123. WARTH, A. H. in: *Progress in the Chemistry of Fats and Other Lipids*, Holman, R. T. ed., IV, pp. 79–96, Oxford: Pergamon Press (1957).
124. WEDMID, Y. and LITCHFIELD, C. *Lipids* **10,** 145–151 (1975).
125. WEDMID, Y. and LITCHFIELD, C. *Lipids* **11,** 189–193 (1976).
126. WEDMID, Y., LITCHFIELD, C., ACKMAN, R. G. and SIPOS, J. C. *J. Am. Oil Chem. Soc.* **48,** 332 A (1971).
127. WEDMID, Y., LITCHFIELD, C., ACKMAN, R. G., SIPOS, J. C., EATON, C. A. and MITCHELL, E. D. *Biochim. Biophys. Acta* **326,** 439–447 (1973).
128. WILLIAMS, N. V. and MASLOV, N. Y. *Schrift. Zentr. Forsch. Lebensmitt. USSR* **4,** 150 (1935).
129. WOOD, F. G., QUOTED IN LILLY, J. C. *Man and Dolphin*, p. 236, Garden City, New York: Doubleday (1961).

Prog. Chem. Fats other Lipids. Vol. 16, pp. 279–308. Pergamon Press, 1978. Printed in Great Britain

MASS SPECTROMETRY OF FATTY ACID PYRROLIDIDES

BENGT Å. ANDERSSON

Department of Structural Chemistry, Faculty of Medicine, University of Göteborg, S-400 33 Göteborg 33, Sweden

CONTENTS

I. INTRODUCTION 279
II. SYNTHESIS, GAS LIQUID CHROMATOGRAPHY AND MASS SPECTROMETRY OF
FATTY ACID AMIDES 279
III. MASS SPECTROMETRIC ANALYSIS OF FATTY ACID PYRROLIDIDES 286
    A. Normal-chain acids 286
    B. Methyl-branched acids 288
    C. Unsaturated normal-chain acids 291
    D. Acetylenic normal-chain acids 295
    E. Cyclic acids 296
    F. Acids containing oxygen 298
    G. Dicarboxylic acids 300
    H. Isotope-labelled acids 300
IV. CONCLUDING REMARKS 307
ACKNOWLEDGEMENTS 307
REFERENCES 307

I. INTRODUCTION

Mass spectrometry (MS) of long chain carboxylic acids has, since the pioneering work of Ryhage and Stenhagen,[33] been a powerful tool in the arsenal of analytical methods available for determination of structure. However, MS formerly has been unable to distinguish between positional unsaturated isomers, *cis–trans* isomers or optical isomers of fatty acids, for which complementary techniques such as nuclear magnetic resonance or infrared spectroscopy must be used. Therefore, efforts have been made to overcome such limitations in order to increase the power of MS. The greatest advances in this direction have been the coupling of a gas chromatograph to the mass spectrometer (GLC–MS) and use of suitable derivatives of the fatty acids to increase their volatility.

An examination of the literature about structural analysis by MS of derivatives of naturally occurring fatty acids reveals that most of the work was done using methyl esters. Other modifications of the carboxylic acid group include esterification with iso-propanol[13] and phenyldiazomethane,[24] conversion to diastereoisomers,[23] and reduction to aldehydes[26] and alcohols. The alcohols have also been converted to methyl ethers[27] and trimethylsilyl ethers.[29,40] All these derivatives for MS or GLC–MS increase the possibility of determining the structure of a fatty acid. The present article is concerned with illustrating the potential of another derivative, in which the carboxyl group of the fatty acid is converted to a pyrrolidide, a cyclic tertiary amide.

Although amide derivatives of fatty acids have long been used for melting point determinations, Gilpin[21] was the first to use amides for MS analyses. Later, Pelah *et al.*,[31] Duffield *et al.*,[18] and Richter *et al.*[32] continued the study, but their primary goal was to understand the fragmentation mechanisms of these amides under electron impact. Vetter *et al.*[39] drew attention to the possibilities of using pyrrolidides in the MS analysis of complicated alicyclic and aliphatic compounds, and later Andersson *et al.*[10] investigated a number of amides of oleic acid and expanded these studies to include different types of fatty acid pyrrolidides. The results of these investigations together with literature reports up to July 1976 will be covered.

II. SYNTHESIS, GAS LIQUID CHROMATOGRAPHY AND MASS SPECTROMETRY OF FATTY ACID AMIDES

The mass spectra of long chain primary, secondary and tertiary amides are well known since the works of Gilpin[21] and others.[18,31] Of these derivatives, the spectra of tertiary

TABLE 1. Relative Retention Times (RRT) and Key Fragments in Mass Spectra of Amides of Oleic Acid (cf. Andersson et al.[10])

Amide	RRT[a] SILAR 10C 240 C	RRT[a] DEGS-PS 220 C	Base peak m/e	Base peak Relative intensity	McLafferty rearr. peak m/e	McLafferty rearr. peak Relative intensity	C7 m/e	C7 Relative intensity	C8 m/e	C8 Relative intensity	C9 m/e	C9 Relative intensity	C10 m/e	C10 Relative intensity	Molecular peak m/e	Molecular peak Relative intensity	Other prominent peaks (m/e)
Primary Octadec-9-enamide	4.4	4.7	41	100	59	71.0	128	2.9	140 / 142	2.6 / 1.4	154 / 156	2.1 / 1.1	168	1.2	281	4.5	83, 263
Secondary N-n-Propyl-octadec-9-enamide	5.1	2.9	43	100	101	78.0	170	9.5	182 / 184	3.0 / 3.2	196 / 198	4.5 / 2.0	210	3.5	323	21.0	114
N-iso-Propyl-octadec-9-enamide	3.5	1.8	101	100	101	100	170	12.0	182 / 184	3.7 / 3.6	196 / 198	5.1 / 1.7	210	4.3	323	30.0	43, 114
N-n-Butyl-octadec-9-enamide	6.0	3.7	128	100	115	97.0	184	13.4	196 / 198	4.0 / 4.7	210 / 212	7.8 / 2.0	224	4.7	337	35.3	
N-Cyclopropyl-octadec-9-enamide	9.4	6.1	57	100	99	46.2	168	4.5	180 / 182	2.0 / 1.6	194 / 196	2.5 / 1.0	208	2.1	321	14.2	112, 303
Tertiary N,N-Dimethyl-octadec-9-enamide	2.9	1.6	87	100	87	100	156	11.7	168 / 170	1.7 / 2.7	182 / 184	3.0 / 0.9	196	4.1	309	29.2	72, 100
N,N-Diethyl-octadec-9-enamide	2.6	1.5	115	100	115	100	184	15.3	196 / 198	1.4 / 2.8	210 / 212	3.2 / 1.0	224	4.7	337	39.8	100
N,N-Di-n-propyl-octadec-9-enamide	2.7	1.8	72	100	143	27.0	212	9.8	224 / 226	0.8 / 1.8	238 / 240	2.0 / 0.6	252	2.8	365	29.0	114, 156

N,N-Di-iso-propyl-octadec-9-enamide	2.0	1.3	86	100	143	11	212	4.0	224 / 226	0.9 / 0.7	238 / 240	1.1 / 0.5	252	1.2	365	16.0	322
N,N-Di-n-butyl-octadec-9-enamide	3.3	2.4	86	100	171	2.3	240	8.1	252 / 254	0.8 / 1.4	266 / 268	1.7 / 0.5	280	2.4	393	27.2	129, 156, 184
N-Octadec-9-enoylazetidine	6.9	4.0	99	100	99	100	168	11.0	180 / 182	1.7 / 3.0	194 / 196	2.9 / 1.0	208	3.6	321	19.0	112
N-Octadec-9-enoylpyrrolidine	8.2	4.8	113	100	113	100	182	11.4	194 / 196	1.0 / 2.5	208 / 210	2.3 / 0.7	222	3.6	335	24.8	
N-Octadec-9-enoylpiperidine	6.8	4.2	127	100	127	100	196	12.8	208 / 210	1.2 / 2.7	222 / 224	2.5 / 0.9	236	4.1	349	32.0	140
N-Octadec-9-enoylhexamethyleneimine	8.8	5.8	141	100	141	100	210	22.7	222 / 224	2.0 / 4.4	236 / 238	4.1 / 1.7	250	6.9	363	57.5	126, 154

[a]Relative to methyltetracosanoate = 1.

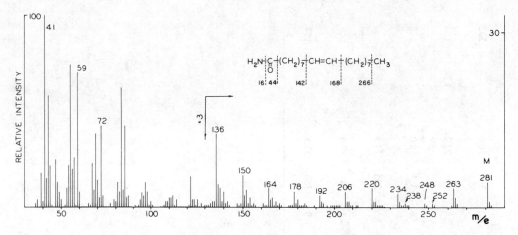

FIG. 1. Mass spectrum of octadec-9-enamide (cf. Andersson et al.[10]).

amides have the advantage of revealing most clearly in their mass spectra the structure of the long chain moiety, because contributions from competing fragmentation pathways are least. Vetter et al.[39] pointed out some advantages of pyrrolidides, Bohlmann et al.[14] used piperidides and Andersson et al.[10] compared a number of amides of oleic acid (9–18:1).* The principal mass spectrometric patterns for these amides will be discussed, followed by their preparation and GLC-behavior. Relative retention times and key fragments of amides of oleic acid are given in Table 1.

The mass spectra of primary amides yield a typical rearrangement ion at m/e 59, which can be correlated with cleavage β to the carbonyl, accompanied by a rearrangement of a hydrogen.[21] Primary amides of higher molecular weight yield m/e 72 corresponding to cleavage γ to the carbonyl.[21] The diagnostically useful fragmentation pattern for the fatty acid chain, where the molecular ion yields fragments containing the amide part of the molecule (M, M–15, M–29,...), is obscured by another series of fragments, which starts with M–18 followed by M–(18 + 15), M–(18 + 29),... The latter series is a dominating pattern when the carboxylic acid chain contains double bonds,[10] as can be seen in Fig. 1.

The spectra of secondary amides show pronounced molecular ions with clear fragmentation of the fatty acid chain. The McLafferty rearrangement ions, which require a hydrogen attached to a gamma carbon, e.g. m/e 115 in N-n-butyloctadec-9-enamide, are dominant[21] (Fig. 2). Another mass spectral feature characterizing the secondary amides, especially acetamides, is the m/e 30 peak $[CH_2NH_2]^+$ which corresponds to

*The abbreviated notation 18:1 indicates a fatty acid chain with 18 carbon atoms and one double bond. The position of the bond is depicted by the prefix 9–. Thus 9–18:1 indicates oleic acid and 9,12–18:2 linoleic acid.[11]

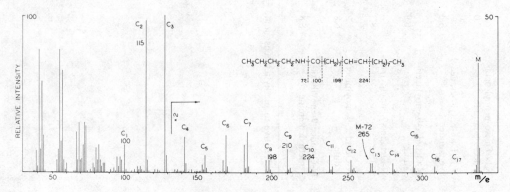

FIG. 2. Mass spectrum of N-n-butyloctadec-9-enamide (cf. Andersson et al.[10]).

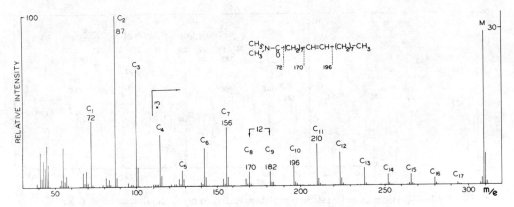

FIG. 3. Mass spectrum of N,N-dimethyloctadec-9-enamide (cf. Andersson *et al.*[10]).

a rearrangement ion resulting from a cleavage of the nitrogen—carbonyl carbon bond and cleavage of the carbon—carbon bond β to the nitrogen atom, accompanied by the transfer of a hydrogen atom.[21] Removal of the amine group occurs and ions of the type $R—C≡O^+$ (m/e 265 in Fig. 2) are formed. This ion is more prominent in the spectra of N-cyclopropylamine derivatives than the acyclic ones,[10] due to decreased electron donation (stabilization) from the nitrogen in these amides.

Tertiary amides have a more pronounced tendency than the other amides to retain the positive charge under electron impact and give intense fragments resulting from cleavage at the carbon–carbon bonds α and β to the carbonyl. This is very similar to the cleavage of long chain esters, described by the following fragments:

$$
\begin{array}{cc}
\text{(a)} & \text{(b)} \\
\text{OH} & \text{OH} \\
| & | \\
\text{amide:} \quad [CH_2{=}C{-}NR_2]^{+\cdot} \quad \text{and} & [CH_2{=}CH{-}C{=}NR_2]^+
\end{array}
$$

$$
\begin{array}{cc}
\text{OH} & \text{OH} \\
| & | \\
\text{ester:} \quad [CH_2{=}C{-}OR]^{+\cdot} \quad \text{and} & [CH_2{=}CH{-}C{=}OR]^+
\end{array}
$$

The fragments of type (a) and (b) are exemplified by m/e 87 and m/e 100 in Fig. 3 and m/e 113 and m/e 126 in Fig. 4. However, when double bonds are present in the chain, the unsaturated center is preferentially ionized in the methyl ester but not in the pyrrolidide. This is well documented by the mass spectra of methyl n-alkylpyrrolida-mates[7] (Fig. 5), where the fragmentation is completely dominated by the pyrrolidine

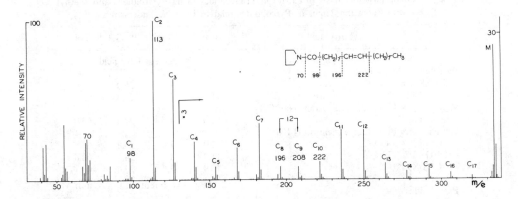

FIG. 4. Mass spectrum of N-octadec-9-enoylpyrrolidine. Experimental conditions same as for Fig. 12.

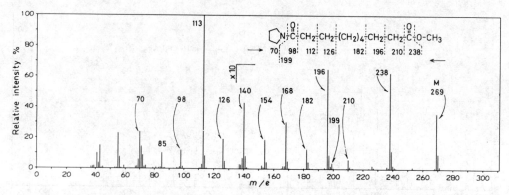

FIG. 5. Mass spectrum of methyl decanpyrrolidamate. Experimental conditions same as for Fig. 29.

containing fragments. As with the secondary amides, there is a rearrangement ion of the following type in the low mass region of the spectra of tertiary amides:

$$[R-CO-NR_1-CH_2-R_2]^+ \rightarrow [HNR_1=CH_2]^+$$

in which the extra hydrogen is extracted from the fatty acid residue,[18,31] exemplified by m/e 45 in Fig. 3. Ions derived from $R-C\equiv O^+$, which is formed by elimination of the amine group, make only a minor contribution to the fragmentation pattern.[10] As a consequence of this, the diagnostically useful fragmentation of the tertiary amides is extremely pronounced. This is exemplified by N,N-dimethyloctadec-9-enamide (Fig. 3) and N-octadec-9-enoylpyrrolidine (Fig. 4).

Because the mass spectral characteristics of the tertiary amides are the most promising, attention was focused on methods for synthesis of these fatty acid amides. A brief investigation[10] of the literature methods indicated that aminolysis is the simplest and most quantitative. Coupling reactions using acid chlorides, carbodiimides and sodium methoxide also gave good results. The use of acid chlorides was recommended if the fatty acid had α- or β-unsaturation or α-branching of the carbon chain.

Compared with the corresponding methyl esters, amides of fatty acids have longer retention times in gas liquid chromatography (GLC) under the same experimental conditions (Table 2). Nonpolar silicone phases such as SE 30[39] and OV-1,[10] the more polar cyanosilicone SILAR 10C[10] and the succinates DEGS[10] and EGS[10] have been used. The silicones are recommended when thermostability is a consideration, and the more polar cyanosilicones when separation of fatty acid amides depending on degree of unsaturation,[10] or position of a branch[2,3,12] are needed. Stationary phase bleeding from

TABLE 2. Quantitative Analysis of Corn Oil Triglycerides using Pyrrolidides and Methyl Esters with Retention Times of Pyrrolidides relative 18:0 (cf. Andersson et al.[10])

Fatty acids	16:0	18:0	9–18:1	9,12–18:2
	Percent fatty acid			
Pyrrolidides by aminolysis				
SILAR 10C, 240 C	11.4	1.7	25.0	61.9
DEGS-PS, 220 C	11.3	1.7	24.4	62.6
Methyl esters by interesterification (NaOCH₃)				
EGS, 170 C	11.0	1.6	24.4	63.0
Methyl esters by interesterification (BF₃)				
SILAR 10C, 170 C	11.0	1.4	23.8	63.8
DEGS-PS, 170 C	11.2	1.7	24.3	62.8
EGSS-X, 170 C	10.9	1.6	24.4	63.1
	Relative retention times, min			
Pyrrolidides				
SILAR 10C, 240 C	0.7	1.0	1.2	1.5
DEGS-PS, 220 C	0.6	1.0	1.1	1.4

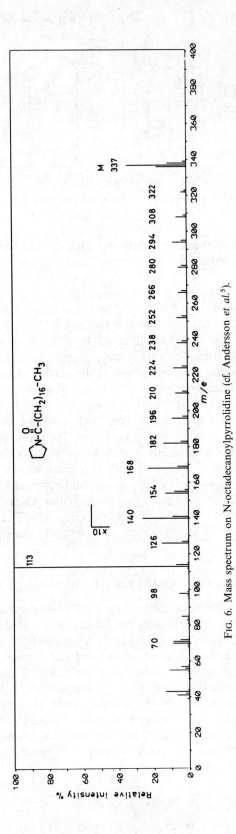

FIG. 6. Mass spectrum on N-octadecanoylpyrrolidine (cf. Andersson et al.[5]).

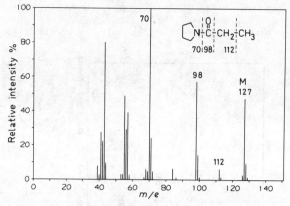

FIG. 7. Mass spectrum of N-propanoylpyrrolidine (adapted from Duffield and Djerassi[18]).

silicones at elevated temperatures gives rise to ions of odd mass numbers. Because the diagnostic ions of amides have even mass numbers, these are not obscured by the column background.

Taking MS, GLC and synthesis into consideration, pyrrolidides were judged to be the most suitable amide investigated.[10] Azetidine is too reactive and difficult to handle, and hexamethylenimine was less reactive and the excess was more difficult to remove from the sample. The N,N-dimethylamide could be a useful derivative because of its volatility, simple mass spectrum and quantitative preparation. However, it is desirable to locate diagnostically useful fragmentations of the carbon chain away from the low mass region because functional groups, such as hydroxyl groups in the fatty acid chain, induce secondary fragmentations which may also occur in the low mass region obscuring diagnostic ions. Piperidides might be preferred both from the MS and GLC point of view, but piperidine is less reactive than pyrrolidine. However, Bohlman et al.[14] used piperidides because they occur naturally in plants, but Vetter et al.[38,39] found pyrrolidides more suitable, because of their reactivity and volatility. In order to test the use of pyrrolidides on naturally occurring fatty acids, Andersson et al.[10] converted a pure fraction of corn oil triglycerides into pyrrolidides. A qualitative and quantitative analysis was performed with different stationary phases on GLC–MS. The analytical data were compared with the analyses of methyl esters and no significant differences were found (Table 2).

III. MASS SPECTROMETRIC ANALYSIS OF FATTY ACID PYRROLIDIDES

Duffield and Djerassi[18] were the first to study the mass spectral fragmentation of the lower homologs (2:0–4:0) of N-acylpyrrolidines. They used high resolution and deuterium labelling and proposed several modes of fragmentation. Richter et al.[32] contributed further knowledge of the electron impact induced fragmentation of these substances. Vetter et al.[39] investigated pyrrolidides of stearic (18:0), oleic (9–18:1) and linoleic acid (9,12–18:2) and suggested their use for determination of the position of double bonds in fatty acids. He expanded these studies to some aliphatic–alicyclic carboxylic acids.[38] Andersson continued these studies in a series of papers[2–7,9–12] to include different types of fatty acids. On the basis of these investigations, several papers[2,3,10,20,22,25] describe the use of pyrrolidides for structural analysis of natural fatty acids.

A. Normal Chain Acids

Because the peaks below m/e 98 in the mass spectra of fatty acid pyrrolidides have no significant influence on the diagnosis of a normal saturated fatty acid chain, they will not be discussed.

A typical mass spectrum of a fatty acid pyrrolidide is represented by N-octadecanoyl-pyrrolidine in Fig. 6 in which the most prominent peaks are due to ions containing nitrogen. Most of these ions contain a pyrrolidine ring and belong to a series of the general formula:

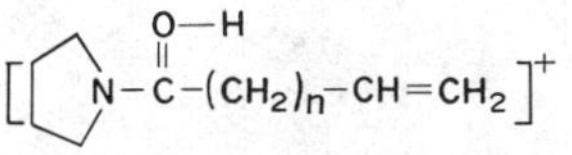

$$\left[\overset{\text{O—H}}{\underset{}{N-\overset{\|}{C}-(CH_2)_n-CH=CH_2}} \right]^{+}$$

where $n = 0,1,2\dots,$.[38] The base peak is the McLafferty rearrangement ion m/e 113, which can be depicted according to route A in Scheme 1.

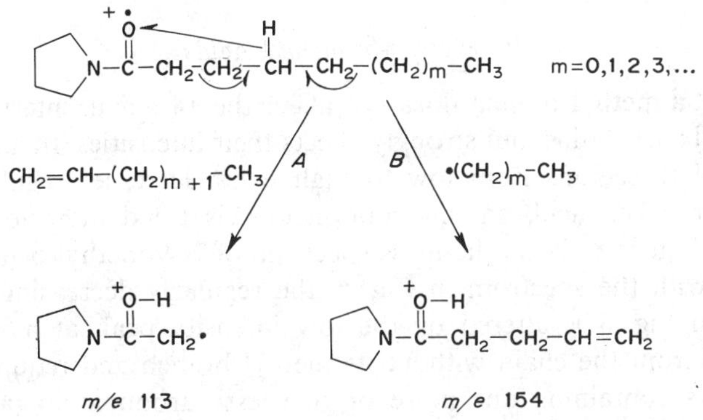

SCHEME 1.

As is observed in Fig. 6, the ions decrease in abundance from m/e 126 to m/e 322 with the exception of m/e 154, m/e 294 and m/e 308. The peak m/e 154 is decreased due to the favorable formation of the dominant McLafferty rearrangement ion, m/e 113, which requires abstraction of the δ-hydrogen, according to route B in Scheme 1 proposed by Vetter et al.[38] This rearrangement of a hydrogen from a carbon atom in the chain to the carbonyl oxygen is analogous to the formation of the m/e 74 ion in the spectra of methyl esters.[35] The disproportionate heights of m/e 294 and m/e 308 are caused by contributions by expulsion of a three carbon fragment and a two carbon fragment from the chain, respectively. This has been substantiated by isotope labelling[5] and is similar to the expulsion observed with methyl esters.[17]

The molecular ion M is always accompanied by an M–1 ion, with an intensity of about 40 per cent of the M ion for the long chain saturated homologs. Although Duffield and Djerassi[18] suggested that the M–1 ion is formed by the loss of a hydrogen from the α-carbon in the pyrrolidine ring and Richter et al.[32] proposed that it is derived from loss of a hydrogen β to nitrogen in the pyrrolidine ring, deuteriation of the fatty acid shows that this loss may also be from the carbon chain.[5]

The base peak in the mass spectra of N-acylpyrrolidines is usually m/e 113 if the acyl group contains a δ hydrogen. In Fig. 7, an exception is illustrated. N-propanoylpyr-rolidine has no δ hydrogen and no McLafferty rearrangement is possible. As the acyl chain increases from 2 to 20 carbon atoms, the intensity of the molecular ion decreases from about half of the base peak to a constant value of about 2 per cent of the base peak.

Palmitic (16:0) and stearic (18:0) acid pyrrolidides were encountered in the analysis of pyrrolidides prepared from corn oil triglycerides[10] (Table 2). The alcohol moieties of waxes, obtained from the free-flowing secretion of swans, have been oxidized and converted to pyrrolidides, and were found to contain the normal chain acids, 16:0, 17:0, 18:0, 19:0 and 20:0.[2,3]

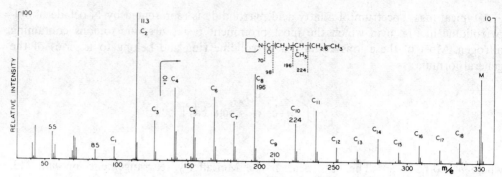

FIG. 8. Mass spectrum of N-9-methyl-octadecanoylpyrrolidine (cf. Andersson and Holman[12]).

B. *Methyl-branched Acids*

The presence of a methyl branch does not affect the 14 a.m.u. intervals characteristic of a saturated acyl pyrrolidine, but strongly affects their intensities. In a normal saturated acid, the series of these ions from low to high mass decreases regularly in intensity. With a methyl branched acid, an interruption of this trend indicates the location of a methyl branch. Figure 8 shows the mass spectrum of N-9-methyl-octadecanoylpyrrolidine. Compared with the spectrum in Fig. 6, the regularly decreasing tendency of the peak intensities in Fig. 8 is altered by the low intensity peak at m/e 210. This peak includes carbon 9 from the chain without its methyl branch and requires two fragmentations. Fragments containing one more or one less carbon atom (m/e 196 and m/e 224) are prominent because cleavage α to a carbon bearing a methyl group is favored,

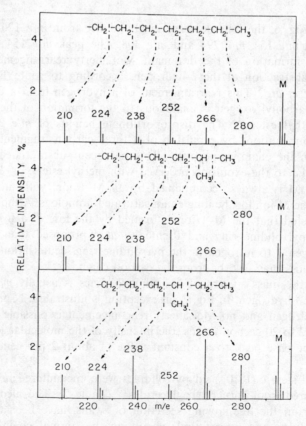

FIG. 9. Partial mass spectra of normal-, iso- and anteiso-N-pentadecanoylpyrrolidines (cf. Andersson and Holman[12]).

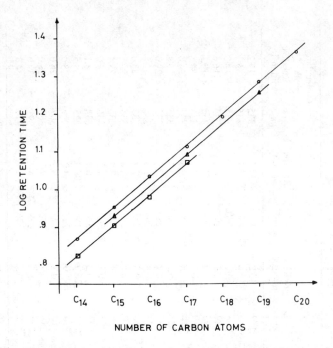

FIG. 10. Logarithmic retention time diagram of a homologous series of saturated normal- (O), iso- (□) and anteiso- (△) fatty acid pyrrolidides (cf. Andersson and Bertelsen[2]).

as is the case with methyl esters and follows route B in Schedule 1.[34] In summary, *a peak of lower intensity than the analogous peak in a straight chain fatty acid pyrrolidide indicates a branched methyl at that position.* This is confirmed by flanking peaks of higher than normal intensity. The mechanisms explaining other prominent peaks in the spectrum have been discussed previously.[38] The variations of peak intensities for methyl branches close to the methyl end of a fatty acid have been used in distinguishing between normal-, iso- and anteiso-structures (Fig. 9). Here, the low intensity peak at m/e 252 flanked by relatively high intensity peaks at m/e 238 and m/e 266 distinguishes anteiso-15:0 from iso-15:0, which has a low intensity peak at m/e 266. A GLC-separation is necessary for the analysis of mixtures of these pyrrolidides, and this can be done on a cyanosilicone column, as is shown in the logarithmic retention time diagram in Fig. 10. The linear relationship between homologous series of pyrrolidides observed in Fig. 10 is also well known for the methyl esters.[19] The diagnostic peaks for localizing iso- and anteiso-methyl groups are more prominent in spectra of pyrrolidides than in spectra of corresponding methyl esters.

Structures of fatty acids with several methyl branches can also be elucidated via their pyrrolidides. In the example shown in Fig. 11, the methyl branching is located

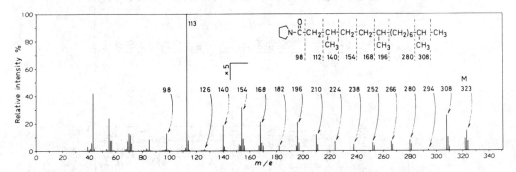

FIG. 11. Mass spectrum of N-3,6,13-trimethyltetradecanoylpyrrolidine. Experimental conditions same as for Fig. 8.

Bengt Å. Andersson

TABLE 3. Key Fragments in Mass Spectra of Monoenoic Fatty Acid Pyrrolidides (cf. Andersson and Holman[11])

Pyrrolidide	m/e	Relative intensity	m/e	Relative intensity	m/e	Relative intensity	m/e	Relative intensity	m/e	Relative intensity	m/e	Molecular peak Relative intensity
cis-4-18:1	124	1.8	126	10.8	138	2.9	139	4.2	152	13.9	166	57.0	335	17.2
cis-5-18:1	126	6.3	138	0.3	140	0.6	152	0.5	166	1.6	180	2.5	335	5.3
cis-6-18:1	140	7.7	152	1.5	154	1.7	166	5.6	168	1.5	180	8.1	335	14.3
cis-7-18:1	154	9.8	166	2.5	168	4.4	180	6.5	182	2.5	194	7.2	335	16.6
cis-8-18:1	168	13.2	180	2.0	182	4.5	194	2.4	196	0.9	208	4.2	335	18.1
cis-9-18:1	182	11.4	194	1.0	196	2.5	208	2.3	210	0.7	222	3.6	335	24.8
trans-9-18:1	182	11.0	194	1.2	196	2.4	208	2.2	210	0.8	222	3.6	335	24.0
cis-10-18:1	196	7.4	208	1.0	210	2.0	222	1.8	224	0.7	236	3.3	335	29.8
cis-11-18:1	210	4.9	222	1.1	224	1.6	236	1.5	238	0.7	250	2.7	335	29.0
cis-12-18:1	224	4.8	236	0.7	238	1.5	250	1.2	252	0.8	264	3.0	335	33.8
cis-13-18:1	238	4.5	250	0.6	252	1.3	264	1.1	266	0.5	278	2.9	335	33.0
cis-14-18:1	252	3.9	264	0.4	266	0.9	278	1.2	280	0.5	292	2.6	335	28.8
cis-15-18:1	266	3.1	278	0.5	280	1.0	292	1.3	306	2.4	320	2.9	335	28.4
cis-16-18:1	266	2.3	278	0.6	280	3.3	292	1.1	294	0.7	306	1.2	335	26.0
cis-17-18:1	280	0.9	292	0.6	294	1.6	306	0.4	308	0.2	320	0.4	335	8.1
cis-4-10:1	124	3.5	126	6.9	138	6.0	139	6.1	152	26.0	166	99.0	223	33.5
cis-9-14:1	182	12.0	194	1.4	196	2.7	208	2.6	210	1.0	222	4.0	279	28.0
cis-9-16:1	182	11.0	194	1.4	196	2.2	208	3.3	210	0.8	222	3.8	307	26.0
cis-11-20:1	210	4.4	222	0.9	224	1.8	236	1.4	238	0.9	250	2.7	363	28.0
cis-13-22:1	238	4.4	250	2.1	252	2.8	264	2.0	266	1.2	278	4.2	391	37.0
cis-15-24:1	266	2.3	278	0.6	280	1.1	292	1.0	294	0.6	306	2.9	419	34.3

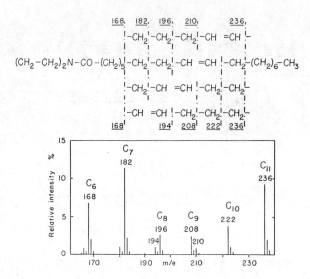

FIG. 12. Postulated fragmentation pathway for N-octadec-9-enoylpyrrolidine (cf. Andersson and Holman[11]).

at carbons 3, 6 and 13, as is indicated by the very low intensity peaks at m/e 126, m/e 172 and m/e 294. Methyl branches have been located in the alcohol moieties of the wax obtained from the free flowing preen gland secretion of swans by this method. The wax was hydrolyzed, the alcohols were separated, oxidized to acids and converted to pyrrolidides.[2,3] Methyl branching is known to occur in the fatty acids of bacterial origin. Gerson et al.[20] identified a methyl branch at carbon 11 of N-11-methyl-octa-dec-11-enoylpyrrolidine in the fatty acids from *Rhizobium*. Vetter and Walther[38] located the ethyl and methyl branching in the long chain moieties of vitamin A analogs at carbons 3 and 7.

C. Unsaturated Normal-chain Acids

Introduction of double bonds into a fatty acid chain disturbs to a minor extent the typical fragmentation pattern of a pyrrolidide.[4,10,39] Following a proposal by Vetter et al.,[39] Andersson and Holman investigated a large number of natural and synthetic monoenoic[11] and methylene-interrupted[4] unsaturated fatty acid pyrrolidides and found that their mass spectra could easily be interpreted to deduce the position of the unsaturation. They concluded that this derivative is superior to the commonly used methyl ester in distinguishing between positional isomers of unsaturated fatty acids. The alternative approach requires further chemical modification at the site of the unsaturation

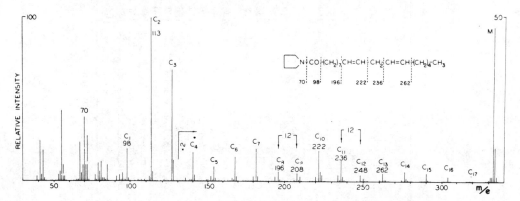

FIG. 13. Mass spectrum of N-octadec-9,12-dienoylpyrrolidine (cf. Andersson et al.[4]).

TABLE 4. Key Fragment in Mass Spectra of Polyenoic Fatty Acid Pyrrolidides (cf. Andersson et al.[4])

Pyrrolidide	m/e	Rel. int.[a]	m/e	Rel. int.	m/e	Rel. int.	m/e	Rel. int.	m/e	Rel. int.	m/e	Rel. int.	m/e	Rel. int.	m/e	Rel. int.	m/e	Rel. int.	Molecular peak m/e	Rel. int.
cis,cis-4,7-18:2	124	1.3	126	3.3	138	3.3	139	2.3	152	11.0	153	7.8	166	4.7	178	2.4	192	3.5	333	16.0
cis,cis-5,8-18:2	140	1.1	152	0.7	154	0.6	166	2.9	168	1.1	178	0.6	180	1.2	192	0.5	206	0.8	333	8.8
cis,cis-6,9-18:2	154	3.8	166	3.9	178	1.0	180	1.0	192	10.0	194	10.0	206	6.3	208	2.4	220	3.8	333	27.0
cis,cis-7,10-18:2	168	4.5	180	4.7	182	3.3	192	1.0	206	13.0	208	13.0	220	6.5	222	2.7	234	4.5	333	28.0
cis,cis-8,11-18:2	182	5.4	194	2.9	196	2.3	206	1.6	208	15.0	222	15.0	234	7.4	236	3.2	248	3.1	333	43.0
cis,cis-9,12-18:2	196	3.4	208	2.8	210	1.9	220	1.9	222	10.0	234	10.0	248	6.8	250	2.4	262	3.1	333	47.0
cis,cis-10,13-18:2	210	2.4	222	1.9	224	1.6	234	1.9	236	7.8	250	7.8	262	6.8	264	1.7	276	2.3	333	40.0
cis,cis-11,14-18:2	224	2.4	236	1.6	238	1.8	248	1.2	250	6.0	264	6.0	276	5.8	278	1.9	290	2.7	333	41.0
cis,cis-12,15-18:2	238	2.4	250	1.4	252	1.5	262	1.2	264	5.9	265	5.9	290	2.2	291	0.7	304	3.0	333	45.0
cis,cis-13,16-18:2	238	5.2	250	1.0	252	1.9	262	0.9	264	1.5	265	1.5	290	3.5	291	1.0	304	1.8	333	43.0
cis,cis-14,17-18:2	252	5.4	260	0.8	262	0.8	264	0.8	276	1.0	278	1.3	290	1.3	292	4.4	304	1.6	333	8.4
cis,cis-5,8-14:2	140	1.2	152	0.9	154	0.8	166	2.2	168	1.0	178	0.5	180	1.1	192	0.5	206	0.8	277	7.1
cis,cis-7,10-16:2	168	4.9	180	4.0	182	2.7	192	0.8	206	12.0	208	1.0	220	5.2	234	2.4	234	3.0	305	16.0
cis,cis-9,15-18:2	196	2.6	208	2.2	210	1.8	222	3.4	236	7.7	264	1.0	276	1.5	290	1.8	290	2.5	333	19.0
cis,cis,cis-6,9,12-18:3	154	3.9	166	3.4	180	6.5	194	8.7	208	8.8	222	2.5	246	3.1	260	3.8	260	3.4	331	23.0
cis,cis,cis-9,12,15-18:3	196	3.3	208	2.6	222	5.3	236	5.1	250	7.7	262	2.7	276	2.9	288	3.5	302	3.1	331	40.4
cis,cis,cis-8,11,14-20:3	182	4.5	194	3.2	208	9.2	222	6.3	236	8.6	248	3.2	262	3.5	274	4.1	274	3.2	359	38.0
trans,trans-8,10-18:2	180	1.9	192	2.1	206	0.5	194	1.5	208	1.1	210	1.0	220	1.3	224	0.7	220	2.1	333	23.0
trans,trans-9,11-18:2	194	0.7	206	1.1	220	0.6	208	1.0	222	0.8	224	0.5	234	0.7	238	0.6	234	1.4	333	10.0
trans,trans-10,12-18:2	208	2.0	220	3.1	234	1.8	220	2.1	234	3.4	238	2.2	248	2.1	248	1.6	248	2.1	333	23.0
cis,cis,cis,cis-5,8,11,14-20:4	140	1.6	152	0.8	166	1.5	180	2.1	194	2.4	232	0.5	246	0.7	258	1.0	286	0.9	357	6.5

[a]Rel. int. = relative intensity.

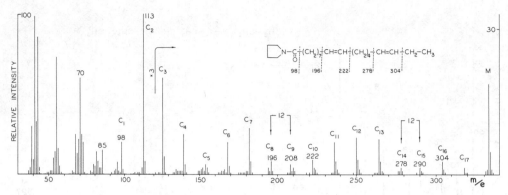

FIG. 14. Mass spectrum of N-octadec-9,15-dienoylpyrrolidine (cf. Andersson et al.[4]).

for potential analysis by MS and no derivative of this type has been completely satisfactory.[30]

Figure 5 shows the mass spectrum of N-octadec-9-enoylpyrrolidine. The base peak is the typical McLafferty rearrangement ion, at m/e 113. The molecular ion at m/e 335 yields a series of ions at m/e 320, 306, 292, 278, 264, 250, 236, 222, 208, 196, 182, 168, 154, 140, 126, 113 and 98. The main peak in each cluster in this series of ions is regularly spaced by 14 a.m.u., except in the vicinity of the double bond, where the interval is 12 a.m.u. occurring between m/e 196 and m/e 208. These fragments correspond to carbons 8 and 9 of the carbon chain. From the mass spectra of isomers from 5–18:1 to the 15–18:1, a rule was formulated: *If an interval of 12 atomic mass units, instead of the regular 14, is observed between the most intense peaks of clusters of fragments containing n and n-1 carbon atoms in the acid moiety, a double bond occurred between carbon n and n + 1 in the molecule.* Other isomers of 18:1 with unsaturation in positions 3, 4, 16 and 17 have mass spectra that also distinguish them from each other, but these do not fit the rule. The characteristic ions are listed for each isomer in Table 3. The 3–18:1 has the base peak m/e 139 that is unique for this isomer.[25] The 4–18:1 isomer has its characteristic series of fragments m/e 126, 139, and 152. The 15– and 16–18:1 isomers have similar key fragments, the only difference being that m/e 280 is less intense than m/e 266 for the 15–18:1 isomer but more intense for the 16–18:1 isomer. The 17–18:1 has its series of fragments m/e 294, 306 and 320 as a unique combination. Andersson and Holman investigated a number of other monounsaturated acids with different positions of the unsaturation and varying chain lengths and found that the rules developed for the isomers of 18:1 also could be applied to other acids. However, they could not distinguish between *cis* and *trans* isomers. A fragmentation model for the monounsaturated fatty acid pyrrolidides was also postulated, involving

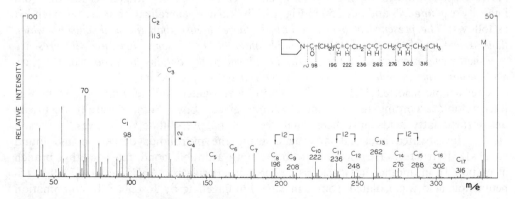

FIG. 15. Mass spectrum of N-octadec-9,12,15-trienoylpyrrolidine (cf. Andersson et al.[4]).

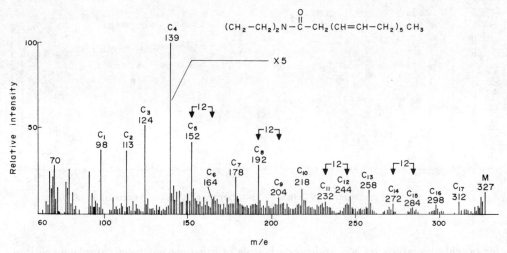

FIG. 16. Mass spectrum of N-octadec-3,6,9,12,15-pentaenoylpyrrolidine (cf. Joseph[25]).

a migration of the double bond, preferentially towards the amide part of the molecule, as is shown in Fig. 12.

Methylene interrupted isomeric polyunsaturated acids were converted to pyrrolidides and their mass spectra were studied by Andersson *et al.*[4] They extended the generalizations developed for the monoenoic acids to include polyenoic acids. The rule formulated for 5–18:1 through 15–18:1 could be applied to most of the compounds listed in Table 4, the exceptions being the dienoic pyrrolidides which have a double bond in positions 4, 15, 16 or 17 in a methylene interrupted system. Figure 13 shows the mass spectrum of N-octadec-9,12-dienoylpyrrolidine, in which the characteristic intervals of 12 a.m.u. are indicated by arrows. The 4,7–18:2 isomer had characteristic peaks at m/e 126, m/e 139 and m/e 152, indicating a double bond in position 4. The 12,15–18:2 and 13,16–18:2 isomers had key fragments at the same mass numbers (264, 278, 290, 304) indicating a double bond in position 15. However, they could be distinguished from each other by peak intensity variations, specific for each isomer. The 14,17–18:2 isomer had the characteristic interval of 12 a.m.u. between m/e 292 and m/e 304. The mass spectra of other dienoic acids also follow this generalization as can be seen for the 5,8–14:2 and 7,10–16:2 isomer in Table 4 and the tetramethylene interrupted 9,15–18:2 isomer in Fig. 14.

The mass spectra of methylene interrupted pyrrolidides of trienoic acids, such as 6,9,12–18:3, 9,12,15–18:3 and 8,11,14–20:3, were also interpretable following the rule developed for the monoenoic acids. Figure 15 shows the mass spectrum of 9,12,15–18:3, in which double bond positions are indicated by the characteristic 12 a.m.u. intervals. However, it is difficult to find the 12 a.m.u. interval when key fragments are of equal intensity, e.g. m/e 248 and m/e 250 in Fig. 15. The earlier generalization must be extended as follows: *The presence of a peak relatively more intense than the peak clusters which flank it and which are involved in probable intervals of 12 a.m.u. indicates the presence of a methylene interrupted system. If the prominent peak contains m carbons of the fatty acid residue, the methylene carbon in the molecule is at position m + 1.*

A tetraenoic acid, 5,8,11,14–20:4, was also investigated and its mass spectrum was interpretable according to the generalizations given above. Some naturally occurring unsaturated fatty acids have been analyzed via the pyrrolidide derivatives.[10] Gerson *et al.*[20] investigated bacterial fatty acids using the pyrrolidide derivative and found an 11–18:1 structure. He was able to locate both a double bond and a methyl branch at carbon 11 in another fatty acid using the same derivative. An N-octadec-3,6,9,12,15-pentaenoic acid was isolated from a marine dinoflagellate by Joseph.[25] In combination with other analytical methods, she used mass spectrometry of the pyrrolidide derivative

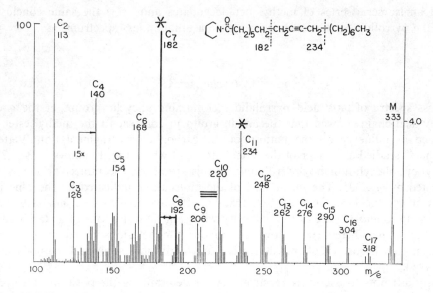

FIG. 17. Mass spectrum of N-octadec-9-ynoylpyrrolidine (courtesy of Drs A. J. Valicenti, W. H. Heimermann and R. T. Holman[37]).

and confirmed the positions of double bonds (Fig. 16). The double bond positions at carbons 6, 9, 12 and 15 are indicated by the 12 a.m.u. interval. The large peak at m/e 139 locates the double bond in position 3.

Conjugated unsaturated fatty acids, such as 8,10–18:2, 9,11–18:2 and 10,12–18:2, also gave spectra different from each other. The double bond in the conjugated system closest to the carbonyl was indicated by the interval of 12 a.m.u., according to the rule, and the second double bond by another interval of 12 a.m.u. immediately following the first interval of 12 a.m.u. Bohlmann et al.[14] studied piperidides of fatty acids with conjugated unsaturation close to the amide group. The characteristic fatty acid chain fragmentation for these substances was obscured by other competing fragmentation pathways and is therefore more difficult to interpret.

D. *Acetylenic Normal-chain Acids*

The monoynoic fatty acid pyrrolidides show mass spectra that somewhat resemble those of the monoenoic pyrrolidides. The molecular ion, exemplified by m/e 333 for N-octadec-9-ynoylpyrrolidine in Fig. 17, is twice as high as the comparable ion in the mass spectrum of methyl octadec-9-ynoate,[37] and of much lower intensity than M (m/e 335, 24 per cent[11]) of the corresponding monoenoic pyrrolidide. A rule, similar to the one developed for the monoenoic acids, has also been developed for the monoenoic fatty acids by Valicenti et al.[37] and that rule is summarized as follows: *If a triple bond occurs between carbon atoms n and n + 1 in the mass spectrum of the pyrrolidide derivative, there will be an interval of 10 a.m.u. between the maxima of the clusters of peaks representing fragments containing n − 1 and n − 2 carbon atoms. In addition, the maxima of clusters indicating carbon atoms n − 2 and n + 2 will have higher intensity values than the maxima of neighboring clusters.* In Fig. 17, the first part of this rule is indicated by triple bonds and the arrows show the interval of 10 a.m.u. The second diagnostic feature is illustrated by asterisks at m/e 182 and m/e 234. This rule is valid for C_{18} monoynoic acids with the triple bond located 5–18:1 to 14–18:1. However, the farther the acetylenic bond is moved from the middle of the fatty acid chain, the more the interval of 10 a.m.u. is obscured by other peaks in the diagnostic clusters. When difficulties occur in application of the rules, reference substances should be consulted, as positional isomers of monoynoic fatty acid pyrrolidides always give mass spectra which distinguish them from each other. Recently, Kleiman et al.[28] investigated an isomeric series of methyl octadecynoates and drew the same conclusion as with the pyrrolidides, that each isomer gave a unique mass spectrum.

E. *Cyclic Acids*

Mass spectra of fatty acid pyrrolidides containing alicyclic groups at the ω-end of a fatty acid chain showed that the amide group in contrast to the methyl ester group retained its influence on the pattern of the carbon chain fragmentation. Vetter and Walther[38] studied a number of these derivatives, of which Fig. 18 shows N-9-(2′,2′,6′-tri-methyl-cyclohexyl)-nonanoylpyrrolidine. In this spectrum, the molecular ion is clearly indicated by m/e 335. The linear part of the molecule is indicated by the nine intense peaks at m/e 98, 113, 126, 140, 154, 168, 182, 196 and 210. Above m/e 210, the peak clusters up to m/e 320 are of low intensity value and indicate ions of a stabilized struc-ture, the six-membered ring. However, no direct conclusions based on only a mass spectrum can be drawn of this derivative, and additional analytical methods must be employed.

Early attempts to use mass spectrometry for determining the position of a cyclopro-pane ring were unsatisfactory because this group is unstable under electron impact. Recently Andersson *et al.*[9] prepared a series of long chain cyclopropane fatty acid pyrro-lidides and investigated their mass spectral properties. They found that *trans* isomers of monomethylene-octadecanoylpyrrolidine gave mass spectra that could distinguish them from each other. A typical representative from this series is the mass spectrum of N-*trans*-9,10-methylene octadecanoylpyrrolidine in Fig. 19. The fragmentation pattern of this derivative is closely related to that of monounsaturated fatty acid pyrrolidides, as is also the case with the methyl esters.[15] The most intense peaks from each cluster differ from each other by 14 a.m.u. except around the site of the unsaturation or the cyclopropane ring. In Fig. 19, the position of the cyclopropane ring is indicated by the peaks at m/e 196 and m/e 208. These peaks are the most dominant in the cluster of ions containing 8 and 9 carbon atoms of the fatty acid chain and are separated by 12 a.m.u. instead of the ordinary 14 a.m.u. The data on other isomers given in Table 5 indicate that the cyclopropane group can be located within one carbon position of its true location if the ring lies between carbons 6,7 and 16,17. Nevertheless, the mass spectrum of each isomer is unique and may be used as a means of identification if spectra of model compounds are available. In summary, *if an interval of 12 a.m.u. instead of the regular 14 is observed between the most intense peaks of clusters of fragments containing n − 2 and n − 1 and n or n and n + 1 carbon atoms of the acid moiety, a cyclopro-pane ring occurs between carbon n and n + 1 in the molecule.* However, the difference of 12 a.m.u. can be obscured by the occurrence of a pair of peaks, 2 a.m.u. apart and of equal intensities, in the peak cluster representing carbons n − 1, n or n + 1. Therefore, other unique peaks in their mass spectra must be considered. Cleavage β to the cyclopro-pane ring produces a stable fragment exemplified by m/e 250 in Fig. 19. If the ring lies between carbons n and n + 1, this fragment includes the ring and one methylene group (n + 3 carbons). It is unique for all isomers with the ring in positions 3,4 up to 14,15. A competing fragmentation pattern that occurs in all pyrrolidides is elimination of the pyrrolidine ring with subsequent creation of the acylium ion RCO^+. In the series of cyclopropanes discussed, this yields a peak at m/e 279 in spectra from the 2,3 through the 7,8 isomers. The contribution from this peak increases drastically for the 8,9 isomer and reaches a maximum in the mass spectrum of the 15,16 isomer (15 per cent of the base peak).

In Fig. 19, this peak is of almost equal intensity to the pyrrolidide-containing m/e 278. Table 5 shows characteristic peaks for each isomer discussed above. Because the 14,15, 15,16, 16,17 and 17,18 isomers have similar mass spectra, they must also be characterized by GLC, in which they have different retention times. The presence of the cyclopropane group can be confirmed by mild hydrogenation and GLC.

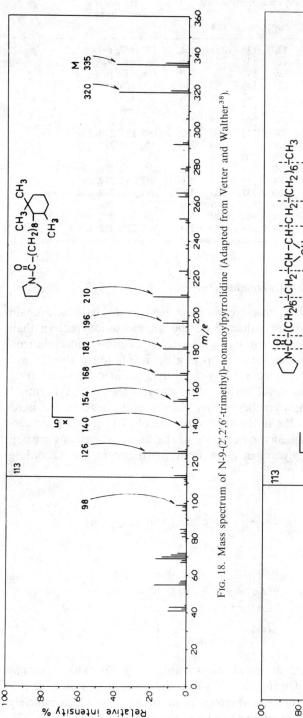

FIG. 18. Mass spectrum of N-9-(2',2',6'-trimethyl)-nonanoylpyrrolidine (Adapted from Vetter and Walther[38]).

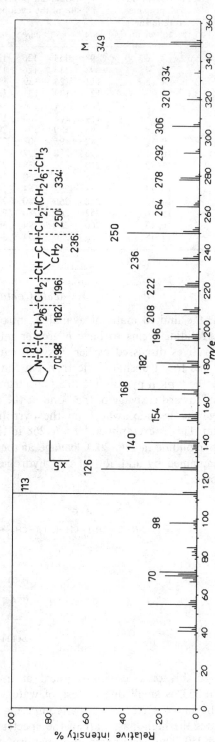

FIG. 19. Mass spectrum of N-trans-9,10-methylene-octadecanoylpyrrolidine (cf. Andersson et al.[9]).

TABLE 5. Key Fragments in Mass Spectra of Monomethylene-octadecanoylpyrrolidines
(Andersson et al.[9])

Position of the cyclopropane ring in the carbon chain	Dominant peaks around the site of the cyclopropane ring						Other peaks typical for each isomer
	m/e	m/e	m/e	m/e	m/e	m/e	m/e
2, 3	98	113	126	138	153	166	139, 152, 167
3, 4	98	113	126	138	152	166	98 (base peak)
4, 5	113	126	139	152	166	180	138, 153, 167
5, 6	126	139	153	166	180	194	138, 140, 152, 167
6, 7	140	154	166	180	194	208	152, 153
7, 8	154	166	180	194	208	222	152 168
8, 9	168	180	194	208	222	236	166, 182, 279
9, 10	182	196	208	222	236	250	180, 194, 210, 279
10, 11	196	210	222	236	250	264	208, 224, 279
11, 12	210	224	238	250	264	278	222, 236, 252, 279
12, 13	224	238	250	264	278	292	236, 252, 266, 279, 280
13, 14	238	252	264	278	292	306	250, 264, 279, 294
14, 15	252	266	280	292	306	320	264, 278, 279, 294
15, 16	252	266	280	292	306	320	278, 279, 294
16, 17	266	280	294	306	320	334	278, 279, 292, 308
17, 18	266	280	294	306	320	334	278, 279, 292, 308

F. Acids Containing Oxygen

From the limited material available, introduction of oxygen in the fatty acid chain of a pyrrolidide seems to have a greater influence on the fragmentation pattern than the derivatives discussed earlier. Oxygen can easily donate an electron under electron impact due to an unshared electron pair, and competing fragmentation pathways will obscure the typical pyrrolidide mass spectrum. Long-chain hydroxy acids show a strong tendency toward cleavage of the bond at the carbon atom carrying the hydroxyl group.[1] Cleavage on the side away from the pyrrolidine ring gives a large peak in the mass spectrum. This is exemplified by m/e 268 in the mass spectrum of N-12-hydroxy octadecanoylpyrrolidine in Fig. 20. Cleavage at the near side of the hydroxyl-carrying carbon is accompanied by addition of one hydrogen atom from the fragment lost according to Scheme 2.

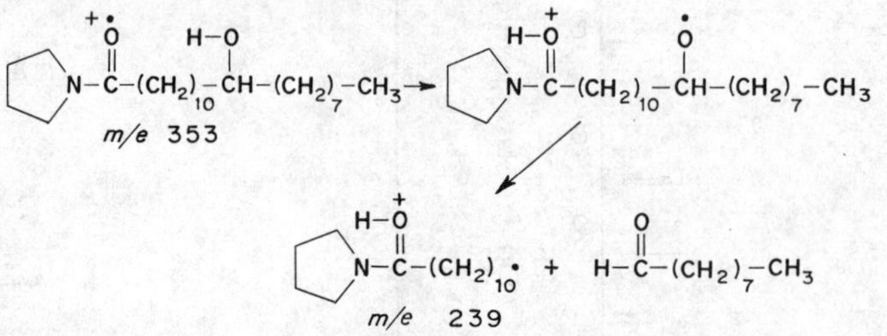

SCHEME 2.

In Fig. 20, this ion gives rise to a peak at an odd mass number, m/e 239. The molecular ion, m/e 353, is small due to loss of water, which yields the more intense ion at m/e 335. The latter gives rise to a competing fragmentation pattern, which completely dominates the high mass part of the spectrum yielding ions at m/e 320, 306, 292, 278, 264 and 250. The basic pyrrolidide pattern starts with the ions at m/e 224, 210, 196.... The ion m/e 250 includes 12 carbons of the fatty acid chain after water elimination, creating a double bond at carbon 12.

In comparison with hydroxy acids, long-chain oxo acids have another characteristic mode of fragmentation.[1] These acids give typical ions derived from β-cleavage at both sides of the carbonyl carbon. In Fig. 21, showing the mass spectrum N-11-oxo octade-

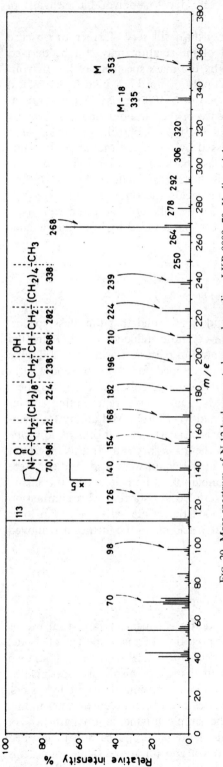

FIG. 20. Mass spectrum of N-12-hydroxy-octadecanoylpyrroline. LKB 9000, 70 eV, direct inlet.

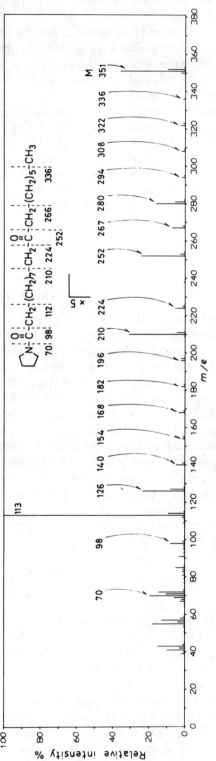

FIG. 21. Mass spectrum of N-11-oxo-octadecanoylpyrrolidine. Experimental conditions same as Fig. 20.

canoylpyrrolidine, this gives rise to a very intense peak at m/e 210 and, due to extraction of a hydrogen by a McLafferty type mechanism, a peak at an odd mass number at m/e 267. The position of the oxo group is also indicated by the missing peak in the typical pyrrolidide fragmentation pattern at m/e 238, due to unfavorable cleavage of a carbonyl.

The presence of an unusual oxygen-containing group will serve further to obscure the usual pyrrolidide ions, although a great deal of information may still be derived from the mass spectrum. An example is provided by the mass spectrum of the pyrrolidide[8] of a furanoid fatty acid isolated from the lipids of northern pike (*Esox lucius*) and other fish species.[22] Figure 22 shows the spectrum of N-10,13-epoxy-11-methyl octadec-10,12-dienoylpyrrolidine. The stabilizing influence of the furanoid ring is shown by the strong molecular peak at m/e 361. Virtually all the typical pyrrolidide peaks, such as m/e 113, 126, 140, 154, 168 and 182, are suppressed by ions from the large furanoid-containing fragment m/e 165, which is also the base peak. The latter ion is derived from cleavage between carbons 8 and 9, which is β-cleavage induced by the furane ring. Cleavage on the far side of the furanoid ring yields the series of ions m/e 290, 304, 318, 332 and 346, typical for an alkyl chain of a pyrrolidide. The furan ring itself is represented by m/e 109, which incorporates the carbon atoms 9 to 14 of the octadecanoyl chain.

G. *Dicarboxylic Acids*

Andersson et al.[6,7] studied dicarboxylic fatty acids and found that the mass spectra of long-chain methyl alkyl-pyrolidamates showed very little influence of the ω-methyl ester group on the characteristic pyrrolidide fragmentation pattern. Location of a double bond could be performed according to the rule developed for the monoenoic fatty acids. Figure 5 shows the mass spectrum of methyl-decane-pyrrolidamate. In this spectrum, the molecular ion at m/e 269 is followed by M–31 and M–59, representing elimination of the methoxy and the methyl ester groups, and M–73 from McLafferty fragmentation of the ester linkage. The base peak is the McLafferty rearrangement ion, m/e 113, and each carbon in the chain is represented in the mass spectrum. In this investigation, the homologs with 4, 5 and 6 carbon atoms in the chain were also studied. Here the amide loses its influences with decreasing number of carbon atoms in the chain. At the same time, a competing fragmentation pattern occurs starting with elimination of the pyrrolidine ring. From this, it can be concluded that a methyl ester group loses its influence upon the fragmentation pattern of a fatty acid chain the further it is moved from the pyrrolidide group.

H. *Isotope-labelled Acids*

The incorporation of stable isotopes in fatty acids by chemical or biological means has previously been used to (a) establish the interpretation of mass spectra of model compounds, (b) facilitate the elucidation of certain basic structural features of unknown compounds, and (c) elucidate metabolic pathways in living organisms. Isotope-labelled fatty acids are investigated according to (a), (b) and (c) above primarily as methyl esters. However, in some cases with methyl esters, difficulties are encountered in determining the exact position of the label. Encouraged by the simple fragmentation pathways of pyrrolidides, Andersson et al.[5–7] analyzed synthetic deuteriated and carbon-13 labelled fatty acids to improve the MS technique under (a), (b) and (c) above. They found this derivative far superior to the methyl ester for location of the isotope label. Figure 23 shows the mass spectrum of N-16,16-dideuteriooctadecanoylpyrrolidine. The molecular ion, m/e 339, appears 2 a.m.u. above that in the naturally occurring N-octadecanoyl-pyrrolidine (Fig. 7), which indicates replacement of two hydrogen atoms with two deuterium atoms. The significant peak for the fragment retaining 15 carbon atoms of the chain appears at m/e 294 and is followed by the next fragment 16 a.m.u. higher at m/e 310. This indicates a *gem*-dideuteriated carbon in position 16 of the carbon chain. The satellite peak $[M-43]^+$, 2 a.m.u. above m/e 292, suggests the elimination of a fragment containing carbons 2, 3 and 4 in the fatty acid chain, as was found in methyl esters.[16] The *vic*-dideuteriated N-9,10-dideuteriooctadecanoylpyrrolidine in Fig. 24 indicates the positions of the label at carbons 9 and 10, as the fragments keeping the deuterium atoms occur at m/e 211 and m/e 225 instead of m/e 210 and m/e 224 (Fig. 7). In Fig. 25, the incorporation of 53 per cent carbon-13 at carbon 2 in N-octadecanoyl-pyrrolidine gives a striking example of the unique mass spectrometric fragmentation pattern of pyrrolidides. The site of the label is immediately given by the partial shift of m/e 113 to m/e 114, m/e 126 to m/e 127, etc. The previously discussed elimination of carbons 2, 3 and 4 of the fatty acid chain is supported by the peak at m/e 294. The elimination of carbons 2 and 3 of fatty acid methyl esters[16] is also shown in Fig. 25 by the increased height of m/e 308. The mass spectrum of a perdeuteriated N-octadecanoylpyrrolidine is shown in Fig. 26. The peaks in the high mass region are of very low abundance compared with the corresponding ones of the non-deuteriated analog (Fig. 7). This phenomenon is probably due to a reinforcement of the carbon–carbon bonds of the fatty acid chain induced by the isotope effect of deuterium.[36] The well documented McLafferty rearrangement ion is shifted from m/e 113 to m/e 116 and the regularly spaced ions are separated by 16 a.m.u. due to the deuterium content.

To give further proof for the postulated fragmentation pattern of ethylenic fatty acid pyrrolidides, a number of normal chain dideuteriated monoethylenic compounds were investigated.[6] The mass spectrum of N-11,12-dideuteriooctadec-11-enoylpyrrolidine in Fig. 27 is typical for these compounds. The spectrum is very similar to that of a saturated compound with no deuterium atoms, but, compared with N-octadecanoylpyrrolidine in Fig. 7, the evenly decreased tendency of the pyrrolidide-containing fragments is disturbed by the peak clusters which include carbons 10 and 11. The most intense peaks in these clusters, m/e 224 and m/e 237, are separated by 13 a.m.u. and followed by another peak at m/e 252, 15 a.m.u. higher. This suggests a double bond which contains two deuterium atoms in position 11,12, following the rule formulated by Andersson and Holman.[11] This rule can be further explained by the postulated fragmentation pathway of N-9,10-dideuteriooctadec-9-enoylpyrrolidine in Fig. 28. Because no cleavage is likely to occur within a double bond, this bond must migrate. The series of ions at m/e 337, 322, 308, 294, 280, 266, 252, 238, *224*, *209*, *196*, 182, 168, 154, 140, 113 and 98, with the diagnostic ions in italics, suggests a migration of the double bond preferentially towards the amide part of the molecule. The mass spectrum of methyl 6,7-dideuterio-dodec-6-enpyrrolidamate in Fig. 29 is an example of a dideuteriated ethylenic fatty acid pyrrolidide, in which the double bond can be located, although an ω-methyl ester group is present. The site of the ethylenic bond is indicated by the peaks at m/e 154, 167 and 184, in accordance with the discussion above, and is located between carbons 6 and 7.

The mass spectra of perdeuteriated methyl alkylpyrrolidamates have also been investigated.[7] The dicarboxylic acids were synthesized via oxidative degradation of a perdeuterio-alkanoic acid with permanganate and were characterized by analytical methods. Because the mass spectra of dimethyl dicarboxylates are very complex and, in most cases, do not have detectable molecular ions, the pyrrolidide derivative was studied. Figure 30 shows the mass spectrum of methyl perdeuteriodecanpyrrolidamate, which exhibits a molecular ion of m/e 285 of about 5 per cent of the base peak, m/e 116, which is the well known McLafferty rearrangement peak. From the molecular ion, it is possible to calculate the total deuterium content by comparing with non-deuteriated analogs. As reported for the non-deuteriated analogues, exemplified by methyl decan-pyrrolidamate in Fig. 6, there is no visible influence of the ω-methyl ester group upon the pyrrolidide fragmentation pattern.

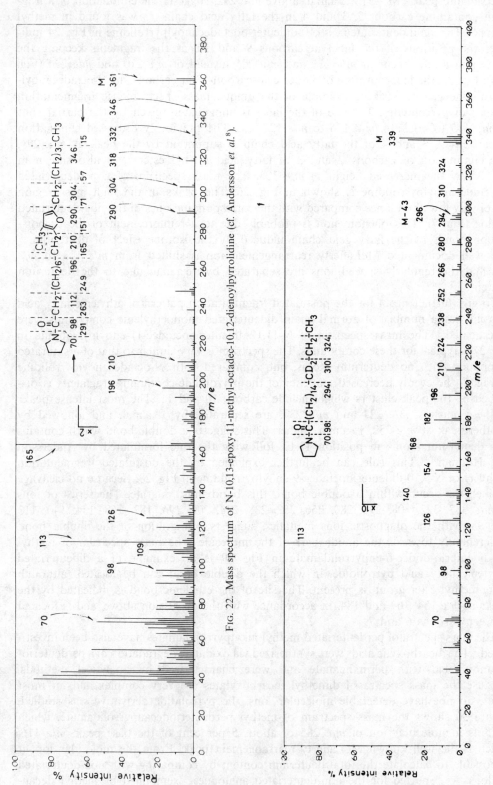

FIG. 22. Mass spectrum of N-10,13-epoxy-11-methyl-octadec-10,12-dienoylpyrrolidine (cf. Andersson et al.[8]).

FIG. 23. Mass spectrum of N-16,16-dideuterio-octadecanoylpyrrolidine (cf. Andersson et al.[5]).

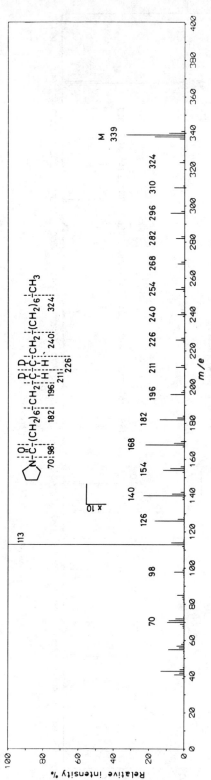

FIG. 24. Mass spectrum of N-9,10-dideuterio-octadecanoylpyrrolidine (cf. Andersson et al.[5]).

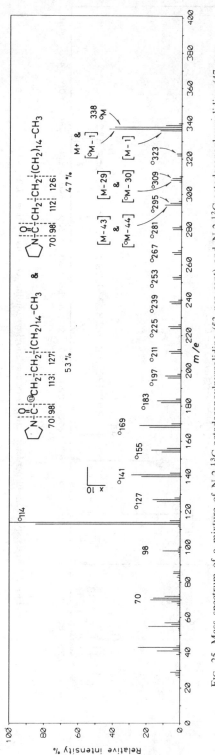

FIG. 25. Mass spectrum of a mixture of N-2-^{13}C-octadecanoylpyrrolidine (53 per cent) and N-2-^{12}C-octadecanoylpyrrolidine (47 per cent) (cf. Andersson et al.[5]).

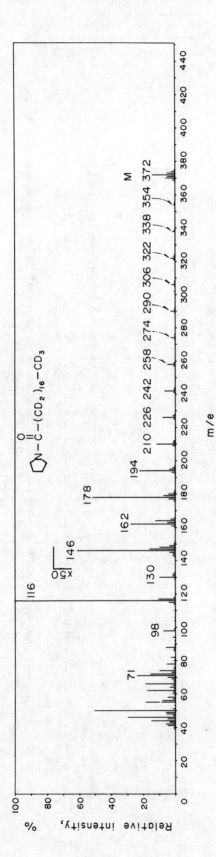

FIG. 26. Mass spectrum of perdeuteriated N-octadecanoylpyrrolidine (cf. Andersson et al.[5]).

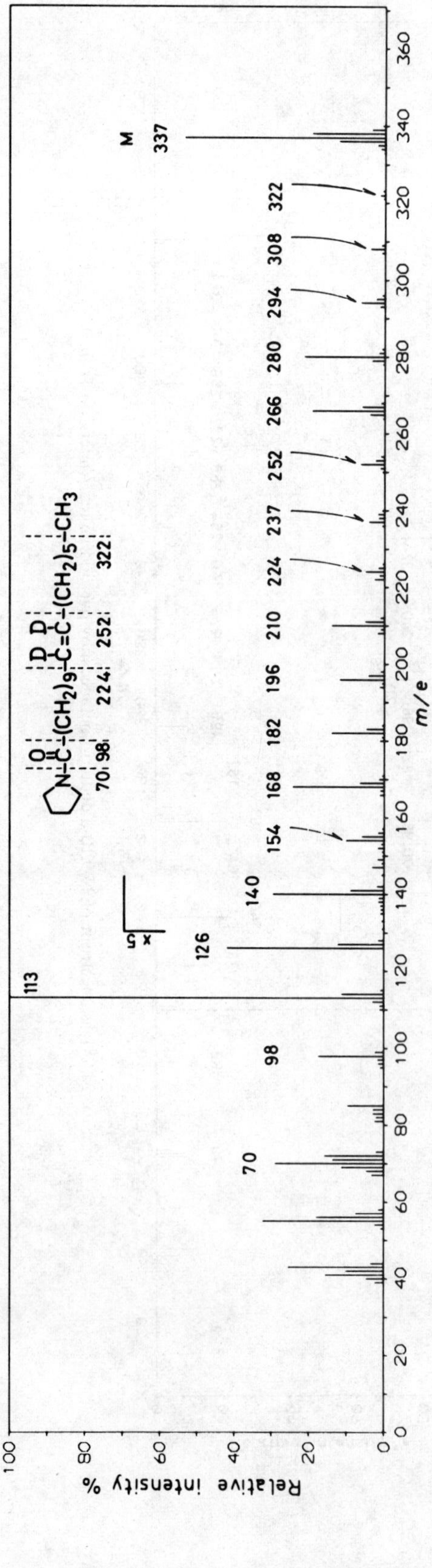

FIG. 27. Mass spectrum of N-11,12-dideuterio-octadec-11-enoylpyrrolidine (cf. Andersson et al.[6]).

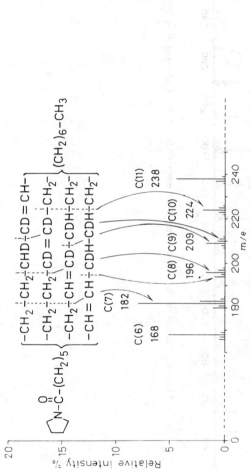

FIG. 28. Postulated fragmentation pathway for N-9,10-dideuterio-octadec-9-enoylpyrrolidine (cf. Andersson et al.[6]).

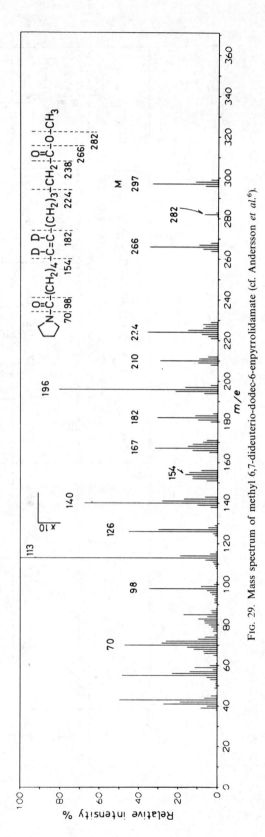

FIG. 29. Mass spectrum of methyl 6,7-dideuterio-dodec-6-enpyrrolidamate (cf. Andersson et al.[6]).

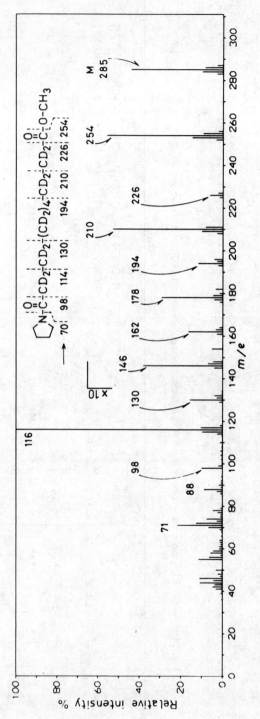

FIG. 30. Mass spectrum of perdeuteriated methyl decanpyrrolidamate (cf. Andersson et al.[7]).

IV. CONCLUDING REMARKS

In the introduction, it was reported that the most common used derivative for GLC–MS and MS analysis of various types of fatty acids is the methyl ester. Therefore, a short comparison between this derivative and the pyrrolidides is appropriate.

The synthesis of either of the two derivatives can be performed in a one-step reaction. However, when the fatty acid occurs with a free carboxyl group, the conversion to pyrrolidides requires a two-step reaction. Very few efforts have yet been made, however, to improve the synthesis of pyrrolidides. The GLC properties favor the methyl esters, because they are more volatile than the corresponding pyrrolidides. This can be compensated by increasing the operating temperatures for the pyrrolidides, for amides are very thermostabile. Pyrrolidides have the advantage in MS because they yield ions at even mass numbers, and these are not obscured by the odd numbered ions from stationary GLC phases, e.g. the commonly used silicones.

Four functional groups found in fatty acids are more easily located by MS of the pyrrolidides. The pyrrolidides of monocyclopropanoic, acetylenic, ethylenic and isotope labelled fatty acids are superior to the methyl esters for location of the functional group. In most cases, the structure of an unknown fatty acid can be deduced from a study of the mass spectrum of its pyrrolidide. Even for the few exceptions, the mass spectrum of the pyrrolidide will lead to the correct structure if mass spectra of model compounds are available because the mass spectra of the isomers are unique. Therefore, with enough suitable reference substances in a library, the structures of most fatty acids can be found by computer search. The use of this derivative in MS will be very attractive to analytical chemists and biochemists, because there is an increasing demand for stable isotope-labelled fatty acids in biological work, and fast and accurate analytical methods are necessary.

Acknowledgements—Grants in support of this Department were obtained from the Swedish Medical Research Council, the Swedish Board for Technical Development, the Wallenberg Foundation and the United States Public Health Service (G.M.–11653).

REFERENCES

1. ANDERSSON, B. Å., unpublished results.
2. ANDERSSON, B. Å. and BERTELSEN, O. *Chem. Scr.* **8,** 91 (1975).
3. ANDERSSON, B. Å. and BERTELSEN, O. *Chem. Scr.* **8,** 135 (1975).
4. ANDERSSON, B. Å., CHRISTIE, W. W. and HOLMAN, R. T. *Lipids* **10,** 215 (1975).
5. ANDERSSON, B. Å., DINGER, F. and DINH-NGUYEN, NG. *Chem. Scr.* **8,** 200 (1975).
6. ANDERSSON, B. Å., DINGER, F. and DINH-NGUYEN, NG. *Chem. Scr.* **9,** 155 (1976).
7. ANDERSSON, B. Å., DINGER, F., DINH-NGUYEN, NG. and RAAL, A. *Chem. Scr.* **10,** 114 (1976).
8. ANDERSSON, B. Å., GLASS, R. L. and SCHLENK, H., unpublished results.
9. ANDERSSON, B. Å., GUNSTONE, F. and STENHAGEN, G., to be published.
10. ANDERSSON, B. Å., HEIMERMANN, W. H. and HOLMAN, R. T. *Lipids* **9,** 443 (1974).
11. ANDERSSON, B. Å. and HOLMAN, R. T. *Lipids* **9,** 185 (1974).
12. ANDERSSON, B. Å. and HOLMAN, R. T. *Lipids* **10,** 716 (1975).
13. BLOMBERG, J. *Lipids* **9,** 461 (1974).
14. BOHLMANN, F. and ZDERO, C. *Chem. Ber.* **106,** 1328 (1973).
15. CHRISTIE, W. W. and HOLMAN, R. T. *Lipids* **1,** 176 (1966).
16. DINH-NGUYEN, NG. *Ark. Kemi* **28,** 289 (1968).
17. DINH-NGUYEN, NG., RYHAGE, R., STÄLLBERG-STENHAGEN, S. and STENHAGEN, E. *Ark. Kemi* **18,** 393 (1961).
18. DUFFIELD, A. M. and DJERASSI, C. *J. Am. Chem. Soc.* **87,** 4554 (1965).
19. GERSON, T. *J. Chromatogr.* **6,** 178 (1961).
20. GERSON, T., PATEL, J. J. and NIXON, L. N. *Lipids* **10,** 134 (1975).
21. GILPIN, J. A. *Anal. Chem.* **31,** 935 (1959).
22. GLASS, R. L., KRICK, T. P., SAND, D. M., RAHN, C. H. and SCHLENK, H. *Lipids* **10,** 695 (1975).
23. HAMBERG, M. and BJÖRKHEM, E. *J. Biol. Chem.* **246,** 7411 (1971).
24. HINTZE, U., RÖPER, H. and GERCKEN, G. *J. Chromatogr.* **87,** 481 (1973).
25. JOSEPH, J. D. *Lipids* **10,** 395 (1975).
26. KARLSSON, K.-A., SAMUELSSON, B. E. and STEEN, G. O. *Acta Chem. Scand.* **22,** 1361 (1968).
27. KARLSSON, K.-A., SAMUELSSON, B. E. and STEEN, G. O. *Chem. Phys. Lipids* **11,** 17 (1973).
28. KLEIMAN, R., BOHANNON, M. B., GUNSTONE, F. D. and BARVE, J. A. *Lipids* **11,** 599 (1976).
29. KOLATTUKUDY, P. E. and WALTON, T. J. *Biochemistry* **11,** 1897 (1972).
30. McCLOSKEY, J. A. *Topics in Lipid Chemistry,* Vol. I, p. 416, Logos, London (1970).

31. PELAH, Z., KIELCZEWSKI, M. A., WILSON, J. M., OHASHI, M., BUDZIKIEWICZ, H. and DJERASSI, C. *J. Am. Chem. Soc.* **85**, 2470 (1963).
32. RICHTER, W. J., TESAREK, J. M. and BURLINGAME, A. L. *Org. Mass Spectrom.* **5**, 531 (1971).
33. RYHAGE, R. and STENHAGEN, E. *J. Lipid Res.* **1**, 361 (1960).
34. RYHAGE, R. and STENHAGEN, E. *Ark. Kemi* **15**, 291 (1960).
35. SPITELLER, G., SPITELLER-FRIEDMANN, M. and HOURIET, R. *Monatsh. Chem.* **97**, 121 (1966).
36. TURKEVICH, J., FRIEDMANN, L., SALOMON, E. and WRIGHTSON, F. M. *J. Am. Chem. Soc.* **70**, 2638 (1948).
37. VALICENTI, A. J., HEIMERMANN, W. H. and HOLMAN, R. T. *J. Am. Oil Chem. Soc.* **54**, 147A (Abstract) (1977).
38. VETTER, W. and WALTHER, W. *Monatsh. Chem.* **106**, 203 (1975).
39. VETTER, W., WALTHER, W. and VECCHI, M. *Helv. Chim. Acta* **54**, 1599 (1971).
40. WALTON, T. J. and KOLATTUKUDY, P. E. *Biochemistry* **11**, 1885 (1972).

INDEX

3-Acetoxyglutaric acid 190
2-Acetoxysuccinic acid 190
Acetylenic normal-chain acids 295
Acholeplasma laidlawii 38, 157
Acide delphinique 258
Acide phocenique 258
Acyclic diterpene alcohols 231, 234
Acyl esters 253
Acyl-sugars 82
Acyl-trehaloses 85
  metabolism 83–85
  subcellular localization 83
Agaricus bisporus 67
Air/sea interface 31, 43
Amides of oleic acid, relative retention times and key
  fragments of 282
Amphiphile-water systems 149
Amphiphiles, soluble 146, 155, 156
Aoyama B 74
Aquatic lipid surface microlayer 31–44
Arthrobacter 77
Arthrobacter paraffineus 77, 85, 86
Arthroderma 171
Arthroderma uncinatum 176
Ascomycetes 175
Asialoganglioside 210

Bacterioneuston 33
Bacterioneuston collector 33
BCG, cord factor from 92
Bilayers
  lipid 125
  membrane, structural features of 125
Bile acid salt/lipid/water systems 159
Biological membranes 125
Birds, waxes from preen glands of 25
Blood-group fucolipids 221
Bombus hortorum 240
Bombus hypnorum 235
Bombus laponnicus 235–36
Bombus lapponicus lapponicus 236
Bombus lapponicus scandinavicus 236
Bombus terrestris 234, 235
Brassica 22
Brevibacterium 77
Brevibacterium thiogenitalis 78
Brevibacterium vitarumen 78
Bubbles, dynamics of 35
N-*n*-Butyl-*n*-octadec-9-enamide 282

Caesalpinioideae 22
Carboxylic acids 82, 279
  conversion to perdeuteriated analog 196
  perdeuteriated, separation of 197
Carotenoids 20
Cedrela toona 231
Cerebroside 141
Chiral compounds 179
Chlorella pyrenoidosa 41
Chlorella pyrenoidosa chick 40
Chlorobium limicola 248
Chlorophyll 245, 247
  in photosynthetic membrane 248–49

Chlorophyll *A*, biosynthesis of 250–53
Chlorophyllide *a*, esterification of 251–53
Chloroplasts 248
Cholera toxin 214–20
Cholesterol and hydrocarbon chains 139–41
Cholesterol ester 134, 139
Cholesterol skeleta 139, 140
Cholesteryl sulphate 140
Cholesteryl-17-bromoheptadecanoate 134, 140
Chromatic acid 70
Claviceps purpurea 67
Conjugated unsaturated fatty acids 295
Continuous flow technique 43
Cord factor 65, 83, 92
  BCG 92
  immunostimulation 92–94
  in oil-in-water emulsion 92–94
  infrared spectrum of 74
  injection to mice 87, 92–94
  isolation and structure of 67–68
  isolation from various species of mycobacteria
    72–74
  isolation of lower homologs of 74
  purification and structure 72–74
  role in pathogenicity of bacteria 94
  toxic properties 87
Cord factor-MBSA complex 92
Corn oil triglycerides 284, 287
Corynebacteria 74
Corynebacterium diphtheriae 74, 75, 84
Corynebacterium fasciens 77
Corynebacterium ovis 74
Corynebacterium pseudodiphtheriae 77
Corynebacterium rubrum 76
Corynebacterium ulcerans 75
Coryno-cord factor 75
Corynomycolenic acid 74, 76, 84
Cr. laurentii 25
Cr. neoformans 25
Cyanolipids 25, 28
Cyclic acids 296
Cytolipin S 209–14

DDT 41
Delphinoidea 276
Delphinus delphis 267
Dermatophyte cell wall lipids 173
Dermatophyte lipids 171–77
Dermatophytes
  fatty acids of 173–75
  lipids in 171–73
  phospholipids in 176
  steroils in 175
Dermatophytosis 171
Desulfolipid-I 81, 90
Desulfolipid-III 94
Deuterated mono- and dicarboxylic acid pyrro-
  lidides 206
Deuteriocarbons 195
Deuterium in organic compounds 195
Dicarboxylic acid 70, 300
  perdeuteriated 195

N-9, 10-Dideuteriooctadecanoylpyrrolidine 301
N-16, 16-Dideuteriooctadecanoylpyrrolidine 300
N-9, 10-Dideuteriooctadec-9-enoylpyrrolidine 301
N-11, 12-Dideuteriooctadec-11-enoylpyrrolidine 301
Digalactosyl diglyceride 157
1,2-Dilauroyl-DL-phosphatidylethanolamine 134
N-N-Dimethylamide 286
N,N-Dimethyloctadec-9-enamide 284
N,N-Dimethyl-n-octadec-9-enamide 283
Dimethyl perdeuterio-n-dicarboxylates 198
Dimethyl perdeuterio-n-heptanedioate 203
Dimethyl perdeuterio-n-hexadecanedioate 204
Dimethyl perdeuterio-n-tricosanedioate 204
L-α-Dimyristoyl lecithin (DML) 164
1,2-Dipalmitoyl-DL-glycero-3-phosphoethanol-
    amine 134
L-α-Dipalmitoyl lecithin 164
1,2 Dipalmitoyl-L-phosphatidylcholine/water system
    148
Dissimilarity Index 14
Diterpenes 231
Diterpenoids 231
Dodecyltrimethyl ammonium chloride/water system
    153
Dolphin head oil lipids 259

Egg-yolk lecithin system 155, 156
Egg-yolk lecithin-water system 148, 156
Electron spectroscopy 2
Electron spin resonance 156
Elephantorrhiza 25
Emulsion capacity 169
Emulsions formed by polar lipids, stability of 163–69
Encyclia 11
Enrichment factor 36, 37, 43
Environmental conditions 18
Epidendrum 11
Epidermophyton 171, 173
Epidermophyton floccosum 171, 172, 173, 175
N-10, 13-Epoxy-11-methyl-octadec-10, 12-dienoyl-
    pyrrolidine 300
Ergosterol 175

Farnesol 233, 234
Fats, composition of 9
Fatty acid amides, synthesis, gas liquid chromato-
    graphy and mass spectrometry of 279–86
Fatty acid pyrrolidides, mass spectrometry of
    279–308
Fatty acids 10, 20, 31, 32, 39, 86, 222
    branched-chain 179
    dermatophytes 173–75
    liver lipids from rats 19
    milk fats 14
    odontocete head oil lipids 261
    odontocete melon 268
Fatty alcohols from odontocete head oil lipids 263
Floral odors of Magnolia 10, 13
Food additives 163
Food emulsifiers 163–65
Formica nigricans 239
Formica polyctena 239
Formica rufa 239
Fortuitin 207
Fucolipids, blood-group 221

Galactodiglycerides 248
β-D-Galactosyl-N-(2-D-hydroxyoctadecanoyl)-D-
    dihydrosphingosine 135
Gangliosides 214
Gas chromatography 10, 12, 241, 253, 279
Gas-liquid chromatography 1, 2, 46, 198

Gel phases 165
Geranyl chloride 232
Geranylcitronellal 231
Geranylcitronellol 231
    characterization of product 234
    comparison between synthetic and natural 234
    natural 234–40
    occurrence in nature 235–40
    synthesis of 232–34
    synthetic procedure 233
Geranylgeranial 231
Geranylgeraniol 231
    in biosynthesis of chlorophyll 250–53
Geranyllinalol 231
Globicephala macrorhynca 258
Globicephala melaena 258, 259, 267, 273
Globicephala scammoni 258
Globicephalus sieboldii Gray 259
Glycerol ethers. See Substituted glycerol ethers
Glycerolipids 45
Glycolipids, trehalose-containing. See Trehalose-
    containing glycolipids
Glycosphingolipids 208, 209, 214, 227
Greenland shark liver oil 45, 51
Gullmaren 42, 43
Gymnosperms 20

Heavy metals 32
Heptaglycosylceramide 223
Hexaglycosylceramides 222
Homofarnesenic acid 233
Hydrocarbon chain matrix, functional groups in
    137–39
Hydrocarbon chains 169
    effect of cholesterol 139–41
    interaction between 125
    lateral packing of 125–43
    see also Subcells
Hydrocarbons 20
    from odontocete head oils 264
Hydroxy compounds, long-chain 179
3-Hydroxybutanoic acid 179
12-D-Hydroxyoctadecanoic acid methyl ester 138
N-12-Hydroxy-octadecanoylpyrrolidine 298
Hydroxy-phthioceranic acids 82
2-Hydroxy-10-undecenoic acid 187

Index of relationship 13, 18–20
Infrared spectroscopy 241
Isophytol 232
Isotope-labelled acids 300–1
Isovaleric acid 259, 276

Juniperus 17
Juniperus californica A 17
Juniperus californica B 17

Leaf oils 19
Lepidium 22
Lepidoptera 25
Linoleate in rats 20
Lipid bilayers 125
Lipid class pattern, odontocete head oil lipids
    259–61
Lipid phases, liquid crystalline 145
Lipid systems of biological interest 155–57
Lipid–water systems, liquid crystalline behavior in
    145–62
Lipids
    in dermatophytes 171–77
    in sound processing 271–75
    insoluble amphiphilic 146
    nonpolar 146

Lipids—cont.
  polar 146, 163–69
  quantitative chemical taxonomy based upon
      composition of 9–29
  soluble amphiphilic 146
  taxonomic relationships based upon 20
  water-soluble amphiphilic 146
  see also under specific lipid systems
Lipopolysaccaride 37
Liriodendron 13
Liriodendron tulipifera 14
Lobularia 22

Magnolia 20
  display of relationships in 13
  floral odors of 10, 13
Magnolia acuminata 14
Marking compounds 10
Mass spectrometry 1, 2, 10, 46, 81, 198, 207, 241, 279–
      308
Membrane bilayers, structural features of 125
Membrane structures 157
Methionine 84
Methyl alkylpyrrolidamates 301
Methyl *n*-alkylpyrrolidamates 283
Methylated bovine serum albumin (MBSA) 91
Methyl-branched acids 288–91
3-Methylbutanoic acid 259
Methyl-decane-pyrrolidamate 300
Methyl 6,7-dideuterio-*n*-dodec-6-enpyrrolidamate
      301
Methylene interrupted isomeric polyunsaturated
      acids 294
N-trans-9,10-Methylene-*n*-octadecanoylpyrrolidine
      297
Methyl esters 39
(—)-2-Methyl-2-ethyleicosanoic acid 131
2D-Methyloctadecanoic acid 129
N-9-Methyl-*n*-octadecanoylpyrrolidine 288
DL-2-Methyl-7-oxododecanoic acid 137
Methyl perdeuteriodecanpyrrolidamate 301
Methyl perdeuterio-*n*-heptanoate 202
Methyl perdeuterio-*n*-monocarboxylates 198
Methyl perdeuterio-*n*-pentadecanoate 202
Methyl perdeuterio-*n*-tricosanoate 203
Methyl phleates 79
Micrococci 20
Micrococcus luteus 22
Micrococcus mucilaginosus 22
Micrococci roseus 22
Micromonospora 65
Microorganisms and model surface layer interac-
      tions 36
Microsomes, trehalose-containing lipid action on
      87–91
Microsporum 171, 173
Microsporum gypseum 172
Microsporum quinceanum 173
Microwave spectroscopy 2
Middle soap 149
Milk fats, fatty acid composition of 14
Mimosoideae 25
Mitochondria, trehalose-containing lipid action on
      87–91
Mnium cuspidatum 14
Mnium medium 14
Mnium punctatum 14
Molecular arrangement in monolayers 101–24
Monocarboxylic acid
  conversion to lower homologs and alkanes 196
  conversion to lower homologs and dicarboxylic
      acids 196

perdeuteriated 195
Monoglyceride/water systems 158
Monoglycerides 166
1-Monolaurin 166
Monolayer data correlation with data for three-
      dimensional crystalline forms 112
Monolayer phases 107, 165
  terminology of 120
  with tilted chain arrangement 122–24
  with vertical chain arrangement 120–22
Monolayer polymorphism 102
Monolayers
  n-aliphatic acids 116–18
  n-alkyl acetates 105–14
  condensed 101
  ethyl esters of *n*-aliphatic acids 114–16
  formation of racemic compounds in 118
  insoluble, general theory of 102–4
  molecular arrangement in 101–24
  molecular structure analysis materials and experi-
      mental techniques 104–5
  molecular structure analysis methods 104
  phase transitions 101, 103
  polymorphism in three-dimensional state 105
Monolinolein 166
Monomethylene-*n*-octadecanoylpyrrolidines 298
Monoynoic acids 295
Multifilm formation at air-water interface 120
Mycobacteria 69–71, 83
  cord factor isolated from various species of 72–74
Mycobacteric acids 84
Mycobacterium avium 71, 83, 84
Mycobacterium bovis 69, 70, 71
Mycobacterium fortuitum 64, 65, 71
Mycobacterium kansasii 71, 92
Mycobacterium marianum 84
Mycobacterium paratuberculosis 70
Mycobacterium phlei 71, 73, 75, 78, 83, 84, 86
Mycobacterium smegmatis 69–71, 76, 84, 85
Mycobacterium tuberculosis 60, 69–71, 74, 83
Mycolic acids 68–72, 83–84

NADase activity 87
Nannizzia 171
Neat soap 149
Neuston 32
Nocardia 76, 77
Nocardia asteroides 76, 77, 84
Nocardia coelica 77
Nocardia corallina 76
Nocardia erythroplis 76
Nocardia kirovani 76
Nocardia polychromogenes 77
Nocardia rhodocrous 77
Nocardo-cord factor 77
Nocardomycolic acid 76
n-Nonadecane 134
Nuclear magnetic resonance (NMR) 46, 52, 81,
      154, 156, 157, 241
Nutritional conditions 18

n-Octadecane 131
n-Octadecanoic acid 128
N-octadecanoylpyrrolidine 300
N-*n*-octadecanoylpyrrolidine 287
N-2-^{13}C-*n*-octadecanoylpyrrolidine 303
N-octadec-9, 12-dienoylpyrrolidine 294
N-*n*-octadec-9,15-dienoylpyrrolidine 293
n-octadec-9-enamide 282
N-octadec-9-enoylpyrrolidine 283, 284, 291, 295
N-*n*-octadec-3,6,9,12,15-pentaenoyl-pyrrolidine 294
N-*n*-octadec-9,12,15-trienoylpyrrolidine 293

Odontocete head and blubber fat differences 266
Odontocete head oil lipids 257–78
    fatty acids from 261
    fatty alcohols from 263
    functions of 275
    hydrocarbons from 264
    lipid class pattern 259–61
    triglycerides from 265
    waxes from 264
Odontocete lipids, sonic properties of 271
Odontocete melon
    as sonic lens 273
    fatty acids from 268
Odontocete melon lipids 266–68
Odontocete melon tissue, sound velocity in 273
Odontocetes 258
    echolocation of 269–71
    ultrasound emission from 269
    ultrasound reception by 270
Oleic acid 36
    amides of, relative retention times and key frag-
        ments of 282
Optical isotropic viscous phases 151, 152
Öresund 42, 43
13-Oxoisostearic acid 126, 137
11-Oxo-octadecanoylpyrrolidine 298–300
Oxygen, acids containing 298–300
Ozonolysis 217

Palmitic acid 36
PCB-compounds 40–42
N-Pentadecanoylpyrrolidines 288
Pentaenoic acid 294
Peptidolipids 207
Perdeuteriated mono- and dicarboxylic acids 195
Perdeuterio-n-tetracosanoic acid, oxidative degra-
        dation of 197
Phase transitions in monolayers 101, 103
Phleic acids 78, 85
Phocaena phocaena 266
Phosphatidylcholine 140, 141
Phosphatidylethanolamines 140
Phospholipids 25, 141
    in dermatophytes 176
Phthioceranic acid 80, 82, 85
Phycomycetes 175
Physeter 269, 270
Physeter catodon 263
Phytol 231, 232
    characterization of products 242–45
    comparison between synthetic and natural 245
    in biosynthesis of chlorophyll 250–53
    in dark grown leaves 253
    natural 245–48
    occurrence in nature 247–48
    synthesis of optical and geometrical isomers of
        240–45
    synthetic procedure 242
Phytylpyrophosphate 248, 252
Picea abies 231
Picea glauca 18, 19
Plaque-forming cells 58
Plastic crystals 153
Platanista 270
Polar lipids, stability of emulsions formed by 163–69
Polluted water 42
Pollution 43
Polytrichum 14
Pongamia 25
Population studies 17
Potassium octanoate/decan-1-ol/water system 151
Prenols, biosynthesis of esterifying 252

N-n-propanoylpyrrolidine 287
Propionibacterium shermanii 65–67
Protein molecules 157
Protochlorophyll pigments 251
Protochlorophyllide 253
Pseudo three-dimensional matrix 11
Pseudomonas 39
Pseudomonas aeruginosa 85
Pseudomonas fluorescens 38
Psithyrus rupestris 235, 240
Pullularia pullulans 67
Pyrophosphate ester 231

Quantitative chemical taxonomy based upon compo-
        sition of lipids 9–29
Quasi-racemic compounds, formation in mono-
        layers 118–19

Racemic 3-acetoxybutanoic acid 180
Racemic compounds, formation in monolayers
        118–19
Racemic 2-hydroxy-10-undecenoic acid 187
Racemic methyl hydrogen 3-acetoxyglutarate 184
RAO function 271
Refsum's disease 241
Reversed-phase partition chromatography 197
Rh. rubra 25
Rhacomitrium 14
Rhamnolipid 86
Rhizobium 291
Ricinoleic acid 188, 191
Ricinus communis L 188

Saccharide 208, 221
Salmonella typhimurium 37
Sampling devices 33
Sapindaceae 28
Schaaffs' function 272
Sediment model ecosystem 41, 44
Seed oils 25, 28
Sequential order 17
Serratia indica 43
Serratia marinorubra 37, 38, 43
Sheep red blood cells 58
Sialic acid 214, 219
'Soap boiler's net soap' 149
Soaps 146
Soap-water systems 146
    basic work on 147
    methods of investigating 147
Sodium cholate/lecithin/water system 159
Sodium cholate/sodium oleate/water system 159
Sodium dodecyl sulphate 138
Sodium octanoate/cholesterol/water system 154
Sodium octanoate-decan-l-ol-water system 149, 150
α-Sodium stearate hemihydrate 131
Sodium taurocholate/monolaurin/water system 159
Sodium taurodesoxycholate/monolaurin/water sys-
        tem 159
Sodium tetradecanoate/water system 148
Solar radiation 33
Sound absorption 273
Sound processing, lipids in 271–75
Sound velocity
    calculation of 272
    in odontocete melon tissue 273
SPAN-20 164, 168–69
Sphingolipids 140 207
Sphingosine 217, 222
Spirochaeta 20
Ställberg-Stenhagen, Stina Lisa 1
Staphylococci 20

Staphylococcus aureus 22
Staphylococcus epidermidis 22
Staphylococcus saprophyticus 22
Staphylococcus sp. 2429 22
Stenella 272
Stenhagen, Einar August 1
Steroils in dermatophytes 175
Streptococcus faecalis 59
Strongly acidic lipids (SAL) 94
Subcells
    concept of 126–27
    hexagonal packing H 133
    hybrid type 134–37
    hybrid type HS1 134
    hybrid type HS2 135–37
    monoclinic packing M‖ 133
    orthorhombic packing O⊥ 127–29
    orthorhombic packing O'⊥ 129–30
    orthorhombic packing O‖ 131–32
    orthorhombic packing O'‖ 132
    simple type 127–34
    triclinic packing T‖ 130–31
Substituted glycerol ethers 45–58
    acetylenic 56
    biological effects of 57–58
    hydroxy-substituted saturated and monounsaturated 52, 57
    isolation and characterization 45
    isopropylidene derivatives 46
    methoxy-substituted, in lipids from man and animals 52
    methoxy-substituted polyunsaturated 51
    methoxy-substituted saturated and monounsaturated 45, 54
    syntheses of methoxy- and hydroxy-substituted 54
    unsaturated 51
Sulfolipid-I 80–82, 90, 91, 94
Sulfolipid-II 82
Sulfolipid-III 82, 91, 94
Sulfuric acid 80
Surface balance 43, 104
Surface films 164, 166
    in continuous flow systems 40
Surface pressure 38
Swelling amphiphiles 155

Taxonomic relationships based upon lipids 20
Teflon plate 34
Terpenoids 20
Terrestrol 234, 235
Tetraenoic acid 294
3-Thiadodecanoic acid 133
Thin layer chromatography (TLC) 45
Thylakoids 248
Trehalose-containing glycolipids 59–98, 207, 209, 210, 212, 214, 220–24
    biological properties 87–95

detection of 60
determination of structure 62
distribution and chemical structure 60–82
esters of 2-D-trehalose with fatty acids of medium chain length 64–67
esters of 2-D-trehalose with β-hydroxy α-branched long-chain fatty acids 67–78
esters of 2-D-trehalose with phleic acids 78–79
esters of 2-D-trehalose with sulfuric acid and phthioceranic acids 80–82
immunostimulation 92–94
in life of bacteria 82–86
linkages with sugars 60
metabolism 83
possible roles 85–86
production of circulating antibodies 91–92
role in pathogenicity of bacteria 94
subcellular localization 83
synthesis 63
toxicity and action on mitochondria and microsomes 87–91
Treponema 20
Trichophyton 171, 173
Trichophyton mentagrophytes 175
Trichophyton rubrum 175, 176
Trienoic acids 294
Triglycerides from odontocete head oil lipids 265
Tri-isotridecanoin 39, 40
N-9-(2',2',6'-Tri-methyl-cyclohexyl)-nonanoylpyrrolidine 296
N-3,6,13-Trimethyl-*n*-tetradecanoylpyrrolidine 289
Triolein 36
Triolein-water emulsion 166
n-Tritriacontane 128
Tropaeolum 22
Tubercle bacillus 1, 67, 80, 83, 92, 94
Tursiops gilli 263, 264, 266
Tursiops truncatus 259, 265–67
TWEEN-80 164, 166–68, 169

Ultrasound emission from odontocetes 269
Ultrasound reception by odontocetes 270
Unsaturated glycerol ethers 51
Unsaturated normal-chain acids 291

van der Waals interaction 128
Variable threshold values 14
Viscous isotropic phases 151, 153, 154, 157, 159

Water. *See* Lipid–water systems
Waxes
    from odontocete head oils 264
    from preen glands of birds 25
Whale head oil lipids 257–78
Whales
    pilot 258, 259, 261, 265
    sperm 263
    toothed 257

CONTENTS OF PREVIOUS VOLUMES

VOLUME 1

The Molecular Structure and Polymorphism of Fatty Acids and their Derivatives. T. MALKIN
Sterols. WERNER BERGMANN
Structure and Properties of Phosphatides. P. DESNUELLE
Chromatography of Fatty Acids and Related Substances. RALPH T. HOLMAN
Derivatives of the Fatty Acids. H. J. HARWOOD

VOLUME 2

The Polymorphism of Glycerides. T. MALKIN
Autoxidation of Fats and Related Substances. RALPH T. HOLMAN
Nutritional Significance of the Fats. HARRY J. DEUEL, JR.
The Surface Properties of Fatty Acids and Allied Substances. D. G. DERVICHIAN
Urea Inclusion Compounds of Fatty Acids. H. SCHLENK
Infrared Absorption Spectroscopy in Fats and Oils. D. H. WHEELER
Counter-current Fractionation of Lipids. H. J. DUTTON

VOLUME 3

Parenteral Administration of Fats. SMITH FREEMAN
Solutions of Soap-like Substances. G. S. HARTLEY
Applications of Low Temperature Crystallization in the Separation of the Fatty Acids and their Compounds.
    J. F. BROWN & DORIS K. KOLB
Synthetic Detergents. W. BAIRD
The Biochemistry of Fat-soluble Vitamins. HENRIK DAM
Oxygenated Fatty Acids. DANIEL SWERN
Low Pressure Fractional Distillation and its Use in the Investigation of Lipids. K. E. MURRAY
Formation of Animal Fats. F. B. SHORLAND
The Triglyceride Composition of Natural Fats. R. J. VANDER WAL
Some Aspects of the Intestinal Absorption of Fats. SUNE BERGSTRÖM & BENGT BORGSTRÖM
Metabolism of the Steroid Hormones. LEO T. SAMUELS

VOLUME 4

The Constitution and Synthesis of Fatty Acids. F. D. GUNSTONE
The Naturally Occurring Acetylenic Acids. E. M. MEADE
The Synthesis of Glycerides. T. MALKIN & T. H. BEVAN
Lipid Waxes. A. H. WARTH
The Synthesis of Phospholipids. T. MALKIN & T. H. BEVAN
The Lecithinases. D. J. HANAHAN
Lipid Dynamics in Adipose Tissue. B. SHAPIRO
Dilatometry. B. M. CRAIG
Ultra-violet Spectrophotometry of Fatty Acids. G. A. J. PITT & R. A. MORTON

VOLUME 5

Standard Methods in the Fat and Oil Industry. V. C. MEHLENBACHER
Monsavon Continuous Process for Soap Manufacture. F. LECHAMPT & R. PERRON
Progress in the Technology of Soybeans. J. C. COWAN
Oilseed Residues. C. M. LYMAN
The Composition and Properties of Fish Oils. L. A. SWAIN
The Production of Marine Oils. H. N. BROCKELSBY & J. R. PATRICK
The Utilization of Fish Oils. H. N. BROCKELSBY & J. R. PATRICK
Recent Advances in the Technology of Drying Oils. A. E. RHEINECK
Technology of Edible Animal Fats. W. C. AULT
The Technology of Margarine Manufacture. G. B. CRUMP

VOLUME 6

Thomas Malkin: A Biography and List of Publications
Plasmalogens. E. KLENK & H. DEBUCH
The Synthesis of Phospholipids. ERICH BAER
Fatty Acid Oxidation. D. E. GREEN
Fatty Acid Biosynthesis. D. M. GIBSON
Relationship of Dietary Fats to Atherosclerosis. L. W. KINSELL
Lipids, Blood Coagulation and Occlusive Vascular Disease. T. H. SPAET.
The Chemistry of Serum Lipoproteins. N. K. FREEMAN, E. T. LINDGREN & A. V. NICHOLS
Lipoproteins of the Fowl—Serum, Egg and Intracellular. O. A. SCHJEIDE
Ozonolysis of Fatty Acids and the Derivatives. R. G. KADESCH
Recent Developments in the Glyceride Structure of Vegetable Oils. H. J. DUTTON & C. R. SCHOLFIELD

VOLUME 7

The Higher Saturated Branched Chain Fatty Acids. S. ABRAHAMSSON, S. STÄLLBERG-STENHAGEN & E. STENHAGEN
Gas Chromatography of Lipids. E. G. HORNING, A. KARMEN & G. C. SWEELEY
Antioxidant Effects in Biochemistry and Physiology. JOHN G. BIERI
The Coenzyme Q Group (Ubiquinones). F. L. CRANE

VOLUME 8

Phospholipids and Biomembranes. L. L. VAN DEENEN
Recent Progress in Carotenoid Chemistry. SYNNØVE LIAAEN-JENSEN & ARNE JENSEN
Nuclear Magnetic Resonance in Fatty Acids and Glycerides. C. Y. HOPKINS
Conformational Effects in Long Carbon Chains in Relation to Hydrogen Bonding and Polarized Infrared Spectra. JOHN S. SHOWELL
Recent Developments in the Thin-layer Chromatography of Lipids. DONALD C. MALINS
Paper Chromatography of Lipids. JAMES G. HAMILTON
Column Chromatography of Lipids. ROBERT A. STEIN & VIDA SLAWSON

VOLUME 9

George O. Burr: Biography
General Introduction to Polyunsaturated Acids. RALPH T. HOLMAN
Analysis and Characterization of Polyunsaturated Fatty Acids. RALPH T. HOLMAN & JOSEPH J. RAHM
Determination of the Structure of Unsaturated Fatty Acids via Degradative Methods. O. S. PRIVETT
The Synthesis of Naturally-occurring and Labelled 1,4-Polyunsaturated Fatty Acids. J. M. OSBOND
The Metabolism of the Polyunsaturated Fatty Acids. J. F. MEAD
Endocrine Influences on the Metabolism of Polyunsaturated Fatty Acids. R. L. LYMAN
Prostaglandins. P. W. RAMWELL, J. E. SHAW, G. B. CLARKE, M. F. GROSTIC, D. G. KAISER & K. E. PIKE
Essential Fatty Acid Deficiency. RALPH T. HOLMAN
Hydrogenation of Fats. H. J. DUTTON
Peroxidation of Polyunsaturated Fatty Compounds. W. C. LUNDBERG & PENTTI JÄRVI
Preparation of Polyunsaturated Fatty Acids from Natural Sources. O. S. PRIVETT
Biosynthesis of Unsaturated Fatty Acids in Higher Plants. E. M. STEARNS, JR.
The Interrelationship of Polyunsaturated Fatty Acids and Antioxidants in vivo. L. A. WITTING
The Role of Polyunsaturated Acids in Human Nutrition and Metabolism. LARS SÖDERHJELM, HILDA F. WIESE & RALPH T. HOLMAN
Odd Chain Polyunsaturated Acids. H. SCHLENK
Biological Activities of and Requirements for Polyunsaturated Acids. RALPH T. HOLMAN

VOLUME 10

Ernst Klenk: Biography
Fatty Livers and Lipotropic Phenomena. COLIN C. LUCAS & JESSIE H. RIDOUT
Metabolism of Plant Terpenoids. GEORGE R. WALLER
Diol Lipids. L. D. BERGELSON
The Biochemistry of Lipids containing Ether Bonds. FRED SNYDER
The Lipid Biochemistry of Marine Organisms. DONALD C. MALINS and JOHN C. WEKELL
Glycosphingolipids of Animal Tissue. ERIK MÄRTENSSON
On Cerebrosides and Gangliosides. E. KLENK

VOLUME 11

Dedication: Prof. J. B. Brown
Phospholipids, Liquid Crystals and Cell Membrane. R. M. WILLIAMS & D. CHAPMAN
Chemistry and Metabolism of Fatty Aldehydes. V. MAHADEVAN
Occurrence of Unusual Fatty Acids in Plants. C. R. SMITH, JR.
Insect Lipids. PAUL G. FAST
Pesticide Residues in Fats and Other Lipids. A. M. PARSONS
The Chemistry of the Sulfolipids. T. H. HAINES
Lipolytic Enzymes. R. G. JENSEN

VOLUME 12

Biography: Dr. Hendrik Johannes Thomasson
Newer Developments in Determination of Structure of Glycerides and Phosphoglycerides. A. KUKSIS
The Analysis of Fatty Acids and Related Materials by Gas–Liquid Chromatography. R. G. ACKMAN

VOLUME 13

Tumor Lipids. L. D. BERGELSON
Spin-Labeling with Nitroxide Compounds. A New Approach to the in vivo and in vitro Study of Lipid–Protein Interaction. W. T. ROUBAL
The Effects of Ionizing Radiation on Lipids. W. W. NAWAR
The Biochemistry of Plant Cuticular Lipids. P. E. KOLATTUKUDY & T. J. WALTON
Odor and Flavor Compounds from Lipids. D. A. FORSS

VOLUME 14

The Chemistry and Biochemistry of Phytanic, Pristanic and Related Acids. A. K. LOUGH
The Lipids of Fungi. P. J. BRENNAN, P. F. S. GRIFFIN, D. M. LÖSEL and D. TYRREL
Infrared Absorption Spectroscopy of Normal and Substituted Long-chain Fatty Acids and Esters in the
    Solid State. INGRID FISCHMEISTER
Lipid Metabolism and Membrane Functions of the Mammary Gland. S. PATTON and R. G. JENSEN

VOLUME 15 No. 1

Dedication; Walter O. Lundberg
Palm Oil and Palm Kernel Oil. J. A. CORNELIUS
Docosenoic Acids in Dietary Fats. J. L. BEARE-ROGERS
Lipids of Olives. E. FEDELI
Books and Reviews on Lipids. F. D. GUNSTONE

VOLUME 15 No. 2

Cyanolipids. K. L. MIKOLAJCZK
Plant Lipoxygenases. G. A. VELDINK, J. F. G. VLIEGENTHART and J. BOLDINGH

VOLUME 15 No. 3

Jojoba Oil and Derivatives. JAIME WISNIAK